K. Stüwe
Geodynamics of the Lithosphere

D1575161

Springer
Berlin
Heidelberg
New York
Barcelona
Hong Kong
London
Milan
Paris
Tokyo

Kurt Stüwe

Geodynamics of the Lithosphere

with 210 Figures

 Springer

DR. KURT STÜWE
University of Graz
Department of Geology and Paleontology
Heinrichstrasse 26
8010 Graz
Austria

ISBN 3-540-41726-5 Springer-Verlag Berlin
Heidelberg New York

Library of Congress Cataloging-in-Publication Data

Stüwe, Kurt, 1959-
  Introduction to the geodynamics of the lithosphere : quantative description
of geological problems / Kurt Stüwe.
  p.cm.
  Includes bibliographical references.
  ISBN 3540417265 (alk. paper)
  1.Geodynamics--Mathematics. I. Title.

Springer-Verlag Berlin Heidelberg New York
a member of BertelsmannSpringer Science+Business Media GmbH
http://www.springer.de
© Springer-Verlag Berlin Heidelberg 2002
Printed in Germany

Camera ready by author
Cover design: E. Kirchner, Heidelberg
Printed on acid-free paper SPIN 10793778 32/3130/as 5 4 3 2 1 0

I dislike very much to consider any quantita-
-tive problem set by a geologist. In nearly every case, the conditions given are much too
vague for the matter to be in any sense satisfactory, and a geologist does not seem to mind
a few millions of years in matters relating to
time...

John Perry, 1895
(In the paper in which he recalculated the age of
the Earth – previously estimated by Lord Kelvin
to be 90 my – to the still accepted age of 4.5 Gy.)

Preamble

Field geologists typically interpret their data in terms of tectonic models that are consistent with their observations in a given terrain, but that often lack an independent test. Such models can be strengthened considerably if they are supported by independent estimates of the magnitude of the implied geodynamic processes. For example, estimates of the thermal energy budget of a metamorphic terrain are an invaluable aid for the tectonic interpretation of metamorphic isograds mapped in the field; estimates for the orogenic force balance of a certain nappe staking geometry interpreted from structural mapping are a fantastic way to test its mechanical plausibility.

This book was written because there appears to be a strong bi-modality in the nature of text books dealing with such problems. Books that introduce the reader to the modeling of geodynamic processes often require a relatively high maths background. On the other hand, books that deal with basic maths usually lack any connection to geology. This book was written in an attempt to bridge this gap. It is the aim of this book to introduce field based geologists to the power of the quantitative treatment of their field data. Because of this, the emphasis of this book lies on the interpretation of data that are typically collected by structural geologists, petrologists and geochronologists in the field, rather than those collected by seismologists or geophysicists.

As an introductory text, little mathematical knowledge is required. All calculations are discussed in detail without omitting steps in the derivations and an extensive appendix on mathematical tools is provided. All computer codes used to calculate the figures are available from the author. They may also be downloaded from the address http://wegener.uni-graz.at.

Contents

Acknowledgements

This book is a reflection of my enthusiasm for the quantitative treatment of geological problems. After going as far as a Ph.D. doing field based "traditional" geology, it was predominantly Mike Sandiford who helped me to enter the field of geodynamics during my postdoctoral years. I have never looked back. I cannot thank Mike enough for helping me to perform this change. G. Houseman and T. Barr are thanked for deepening my understanding of geodynamics in my years at Monash University. I thank T. Bartosch, B. Bookhagen, D. Coblentz, M. Coffin, K. Ehlers, H. Gibson, B. Grasemann, M. Hintermüller, G. Houseman, M. Raab, J. Robl, R. Rudnick and P. Strauss for reading and correcting various chapters of this book. Particular thanks go to T. Barr who read the whole manuscript in an unparalleled effort and has been a constant helpful companion in many questions throughout the writing. W. Engel, head of the planning section at Springer is thanked for his support and help over the years. The fact that we both agreed to publish this book as our now second project, speaks for the good cooperation I have felt through the years. A. Stasch is thanked for all these little LaTeX hints.

Sadly, the number of errata in a book, F, is never zero. It follows the relationship: $F = F_0 e^{-xt}$, where F_0 is the number of mistakes in the first draft, t is correction time and x is an unknown decay constant. When I wrote the first version of this book F_0 was many thousand. The time t I have taken to edit this book is now already about 3 years. The magnitude of x remains unknown. However, I hope that it is at least of the order of $5 \times 10^{-8} \mathrm{s}^{-1}$. I leave it as the first exercise for the interested reader to determine the number of remaining mistakes. Clearly, these are my own fault. Some of them will be obvious to me within hours of sending this manuscript into print, about others I invite observant readers to inform me. As an excuse I can only say that I have written this book as a field geologist and not as a theoretical geophysicists. I therefore hope, that the advantages of a text book written on the basis of many years of field observation outweigh the disadvantages of potential mathematical shortcomings. Good luck!

January 2002 Kurt Stüwe

1. Introduction

The large scale structure of the earth is caused by *geodynamic processes* which are explained using *energetic*, *kinematic* and *dynamic* descriptions. While "geodynamic processes" are understood to include a large variety of *processes* and the term is used by earth scientists quite loosely, the methods of their *description* involve well defined fields. *Energetic* descriptions are involved with distribution of energy in our planet, typically expressed in terms of heat and temperature. *Kinematic* descriptions describe movements using velocities, strains and strain rates and *Dynamic* descriptions indicate how stresses and forces behave.

As structural and metamorphic geologists we document in the field only the *consequences* of geological processes. The underlying *causes* are much harder to constrain directly. However, it is absolutely crucial to understand these causes or: *"driving forces"*, if we are to explain the tectonic evolution of our planet. This book deals with the dynamic description of geological processes. Our descriptions relate *causes* and *consequences* – *tectonic processes* with *field observations*. In many cases, we will use equations as a concise form to describe processes and observations in nature. As we will be dealing mostly with large scale tectonic questions, the observations that we shall use are also on a large scale. For example, we shall use observations on the elevation (Fig. 1.1, 1.2) and heat flow of mountain ranges, the thickness of continents and the water depth of the oceans.

As the processes we seek to describe are *changing* with time or space, many of the descriptive equations will be *differential equations*. However, it is *not* the aim of this book to blast the reader with high level mathematics. Rather it is the aim to introduce field geologists to the beauty and simplicity of descriptions with equations. Thus, all equations will be explained from basic principles so that an intuitive understanding of them can be achieved. While I prefer the term "description", many colleagues would refer to the contents of this book as "modeling". Thus, it seems useful to commence with an explanation of the term "model".

1.1 What is a Model?

Models are tools that we use to describe the earth around us in a simplified way so that we can understand it better. Sadly, many geologists misunderstand the word model and think of it only as something complicatedly mathematical that has little to do with field work (s. Greenwood 1989). In doing so, many field geologists overlook the fact that field mapping itself is also a form of modeling. Let us explain this, using the example of a geological map.

A map is a transformation of reference frames; for example the projection of the geographic position of field locations onto a piece of paper. However, a geological map that is a mere representation of field data in a new reference frame (i.e. our piece of paper on which the map is drawn) is usually a poor map (Passchier et al. 1990). It may still be useful to find a given outcrop of a given lithology, structure or metamorphic grade, but as scientists we usually want to go beyond that and produce an interpretative map, even if it is on the simple level of inferring where a given lithological contact is. After all, we go out in the field to *clarify* the field relations. We make maps so that we can explain some features of nature to a colleague geologist to the same amount of work we invested to produce the map in the first place. In order to achieve this aim, the geological map must illustrate field relationships in a *simplified* and *interpreted* manner. This forces the field geologist to a constant decision-making process. First of all the geologist has to decide *what* is to be mapped. Is it topography? Is it structure? Is it metamorphic isograds or is it lithology? Which of these (and many others) is to be mapped depends on the question with which we go into the field (Fig. 1.1). Then, the geologist has to decide on the scale on which we are mapping. Once that is decided many more decisions are to be made. Which observations are too small to be mapped and should be neglected. Which ones should be drawn into the map? Which ones are to be emphasized by lines? Can a contact seen in two outcrops be mapped as a line, even in the paddock separating the two outcrops? The geologist is modeling!

If the map is good, then it helps the reader - like any other good model - to understand nature quickly and easily. It also helps to make predictions how the geology may look at different places that were not mapped yet. For example, constructing profiles across our map helps us (to a certain degree) to explain how the geology looks underground. In numerical-, analogue-, or thought-models this process is identical. Mathematical models consist of a series of rules that determine which observations in nature are to be neglected and which ones are to be emphasized. The former will *not* appear in an equation, the latter *will* appear as a parameter in the equation. As such, a mathematical model is no different from the field work of a geologist.

Every model can be considered as a tool that can be used to make predictions about observations in nature. Just as a geological map can be used to construct cross sections and thus predict the geology underneath the surface, a numerical model can be used to make predictions about temperatures,

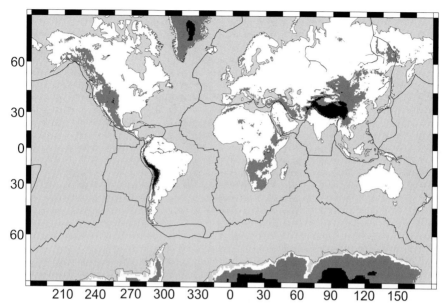

Figure 1.1. Topographic map of the continents showing the large mountain ranges which always have inspired geologists to geodynamic models explaining their origin. Elevations higher than 1 000 m above sea level are shown as gray, elevations higher than 3 000 m are shown in black. The map bears all the characteristics of a good *model*. It captures the essence of the planets distribution of high mountain ranges within only about 0.01 m² of this page, which is only possible by making a range of simplifying assumptions, for example on resolution or choice of cut-off elevation. As such, elevations above 3 000 m in the Azores or Japan escape the scale. Also – while the model largely succeeds in its aim – it is also "wrong" in some places. For example, much of the land surface of (ice covered) Greenland and Antarctica is actually below sea level

forces or velocities which cannot be observed directly because of their enormous time scale or depth.

If the choice of parameters that we consider in our model (and the rules that relate them to each other) are good, then our model is good and it will predict many new observations which will be proven to be correct by future observations. If our choice of parameters and rules is bad, then our model may explain the one or other field observations, but it will predict many other features that will be proven wrong by future observations. Modeling is therefore an iterated back and forth between the choice of parameters and rules that are to be considered or neglected, new observations in nature and finally improvement of the model based on the new observations. Good models are *consistent* with a large number of observations (s. p. 6), but models are hardly ever *unique* in fitting those observations.

The Difference Between "Consistent" and "Unique". The difference between consistent and unique is an important one that is often not rec-

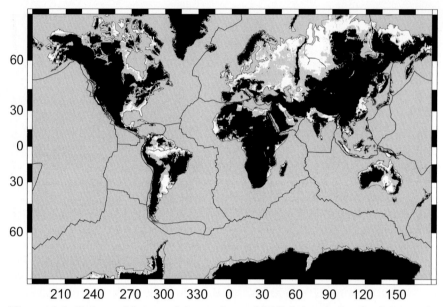

Figure 1.2. Topographic map of the continents highlighting the low lands on earth. Elevations between 100 m and 250 m above sea level are shown gray, elevations below 100 m above sea level are shown in white. Everything above 150 m above sea level is shown in black. See Fig. 1.1 for more information

ognized, even by modelers themselves. *Unique* means - as the word says - that the model is the very only explanation for a set of observations. *Consistent* means that the proposed model *does* explain a set of observations, but that other models may also explain the same set of observations. The largest majority of models are consistent but by no means unique. For example, the heliocentric Copernican model for the solar system (which states that the planets rotate around the sun) is a model which is consistent with our observations of when the sun rises and sets and so on. However, a *geocentric* model (initially designed by Ptolemy) in which the planets and the sun rotate around the earth is possible too. The geocentric model is amazingly more complicated than the heliocentric model and it involves weird planetary motions including epicycles and cycloid curves. However, it also is consistent with our observations on planetary motions. Neither the heliocentric, nor the geocentric model are therefore unique. When viciously defending a model in a discussion it is always sobering to remind oneself that practically all models are only consistent (at best).

The Difference Between "Good" and "Accurate". The difference between *good* (or possibly: "adequate") and *accurate* models is related to the difference between consistent and unique models, but it is not quite the same. Here, it is important to realize that the *best* model must not be the most *ac-

curate model! The best model is the one that finds the best balance between accurate description and simplicity. A good example for this is given by a comparison of Newton's law of gravitation and relativity theory when applied to the description of planetary motion (e. g. Hawkins 1988). Newtonian theory states that the gravitational attraction, F, between two masses is directly proportional to the masses of the two bodies m_1 and m_2, and inverse proportional to the square of the distance r between the two bodies. This model is incredibly simple and may be described by a simple equation:

$$F = G\frac{m_1 m_2}{r^2} \quad . \tag{1.1}$$

The constant of proportionality is called the gravitational constant, G. Its value may be found in Table C.4 in the Appendix of this book. This model describes the elliptic motions of the planets (that were discovered by Kepler in order to improve the Copernican model) extremely well. However, very detailed measurements early this century showed that the motions of some planets differ very marginally from those described by eq. 1.1. These differences may be explained with the model of general relativity, which describes the planetary motions more exactly than Newton's law. Thus, one might consider Newton's model to be superceded by general relativity and use this new model from now on. However, general relativity is much harder to formulate in terms of simple equations than eq. 1.1. It is therefore often not very practical to use. In fact, most earth scientists are familiar with the meaning of eq. 1.1, but few are familiar with the equations of general relativity. Moreover, for the largest majority of purposes – for example to find a planet with a telescope in the sky, or for the interpretation of gravity anomalies by a geologist – Newton's model (eq. 1.1) is sufficient. Thus, for most purposes Newton's model is better (because simpler), albeit less accurate. In short, a good model should find a good balance between *simplicity* of the model and *accuracy* in describing a set of observations.

The Difference Between "Accuracy" and "Precision". In the last paragraph we have used the word *correct* to describe a very good correlation between model description and observation in nature. In general this is the same what is meant by the word *accurate*. However, precision is something different. Precision describes how good a model or an experiment can be repeated with the same result (s. p. 375). Let us illustrate this with an example. A radiochemical analysis may indicate that a rock formed 100 my ago. This analysis is very precise if every time we perform it, we arrive at the same age of 100 my. This applies to errors as well. The analysis is still very precise if we come up with an answer of 100± 50 my, if that answer is reproducible with the identical error limits and we know these error bars very well. However, analyses with huge error bars are not very accurate. In fact, even analyses with very small analytical errors may be not very accurate at all. It could be that our rock actually formed 300 my ago and but was analyzed using the wrong radiometric system – albeit with a high precision.

In conclusion of this section, let us define a good model by a description having the three following properties:

- A good model should describe a *large* set of observations with a comparably *small* set of parameters.
- A good model must be useable as a tool in order to make predictions about observations that can be made in future.
- A good model must be testable by observations.

Note that none of these three requirements includes *accuracy*! The deciding factor for a good model remains the balance between accuracy and simplicity. Accurate description of nature is a virtue that remains reserved to (explanation-free) collections of measured accurate data. (This is very much aside from the fact that many measured data may not be accurate either.)

Geodynamics describes the dynamic evolution of earth through space and time. This dynamic evolution occurs on time scales of up to many hundreds of million years and spatial scales of up to thousands of kilometers. Direct observation is therefore often difficult. Geodynamics is therefore a science which relies to a much larger extent on the model tool than many other sciences. However, mapping in the field, analogue modeling in the laboratory and programming on the computer are all three modeling techniques that are of equal importance in this process. Integrated use of various modeling techniques is the most elegant way to arrive efficiently at a good description of the nature around us.

1.2 Spatial Dimension of Geological Problems

When a geological process is to be described by a simple model, one of the first decisions that has to be made is often the number of spatial dimensions that are to be considered. This decision has to be made according to the requirements of a good model, as discussed above. That is, it should be tried to describe the problem in question with as few spatial dimensions as possible, without loosing the essence of the problem. Fortunately, a very large number of geological problems can be considered one-dimensionally or even without any spatial dimension. Before we go on to discuss some problems that require one-, two-, or three-dimensional consideration, it should be emphasized that "dimension" need not always be the "spatial dimension". For many problems it is important to consider time or some other independent variable as the dimension which may then become an axis on a diagram which we use to illustrate some feature of our description. In this book we will stick to (unless otherwise noted) the meaning of a *spatial* dimension when we talk of the dimension of a problem and use the SI *units*, when we use another variable as our "dimension".

One-dimensional Problems. A simple example of a one-dimensional problem is the description of temperature in the lithosphere as a function of depth (Fig. 1.3a). A one-dimensional description of this problem is sufficient when the lateral extent of the plate is large compared to the thickness of the lithosphere and there is no other lateral changes in physical parameters or structure. Continental lithosphere is often of the order of 100 km thick, but continents are usually many hundreds or even thousands of kilometers in lateral extent. The temperature as a function of depth can therefore be well-illustrated in a diagram in which one axis denotes depth and the other temperature. The depth axis is the dimension of the problem, the temperature axis is the evaluated variable. Similar logic may be applied to the description of temperatures around magmatic dikes.

One-dimensional problems need not be in Cartesian coordinates, but can also occur in spherical or cylindrical coordinates, for example temperatures around a spherical intrusion or compositional zoning profiles in a cylindrical crystal. One-dimensional descriptions are useful when there is little or no variation of the variable of interest in the other spatial directions.

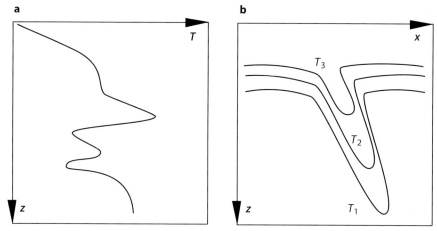

Figure 1.3. Examples of illustrations of one- and two-dimensional models. **a** shows an example of temperature evaluated as a function of a single spatial coordinate, z. The axes of the surface of this page are used up for this one-dimensional description. The strange geotherm is uniquely defined: Every depth corresponds uniquely to a single temperature. **b** shows a two-dimensional model, for example temperatures in a subduction zone as a function of depth z and horizontal distance x. The surface of this page is used up for the two spatial dimensions of the model and the evaluated variable must be portrayed by the contours.

Two-dimensional Problems. An example of a typical two-dimensional problem is the temperature distribution in *subduction zones*. There, the subduction angle as well as the plate thickness are both critical for the shape of

Figure 1.4. A
"three-dimensional" illustration
of the two-dimensional function
$p = \sin(x) \times \sin(y)$ shown on the
two-dimensional paper of this
book page

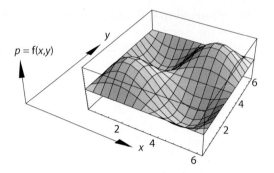

the isotherms and a vertical (z) *plus* a horizontal spatial coordinate (x) are
therefore needed for a meaningful description (Fig. 1.3b). Fortunately, the
third spatial dimension – the direction parallel to the trench – can often be
neglected because trenches are usually long and the subduction direction is
often at roughly right angles to the trench. Calculated temperatures can be
illustrated as contours on a diagram with two spatial dimension of the axes.

The same two-dimensional problem can also be illustrated by using a
perspective sketch where the two spatial coordinates *and* the temperature axis are all shown in one diagram. Fig. 1.4 illustrates such a "three-
dimensional" illustration of a two-dimensional function using the abstract
function $f(x, y) = \sin(x) \times \sin(y)$ as an example. "Three-dimensional" is
shown here in quotation marks, 1st because the third dimension (p, or T
in Fig. 1.3b) is not really a spatial dimension but the evaluated variable, and
2nd because a perspective sketch is still drawn on the two dimensions of this
page. Thus, such drawings remain two-dimensional on paper (unless they are
built of wire or something else) and they only appeal to the three dimensional imagination of the reader. For two-dimensional models that are used
to evaluate temperatures "three-dimensional" illustrations are not common
practice, but for geomorphological models and many geophysical questions
"three-dimensional" illustrations of two-dimensional model results are quite
instructive and are often used.

However, be careful! The fact that two-dimensional models are often shown
as surfaces over a grid of two spatial coordinates should not be mistaken for
a real three-dimensional model. The third dimension is only the evaluated
variable! For Fig. 1.3b we could write $T=f(x, z)$: temperature is a function
only of x and z. In landscape models (e. g. Fig. 1.5) this is often quite confusing as the evaluated variable (surface elevation) has the same units (meters)
as the two spatial coordinates it is a function of (s. p. 173). To stop confusion
it is often useful to use the expression *potential surface*. Just like geophysicists evaluate gravitational- or electromagnetic- potential as a function of two
spatial coordinates, so can surface elevation or temperature be viewed as a
potential surface overlying the plane defined by the two spatial coordinates.

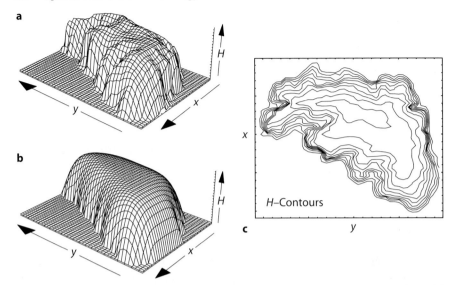

Figure 1.5. a, b "Three-dimensional" illustration of a two-dimensional model shown on the two dimensions of this page. **c** Two-dimensional illustration of the same model results. The coordinates x and y are the spatial coordinates of the model and H is the evaluated variable - or potential surface. In other models the vertical axis may be the gravitational potential (in m s^{-2}), temperature of a metamorphic terrain or the concentration of an element in a crystal. However, in the shown example, H is of the same units as the spatial coordinates - the model is a landscape model. **a** and **c** are illustrations of field data of the shape of Ayers Rock in Central Australia and **b** shows the modeled shape according to Stüwe (1994)

Three-dimensional Problems. Three-dimensional models are not only very difficult to design, their results are also hard to show graphically. For many geological problems – for example those that involve the modeling of stress or strain – three-dimensional models involve tensors algebra, which makes it often hard to follow their results intuitively. Three-dimensional models should therefore only be used if the problem that is to be solved cannot be simplified in its spatial dimensions (s. p. 6).

Examples of important geological problems that are inherently three-dimensional are mantle convection or oblique subduction. Such problems are only solvable with three dimensional models and brave earth scientists use modern methods of calculation and illustration to tackle such difficult problems (e. g. Braun and Beaumont 1995; Platt 1993a). Practically, three-dimensional model results can only be illustrated on a computer screen or with series of contoured diagrams or videos. In this book we refrain from discussing any problems that can only be described using three-dimensional models.

Zero-dimensional Problems. A very large number of geological problems can be solved quite elegantly without considering any spatial dimensions. For

example, the surface elevation of mountain belts in isostatic equilibrium or
the influence of heat production on temperature of rocks can be estimated
without spatial considerations and can still give use enormous insight into the
nature of many tectonic processes. When neglecting all spatial dependence
of a problem, this gives us the great freedom to evaluate the model results
graphically as a function of two variables simultaneously. Both axes that
may be drawn on paper can be used to evaluate the influence of two different
variables against each other.

Zero-dimensional problems should not be confused with *dimensionless
variables*. In many problems of this book we will encounter variables that
are normalized to some standard value and therefore have no units. This is
called a *dimensionless variable*. The main use of dimensionless variables is
that a *single* solution with dimensionless variables can represent *many* solutions of dimensional variables. For example, the elevation of Mt Blanc,
$H_{(MtBlanc)}$, can be either given in meters, or it could be described in terms
of its proportional elevation h (in %) relative to, say, Mt Everest:

$$h = \frac{H_{(MtBlanc)}}{H_{(MtEverest)}} = \frac{4\,807\,\text{m}}{8\,848\,\text{m}} = 0.543 \quad . \tag{1.2}$$

0.543 is the dimensionless elevation of Mt Blanc relative to Mt Everest. The
use of the dimensionless elevation in eq. 1.2 may not be immediately clear.
The next example will illustrate the usefulness of this approach better.

When describing the thermal evolution of contact metamorphic rocks,
(sect. 3.6) we will often encounter dimensionless temperatures of the following
form:

$$\theta = (T - T_\text{b})/(T_\text{i} - T_\text{b}) \tag{1.3}$$

There, T_i and T_b are the temperatures of an intrusion and the background
temperature of the host rock, for example: $T_\text{i} = 900\,^\circ\text{C}$ and $T_\text{b} = 300\,^\circ\text{C}$. T is
the variable temperature which may change as a function of distance from the
intrusion or time. When $T = 600\,^\circ\text{C}$, then this in itself is not very instructive.
However, when expressed as $\theta = 0.5$ we can see that this temperature is exactly
half way between the host rock and the intrusion temperature. For many
questions this is much more instructive.

In other examples we will encounter even more complicated formulations of
dimensionless variables. For example, in diffusion problems it is often useful
to evaluate temperature as a function of the dimensionless variable:

$$T = \text{f}\left(\frac{\kappa t}{l^2}\right) \tag{1.4}$$

This may appear quite confusing, but it has also the purpose of simplifying
the results, similar to what we did in eq. 1.3. Eq. 1.4 shows that the variables
κ (diffusivity), t (time) and l (size) are coupled in the particular form of this
equation ("f()" in eq. 1.4 means "function of"). Using dimensionless variables

is not only useful for the better illustration of the meaning of a result, but is also a great aid in differential calculus.

1.2.1 Reducing Spatial Dimensions

Deformation of lithospheric plates is – in the most general case – a three-dimensional problem (however: s. sect. 2.2.2). In three dimensions, stress, strain and strain rate are described by tensors and any fully three-dimensional description of continental deformation does therefore involve tensor calculations (s. sect. 5.1 and A.3). In order to reduce the complications that arise in such calculations, it is useful to see if the number of considered dimensions can be reduced.

When describing lithospheric deformation, there are various well-established simplifications that allow us to neglect some or even all of the components of a tensor. Of course, whether or not these simplifications should be used depends on the nature of the problem. Two of these simplifications are important enough for plate tectonic modeling so that they are mentioned already here, in this first chapter of the book.

Plane Strain Approximation. The plane strain approximation reduces three-dimensional problems to two dimensions. It assumes that all deformation is strictly two-dimensional so that all strain and displacement occurs in plane and no strain perpendicular to this plane. In plate tectonic modeling we normally consider only the case of no volume change. Then, the total amount of shortening in one spatial direction must be compensated by stretching in the other (Fig. 1.6b; Fig. 1.7). No area change occurs. Tapponier (e. g. 1982) has made great advances in our understanding of continental deformation using this assumption in his descriptions of the India-Asia collision. It should be kept in mind that the plane strain approximation implies that the *normal stresses* on the surface of the plate vary. The plane strain approximation may be viewed as deformation that occurs in a thin film that deforms between two fixed plates. Using this analogy it can be imagined that in the areas where the most deformation occurs, there is also the largest stresses exerted on the two confining plates. It is therefore different from the "thin sheet" or "plane stress" approximation.

Thin Sheet Approximation. When lithospheric shortening in one horizontal direction is compensated both by stretching in the second horizontal direction and by thickening (or thinning) in the vertical, the lithospheric deformation becomes three-dimensional. In order to still be able to deal with this problem in two dimensions the *thin sheet* approximation may be assumed (s. e.g. Houseman and England 1986a; s. however: Braun 1992). The thin sheet model is based on the assumption that the normal stresses at the surfaces of the plate are constant. Thus, the plate may thicken or thin in the vertical direction as to maintain the surface stresses constant. The thin sheet approximation is also called the *plane stress* approximation because

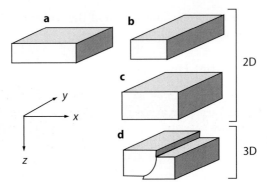

Figure 1.6. Illustration of two- and three-dimensional deformation models (abbreviated with 2D and 3D in the figure). The shortening of the block **a** in the x direction is compensated in **b** only by stretching in the y direction. The thickness of the block remains constant. This is plane strain deformation. In **c** the shortening of the block is compensated by both stretching in the y and z directions. The thickening or stretching in the z direction is homogeneous. This corresponds to the two-dimensional thin sheet approximation. In **d** the shortening in the x direction is compensated by stretching in the y direction and also by inhomogeneous thickening in the z direction. This kind of deformation can only be described with a fully three-dimensional model.

In **b** and **c** the stretching in the y and z directions is no function of x (in contrast to Fig. 1.7). This is no requirement of the plane strain or thin sheet approximations. The illustrated geometries are therefore special cases which could also be described one-dimensionally. Stretching in y and z direction could both be evaluated as a function of the shortening in x only

both stresses and strain rates are averaged in the vertical direction. Because of this assumption there is no vertical strain rate gradient (Fig. 1.6; 1.7) (England and McKenzie 1982, England and Jackson 1989). Using z for the vertical spatial coordinate and $\dot\epsilon$ for strain rate, the thin sheet approximation can be described by:

$$\frac{d\dot\epsilon}{dz} = 0 \quad . \tag{1.5}$$

The thin sheet approximation is a good approximation for the description of lithospheric scale deformation when:

1. The shear stresses at the surface and the base of the lithosphere are negligible.
2. If the topographic gradients at these two surfaces are small.

Both are usually given on the scale of whole lithospheric plates (s. more detailed discussion in sect. 5.3 and 6.2). Note however that this description can only describe homogeneous thickening of the lithosphere. Thus, this thickening strain can be evaluated as a variable without the need for a third spatial dimension.

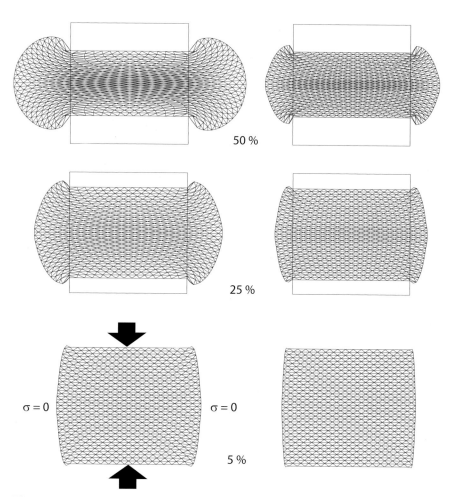

Figure 1.7. A comparison of two-dimensional deformation of a quadratic plate using the plane strain (left row) and thin sheet (right hand row) approximations. The illustrated examples lead to extremely heterogeneous deformation fields and are therefore quite complicated examples. The area of the plates in the left hand row remains constant throughout deformation. The area loss in the right hand row is compensated by thickening in the third dimension. The boundaries along the left and right sides of the plate are stress-free. The top and bottom of the plate are forced to converge at a constant rate (s. sect. 5.3.1 and A.1.1). These boundary conditions are schematically shown in the sketch at the bottom left. The shown time steps are after 5%, 25% and 50% of shortening. The figure was calculated with the computer program BASIL of Houseman, Barr and Evans

2. Plate Tectonics

In this chapter we repeat basic aspects of the theory of plate tectonics. In the first part of the chapter we present a summary of the history of this exiting conceptual model and discuss some basic principles how to describe it. In the second part of this chapter we discuss the layered structure of the earth and the geographic distribution of lithospheric plates. We will also use this chapter to introduce the terminology that is used in the remainder of the book. As such, the chapter is meant as a basis for all following chapters.

2.1 Historical Development

Observations pertaining to the theory of plate tectonics are at least 500 years old. In the late 16th century, Sir Francis Bacon observed that the coast lines of the American and African continents have a matching shape. At Darwin's times the connections between the two continents were already well-established. However, it was only Alfred Wegener who presented the first synthesis explaining these similarities with a theory of plate motion. In part, his synthesis was based on his own observations on the climatological connections between the two continents. While Wegener's synthesis was ultimately proven wrong by the first detailed bathymetric surveys of the oceans in the middle of the past century, his publications (and those of others around his time, e. g. Taylor 1910) are still viewed as the basic foundation of plate tectonic theory (Wegener 1912a,b; 1915). At the time of Wegener, the significance of mid-oceanic ridges and subduction zones were still unknown. However, the deepest point in the world (the Mariana trench with 11.5 km below sea level) and the eastern Pacific rise had already been discovered by the research vessel H.M.S. Challenger around 1875. Also, mantle convection was already established as the driving force for plate motion (Holmes 1929, Griggs 1939). Nevertheless, Wegener had no model for the processes on the ocean floors and thought of plate motion as some "plough-like" motion of the continents through the oceans. From the time of this theory, only the names of the ancient super continents *Gondwana* and *Laurasia* (du Toit 1937) are still being used.

The real break through of plate tectonics came not until the mid-twentiest century when the first detailed bathymetric surveys of the Atlantic were

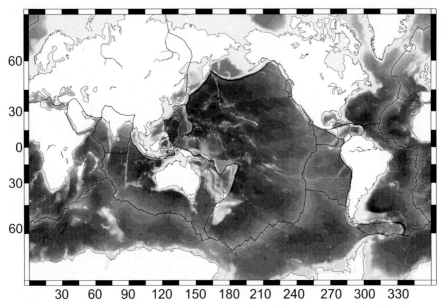

Figure 2.1. Topographic map of the ocean floor. Most regions where the water depth is less than 200 m (nearly white regions) are made of continental lithosphere. Note that in some regions substantial portions of continental lithosphere are actually below sea level, for example east of New Zealand, between North-America and Siberia, in the Mediterranean or east of southern South America

performed around 1950 to 1965 (Heezen 1962, Menard 1964). During these surveys, the gigantic mountain ranges and valleys of the ocean floors were discovered that are now known as mid-oceanic ridges and subduction zones. Earlier on, however, these valleys and ridges were interpreted by all kinds of theories, for example the *expanding earth theory* (Carey 1976, King 1983). However, since the mid-sixties it is well-established that mid-oceanic ridges are areas of lithosphere production, while subduction zones are areas where lithosphere is being consumed (*destructive plate margins*). The chains of volcanoes that produce new oceanic lithosphere along the mid-oceanic ridges had already been predicted by some authors, but they were only discovered by the submarine research vessel Alvin (s. Edmond and Damm 1983). Subduction zones and mid oceanic ridges are in volumetric balance so that there is no need for an expanding earth theory (Hess 1961, Vine and Mathews 1963). Today our knowledge of mid-oceanic ridges and subduction zones is well-established as the basic foundation of plate tectonic theory (sect. 2.4.3) (e. g. Morgan 1968).

The total length of mid-oceanic ridges is of the order of 60 000 km. Most of them are located on the ocean floors (s. sect. 2.4.3). The average rifting rate at these ridges is about 4 cm per year, (Table 2.3) which implies that the total production rate of new surface on earth is about is about 2 km^2 y^{-1}. The total surface of the oceans is about $A_0 = 3.1 \cdot 10^8$ km^2 which means

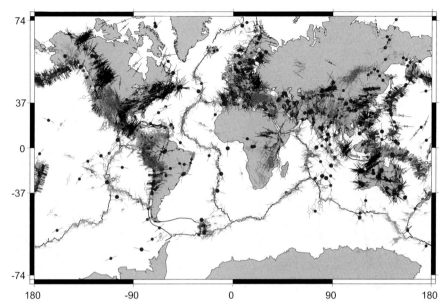

Figure 2.2. The intraplate stress field of the world. The different symbols indicate different methods of determination including earthquake focal mechanisms, borehole breakouts and geological indicators. The different shadings indicate different deformation regimes: darkest are thrust faults, medium gray are normal faults and light shading are strike slip faults. For detailed resolution of this figure see the original CASMO facility on the world stress map home page, which was used to create this map (Mueller et al. 2000; see p. 417)

that all oceanic lithosphere is being renewed about every 155 my. This is a geologically short time span and means that oceanic lithosphere is one of the younger features of this globe. It is therefore ironic that plate tectonic theory – which is now mostly being applied to our very detailed observations in the continents – has its origin *not* on the continents, but at the deepest points of the ocean floors. Cox (1972) has summarized the revolution of plate tectonic theory. He considers the plate tectonic theory to be based on four independent data sets:

- The topographic maps of the ocean floors (Fig. 2.1).
- The magnetic maps of the ocean floors.
- The age dating of the magnetic maps.
- Detailed maps of the epicenters of global earthquakes (Fig. 2.4).

Since the fundamental break through of the sixties, plate tectonic theory has made dramatic and rapid advances. It was soon discovered that many observations can be explained by astoundingly simple physical models, all within the plate tectonic concept. For example, the startling simple quadratic relationship between space- and time scales of diffusion processes (sect. 3.1.4)

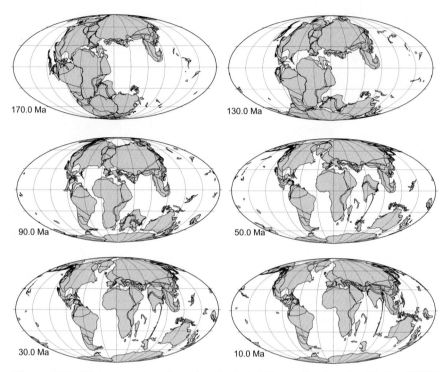

Figure 2.3. Plate reconstructions for the last 170 my. Produced with the ODSN home page (s. p. 415). Maps are in Mollweide projection reconstructed relative to the magnetic reference frame

has now been used to explain an enormous variety of processes including the water depth of the oceans (sect. 4.2.1), the duration of metamorphic events, the magnitude of contact metamorphic aureoles (sect. 3.6.2) or the shape of chemical zoning profiles in minerals (sect. 7.2.2).

Such amazing success of simple physical models has lead in the past 30 years to an unparalleled development of plate tectonic theory. Much of this development has been characterized by the application of simple analytical models to geological problems. Many of these models are introduced in this book.

Recent Developments and Future. The global distribution of earth-quakes shows that *oceanic* lithosphere acts according to the principles of plate tectonic theory: The plates are large and flat and their margins are narrow compared to their size. Seismicity is confined to the rims of these plates. *Continents* behave differently. They are characterized by deformation and seismicity which reaches wide into the continents themselves, their thickness varies dramatically and some seismicity even occurs in their centers. Their plate boundaries are diffuse. It has therefore been suggested to use the term "cheese tectonics" as a superceding term for "plate tectonics", because conti-

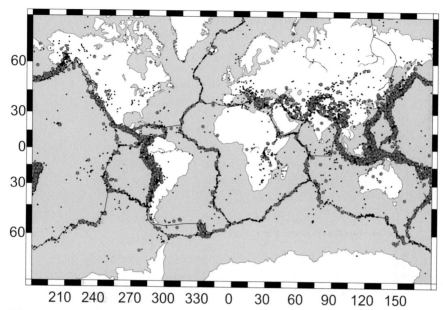

Figure 2.4. Global seismicity. All recorded earthquakes above magnitude 6 since 1973 are shown. Note that most earthquakes along oceanic plate boundaries (e.g. the mid-Atlantic ridge) occur very near the plate boundary, while earthquakes associated with plate boundaries in continental lithosphere cover much more diffuse areas (e.g. the India-Asia collision zone). See Fig. 2.17 for more details about the plates

nents seem to deform rather like soft camembert than like rigid plates. Like many good jokes this one has a true core. It has been known since the turn of the century that rocks behave like fluids on geological time scales (s. summary by Gordon 1965 or England 1996). However, only in the past 20 years geologists have begun to actually use the theory of deformation of viscoelastic materials to describe the dynamic evolution of continents (England and McKenzie 1982, England and Jackson 1989). In part these new descriptions were triggered by studies like those of Goetze, (1978) or Brace and Kohlstedt, (1980) who provided us with the first simple models of lithospheric rheologies.

Because of such models it has now been possible to describe largest scale tectonic processes with simple dynamic models. Even coupled thermomechanical approaches can now be performed with startling simplicity (e.g. Sonder and England 1986) and have found their use even by non-geophysically oriented earth scientists.

The current development of plate tectonics is going more and more towards the use of numerical models and further away from simple analytical models. Digital data sets of global observations make it now possible to tackle problems that can only be solved using large numbers of data. Such models can now be used to explain problems that go beyond single observations and pertain to the whole globe. The future will surely be characterized by an increasing use of large data sets and numerical models.

2.2 Working on a Spherical Surface

The earth is nearly a sphere (s. sect. 4.0.1) and many aspects of the geometry and the mechanics on a sphere are different from its Cartesian equivalent. In this section we discuss some aspects of spherical coordinates that may need to be considered when solving geodynamic problems on very large scales. We also discuss some ball park estimates that can be used to judge whether it is sufficient to neglect the spherical geometry.

2.2.1 ...or is the Earth Flat After All?

By far the largest number of geodynamic problems can be described assuming a flat earth. Fig. 2.5 shows that over a line one thousand kilometers in length, the curvature of the earth causes a 20 km deviation from a straight line. This is less than 4% of the extent of the feature. For geological features of some tens of kilometers extent, the deviation is only some tens of meters, which is about 0.1% of the extent of the feature. Thus, only for observations on the largest scale, the spherical geometry must be considered.

One of the most classical examples of problems that should not be described on a flat earth is the shape of long subduction zones along the surface of the earth. On a flat earth, the trace of a subduction zone should be linear. It should be as straight as the line that is formed by the points of maximum curvature of a sheet of paper hanging off the edge of a table.

However, many subduction zones are curved along the surface of earth. This is often explained with the ping pong ball model. This model compares the curvature of trenches on the surface of the earth with the curvature of the indentation edge on a dented ping pong ball. If the indented part of a ping pong ball is not deformed in itself, then this edge forms a small-circle on

Figure 2.5. The difference between a flat and a curved surface of the earth. The maximum deviation of a curved surface from a flat surface, H, is given by $H = R - \sqrt{R^2 - (r/2)^2}$. For $r = 1\,000$ km as shown here and the value for R from Table C.3 the deviation is $H = 19.6$ km

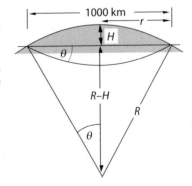

Figure 2.6. Six plates on a flat surface. The relative motions of some plates are shown by the arrows. However, the relative motions of plates C and E, D and E, B and E, B and F as well as that between A and F are completely unconstrained by the shown relative motions

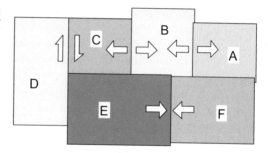

the surface of the ball. Exactly this is observed in subduction zones. In fact, the model can be used to predict the subduction angle ($2\ \theta$), which should be given (according to Fig. 2.5) by:

$$\sin(\theta) = \frac{r}{R} \tag{2.1}$$

where r is the small circle radius and R is the radius of earth. Most of the small circle radii of subduction zones on earth correspond well with the subduction angle predicted by eq. 2.1 (s. Isacks and Barazangi 1977). However, the subduction angles may also depend on a large number of other parameters, for example whether subduction occurs *in* or *against* the direction of convection in the asthenosphere (Doglioni, 1993).

Geometrical problems can be described on a flat earth using the familiar Cartesian coordinate system. In this book we use x and y for the coordinates oriented parallel to the surface of the earth and z for the vertical direction. In section 2.3 we discuss the relationship between those coordinates and those of other coordinate systems. In section 4.0.1 we discuss some confusions that may arise from mixing different reference frames.

Kinematic and *dynamic* problems on a flat earth can be described by motion and forces in the horizontal and vertical directions. Velocities and forces are described by vectors. This means that they have a direction and a magnitude and can be split up into vector components that are parallel to the axes of a Cartesian coordinate system. For combined paths we can therefore use the rues of vector algebra. Be careful to note, however, that all motions (and forces) are relative! This is important to realize as an observed motion in *one* place of the globe must not imply that this motion is elsewhere the same (for example China is moving towards Tibet, but the region *between* China and Tibet is under extension). Fig. 2.6 illustrates some more examples. All this is quite different on a spherical surface, which we deal with in section 2.2.3.

2.2.2 Geometry on a Sphere

On a spherical surface the position of a point is described by its longitude ϕ, and latitude λ (Fig. 2.7). As with time, spherical geometry is one of the few branches in science where the duo-decimal system is still in use. Correspondingly, a right angle has 90 degrees and longitude and latitude around the globe are divided into 360 degrees. (The use of 100 degrees for a right angle was attempted by the introduction of "new degrees" but has not found footing in science). Every degree of longitude is described by a great circle which goes through the geographic poles. These great circles are called *Meridians*. Great circles are lines on the surface of a sphere that are defined by the intersection of a planar surface through the center of the sphere, with the surface of that sphere. Meridians are therefore a special kind of great circle, namely one that goes through the poles. Small circles are defined as intersections of all other planar surfaces with the surface of a sphere. 180 of the 360 Meridians are numbered west of Greenwich and the other 180 east of Greenwich, which has been internationally agreed upon to be the reference for longitude. Each degree of latitude is defined by a small circle parallel to the Equator and at right angles to the axis that connects the poles. 90 degrees of latitude are north of the equator and 90 are south. Note that there is a total of 360 Meridians, but only 180 degrees of latitude. The spacing of the degrees of latitude is chosen so that they divide the Meridians into 360 sections of equal length. Thus, the distance (along the surface of the earth) between degrees of latitude is constant everywhere on the globe, while the distance between degrees of longitude is largest at the equator and zero at the poles. For more detailed description of locations on a spherical surface, every degree is divided into 60 arc minutes and every arc minute into 60 arc seconds. Just to make things worse, the duo-decimal system is often coupled with the decimal system: Geographic locations are often described by degrees and decimals. That is, tenth and hundredths of degrees are given, rather than arc minutes and arc seconds.

Other important lines on spherical surfaces are *rhumb lines* (also called *loxodromes*). These are lines that intersect degrees of latitude and longitude at constant angles. Rhumb lines are easy to follow, for example when setting constant course on a ship, but they form spiral-shaped curves on a sphere and they are *not* the shortest connection between two points (Fig. 2.7). The angle between magnetic north and the lines of longitude (geographical north) is called the magnetic declination. The vertical angle between the normal to the Geoid surface and the magnetic field lines is called the magnetic inclination. The circumference of a great circle on earth is about $2R\pi \approx 40\,000$ km. (If it were exactly $40\,000$ km, then the radius of the earth would have to be $R = 6\,366.2$ km; in reality the equatorial radius is $6\,378.139$ km and the polar radius is $6\,356.75$ km). In fact, one meter was long defined as the $1/40\,000\,000$ part of the circumference of earth. One degree of longitude at the equator (and *all* degrees of latitude) is therefore about $40\,000/360 \approx 111$ km. On small circles north and south of the equator, the distance between full degrees of longitude, l, decreases with the cosine of the latitude:

$$l \approx \cos(\lambda) \cdot 111 \quad . \tag{2.2}$$

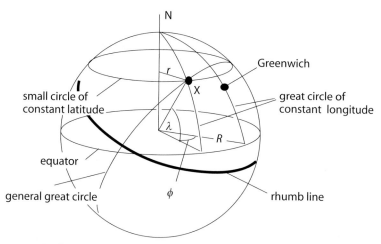

Figure 2.7. Definitions of important lines and angles on a spherical surface. The geographic longitude ϕ of point X is *west* of Greenwich

In eq. 2.2 we have used the approximate value for one degree of longitude at the equator. Correspondingly, the small circle radius of each small circle of constant latitude shrinks with the cosine of the latitude: $r = \cos(\lambda) \times R$, where R is the radius of earth. One arc minute of latitude is defined as one nautical mile which is ≈ 1.8 km. Along the equator, distances between degrees of longitude and latitude are of equal length. All distances and angles on a spherical earth can be calculated with simple combinations of the trigonometric functions. Throughout the book it is assumed that the use of those, as well as some sound ability of spatial imagination is familiar to the reader. As a reminder, some of the relationships we will need are summarized in Appendix B, Tables B.4, B.6, B.7, and Fig. B.1.

2.2.3 Kinematics on a Sphere

On a flat surface, velocity v and speed have the units of m s^{-1}. Velocity is a vector and speed a scalar quantity. For example, the Indian plate has a *speed* of 0.05 m s^{-1}, but a *velocity* of 0.05 m s^{-1} *moving north*. The equivalent to velocity on a spherical surface is the angular velocity w. w has the units of radian per time, which is s^{-1}. The axis that is perpendicular to the planar surface swept over by angular motion is called the pole of rotation or *rotation pole* (Fig. 2.8). The velocity that corresponds to a given angular velocity depends on the distance of the angular motion from the pole of rotation. Acceleration in a straight line is the change of velocity over time and has the units m s^{-2}. Correspondingly, the angular acceleration has the units of s^{-2}. The differences in units between linear velocity and angular velocity has lead to a lot of confusion in the literature. However, they are extremely

Figure 2.8. Illustration showing the meaning of rotation poles. The arrows are vectors showing the direction and magnitude of relative motion of the two plates (shaded regions). The thick line connecting the arrow heads is the new plate margin after some time. The axis of the earth is only shown to emphasize that it has nothing to do with the rotation pole of plate motion

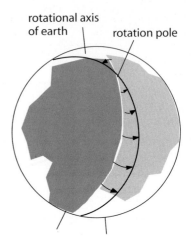

important differences! For example, a *constant* rate of plate motion with a constant angular velocity will cause differences in the rates of relative plate motions along the plate margin. The relative plate motion rate depends on the small circle radius of the velocity vector (Fig. 2.8). In fact, even qualitative changes from divergent plate motion to convergent plate motion may occur along a plate margin because of this (compare Fig. 2.6 and 2.8).

2.2.4 Mechanics on a Sphere

Plate tectonic forces are often described in the literature as "torques". For example, ridge "push" is a force, while many authors rather use the term ridge "torque". Strictly speaking, we should only use torques to understand the mechanics that cause plate motion on the earth's surface as plates do not move in a linear direction but rather around a rotation pole (the center of earth). In calculating a force or torque balance of a mountain belt, where every point in the belt is virtually the same distance from the pole of rotation, the distance to the rotation pole cancels out and torque balances and force balances are practically equivalent (s. p. 203).

Force F is given in Newtons [N] and: $1\,\mathrm{N} = 1\,\mathrm{kg\,m\,s}^{-2}$. Force is a vector with a magnitude and direction. Horizontal forces are therefore tangential to the globe. The equivalent on a spherical surface is torque. Torque (which is different from angular momentum!) is the turning moment which is exerted by a force about an axis. It is given by the product of force and the distance from the axis about which the torque acts. Torque has the units of Nm or $\mathrm{kg\,m}^2\,\mathrm{s}^{-2}$. A force of 10^{12} Newton that acts in direction of a great circle on the earth's surface, corresponds to a torque of $6.37 \cdot 10^{18}$ Nm. The torque changes along a plate margin, as the normal distance of the plate margin to the rotation pole changes. The units of torque can be read as "Newton times meter of leverage", where the "meters of leverage" are the normal distance to the rotation pole. In the literature, "forces" are often given in Newtons per

Table 2.1. Important kinematical and mechanical parameters and their units. Each parameter is given with both, the linear and the spherical equivalents

physical parameter	unit
velocity	$\mathrm{m\,s^{-1}}$
angular velocity	$\mathrm{s^{-1}}$
acceleration	$\mathrm{m\,s^{-2}}$
angular acceleration	$\mathrm{s^{-2}}$
force	$\mathrm{kg\,m\,s^{-2}}$
torque	$\mathrm{kg\,m^2\,s^{-2}}$
mass	kg
moment of inertia	$\mathrm{kg\,m^2}$
linear momentum	$\mathrm{kg\,m\,s^{-1}}$
angular momentum	$\mathrm{kg\,m^2\,s^{-1}}$

meter. It is important not to confuse this with torques, which have the units of Newtons times meters. Such similarities in units can cause confusion.

Force = mass × acceleration ($F = m \times dv/dt$) and similarly torque = mass × angular acceleration. In plate tectonics the changes in velocity and angular velocity occur over very long time periods, so that accelerations and angular acceleration are negligible (s. p. 244). Thus, the common assumption is that the sum of the torques or the net torque acting on a plate is zero or, correspondingly, that the sum of the forces or net force acting on a smaller region such as a mountain belt is zero.

Aside from force and torque, there are some other important mechanical parameters that we will need in this book. The linear momentum I is the product of mass and velocity: $I = mv = \mathrm{kg\,m\,s^{-1}}$. Just as momentum is given by $I = mv$, the *change* of momentum is given by: $\Delta I = m\Delta v$, mass times the *change* in velocity. Considering that force has the units of the product of mass and acceleration (the change in velocity) we can write that: $F = mdv/dt$. Thus, the change of momentum is: $\Delta I = Ft$, force acting over a given time. On a spherical surface, the angular momentum is analogous to the linear momentum. The angular momentum D is the product of the moment of inertia J and the angular velocity w: $D = Jw$. The moment of inertia is the ratio of torque and angular velocity and has the units of $J = \mathrm{kg\,m^2}$. Angular momentum has the units of $D = \mathrm{kg\,m^2\,s^{-1}}$. In plate tectonics, changes in momentum and angular momentum are ignored because changes in velocity and angular velocity of plates takes place over very large time periods (s. sect. 5.3.3). In most problems then, the net forces or torques acting on a given plate must balance or add up to zero (s. p. 203). However, momentum

and angular momentum help to understand why forces *or* velocities can form
the boundary conditions for orogenic processes.

2.3 Map Projections

Map projections are mathematical or geometrical models that portray the
features of the spherical surface of our globe on the two dimensions of paper.
The map projection model serves the purpose to make these features easier
to look at and pays for that by accepting some distortion. Usually, the process of projection involves the conversion of latitude λ and longitude ϕ into
Cartesian coordinates (Robinson et al. 1984). Map projections have become
an increasingly important part of geology since global data sets are used. Map
projections can be performed geometrically or they can be purely mathematical, without any apparent geometrical equivalent. Among projections *with* a
geometrical equivalent we discern three important types (Fig. 2.9):

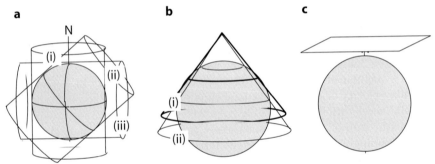

Figure 2.9. Schematic illustration of three examples of commonly used geometrical
map projections. **a** *Cylindrical projection* with examples of (i) normal or equatorial,
(ii) transverse and (iii) oblique orientations of the projection surface; **b** *conical
projection* with examples of (i) tangential orientation (touching the sphere along
the thick line) and (ii) secant orientation (intersecting the globe) of the projection
surface; **c** *Azimuth* projection with polar orientation of the projection surface

- Cylindrical projections,
- conical projections,
- azimuth projections.

These three types of projections can be imagined as projections of the earth's
surface from an imaginary light source (usually assumed to be located at the
center of earth) onto some enveloping surface which is then rolled open to
form a planar surface. In these three projections, the enveloping surface is
a cylinder, a cone or a planar surface. However, there is several hundreds of
known projections that serve a large variety of purposes. Most of these are

a

b

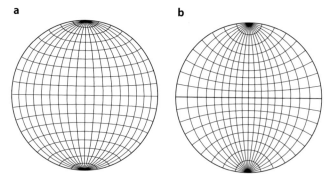

Figure 2.10. a The equal-area net by Schmidt. **b** The conformal net of Wulff

not geometrical but are only described by a *projection function* that relates spherical and planar coordinates. Just about all map projections are either:

– Conformal (orthomorphic) projections,
– equivalent (equal-area) projections,
– azimuthal projections.

Conformal projections represent angular relationships the way they are on the spherical surface. *Equivalent* projections render the same areas as those on the spherical surface. *Azimuthal* projections retain azimuthality, that is, the directional relationships along great circle bearings are the same on the map as they are on the sphere. No projections on two-dimensional paper can retain more than one of these relationships from a sphere at the same time. However, many projections aim to find a useful compromise between different relationships, on the expense of being neither true in angle nor true in area or great circle bearing. Most geologists are familiar with at least one conformal and one equivalent projection, even if they have never used maps of the globe. These are the nets of Schmidt and Wulff that are used by crystallographers for the presentation of angular relationships of crystal faces and by structural geologists for the equivalent illustration of planar structures, respectively (Fig. 2.10).

In the following, we discuss the Schmidt and the Wulff projection and some important others. We confine our discussion to projections that can be used with the software GMT and are described in detail there (Wessel and Smith, 1995ab). A large number of other projections are explained by Snyder (1987) and Snyder and Voxland (1989). All their projections are usable with the software PROJ by Evenden (1990) and there is great overlap with the projections incorporated in GMT.

Cylindrical Projections. The common feature of all cylindrical projections is that the projecting surface is a cylinder which is assumed to envelop the earth. *Pseudocylindrical projections* have no direct geometrical equivalent to cylindrical projections, but they share the feature that lines of constant latitude appear on them as parallel straight lines (if viewed from the equator).

● *Mercator and other cylindrical projections.* The *Mercator projection* is the most commonly used projection for maps of the whole world. It was first used by G. Mercator in 1569 for navigation on the oceans (Fig. 2.11a). For this projection, the projection surface is a cylinder a cylinder that touches the globe along the equator and has its axis therefore parallel to the axis of earth (Fig. 2.9a). Every point of the earth's surface is projected onto this cylinder from an imaginary light source at the center of earth. The imaginary cylinder is then rolled open. The map is ready! There is two great advantages of this projection:

1. There is *no* distortion along the equator and the distortion in equator-near regions is negligible.
2. Lines of constant longitude, latitude and rhumb lines appear as straight lines.

Because rhumb lines are easy to navigate along, Mercator projections used to be common for shipping (before navigation software made more direct routing much easier). However, note that the shortest connection between two points on the globe – a great circle – appears on the Mercator projection as a curved line. The biggest disadvantage of the Mercator projection is that the areal distortion increases with latitude. Areas near the poles appear too large and the poles themselves cannot be represented at all. For example, Greenland appears to be larger than south America although it has only an eight part of its size.

This problem can be circumvented by using the *transversal Mercator projection*. There, the cylinder touches the globe along one of the Meridians and its axis is at right angles to the axis of the earth (Lambert, 1772). Accordingly, there is no distortion along this Meridian and the distortion goes towards infinity on the equatorial points of the Meridian that is 90° from the Meridian where the cylinder touches. Both lines of constant longitude *and* latitude are curved on this projection.

The *universal-transversal-Mercator projection (UTM)* is defined by a total of 60 different Meridians (all separated by 6 degrees of longitude) all of which are touching Meridians for 60 assumed projection cylinders. In other words, it is a combination of 60 transversal-Mercator projections. This avoids serious distortion of any point on the surface of the globe.

The *Cassini-projection* (or Cassini-Soldner projection) is also a cylindrical projection. It is neither equivalent nor conformal, but it forms a good compromise. As for the transversal-Mercator projection, there is no distortion along a central Meridian. Meridians that are 90° from this, appear as straight lines and so does the equator. This projection is useful for areas with a large north-south extent.

The *Plate-Carree projection* looks a bit like a Mercator projection, but the distances between lines of constant latitude are forced to be constant. This avoids (forcefully) the distortion of the Mercator projection near the poles, but it leads to a number of other distortions that have no geometrical equivalent.

Conical Projections. In conical projections, the earth's surface is projected onto a cone, which can then be cut and rolled open like a cylinder. In common conical projections a cone is used which has its axis parallel to that of the globe (Fig. 2.9 b). The cone can touch the globe (tangential orientation), or it can intersect it (secant orientation). Thus, the cone can be defined by one or two lines of longitude where it touches or intersects the surface of the globe, respectively. On those lines the projection has no distortion. Everywhere else, areas appear too *large*, except in secant orientations of conical projections, where areas appears too *small* in the cone section that lies inside the globe, i.e. between the two latitudes of intersection.

The two most important conical projections are the *Albers equivalent projection* (Albers, 1805), and the *Lambert conformal projection* (Lambert, 1772). However, there is also a perspective-conical projection, as well as a series of poly-conical projections. All these have in common that a series of cones with a series of touching lines on the globe are assumed, or that some scaling factor is assumed in addition to the cone-projection. Many poly-conical projections that are used to portray the whole of the globe have heart-shaped outlines.

Azimuthal Projections. Azimuthal projections are direct projections onto a planar surface. This planar surface may be assumed to touch the globe, or it may penetrate it. The net of Schmidt is an azimuthal projection where the projection surface penetrates the earth along the equator. Azimuthal projections are common to portray the areas near the poles as a supplement for Mercator projections. Fig. 2.11 shows four examples of azimuthal projections with circular outlines. Of those, only the equi-distant azimuthal projection (Fig. 2.11b) is useful to portray the entire globe. On the other three, only half of the globe can be portrayed. *Lambert's equivalent projection* (Fig. 2.11c) is familiar to us from the net of Schmidt. However, on Fig. 2.10a the lines of longitude and latitude are distorted relative to the net of Schmidt from Fig. 2.10a because the projection plane is not assumed to be parallel to the equator. The *stereographic projection* (Fig. 2.11d) corresponds to the net of Wulff. Again, the difference to Fig. 2.10b arises from the inclined orientation of the assumed projection plane. Lines of longitude and latitude appear therefore here as strange curves. The *orthographic projection* (Fig. 2.11a) is neither equivalent nor conformal.

Other Map Projections. When we look at the walls in travel agencies, we encounter a huge variety of world maps on which the whole globe fits nicely into an ellipse, into a rectangle, into a circle or into other weird and wonderful shapes (Fig. 2.11, 2.12). Most of these map projections are projections *without* geometrical equivalent. They only use some mathematical function that relates the spherical coordinates to planar coordinates.

The *Robinson projection* is a pseudocylindrical projection and serves the purpose "to look right" (Fig. 2.11b). In truth it is neither conformal nor

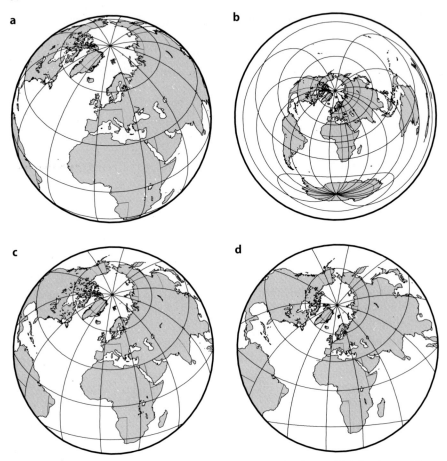

Figure 2.11. Four examples of azimuthal map projections with circular outlines.
a orthographic projection, **b** equidistant azimuthal projection, **c** Lambert's equal
area projection and **d** stereographic projection. All shown examples are for inclined
orientations of the rotation axis of earth relative to the projection direction. If **c**
and **d** were projected parallel to the axis of earth, then lines of constant longitude
and latitude in **c** would correspond to those of Fig. 2.10a and lines of constant
longitude and latitude in **d** would correspond to those of Fig. 2.10b

equivalent. No point of the map is without distortion. The *Winkel Tripel
projection* is similar in that it is nowhere without distortion (Fig. 2.11c). The
Eckert IV projection is a pseudocylindrical equivalent projection (Fig. 2.11d)
which is common for Atlases. The *Mollweide projection* (Mollweide, 1805) is
a pseudocylindrical equivalent projection, in which longitudinal lines appear
as straight lines and Meridians as parts of ellipses with equal distances to
each other (Fig. 2.11e). The *sine projection* can also be used for the whole
globe, but it is often used only for South America and Africa (Fig. 2.11f). It is
equivalent. *Split versions* of the sine-projection (Fig. 2.11h) help to eliminate

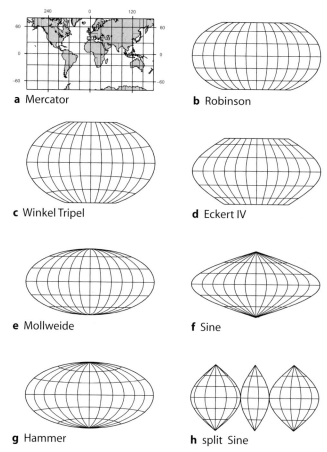

Figure 2.12. Eight examples of projections commonly used to portray the entire globe.

much of the distortion and are common for cut out paper models of the globe. The *Hammer projection* or *Hammer-Aitoff projection* is an equivalent projection in which the circumference is an ellipse and the equator, as well as a central Meridian, are straight lines.

2.4 The Layered Structure of Earth

The layered structure of earth can be considered based on:

− different materials,
− different physical properties.

When considering the layers made of *different materials*, there is three: a crust, a mantle and a core (layers above the surface of solid earth like the

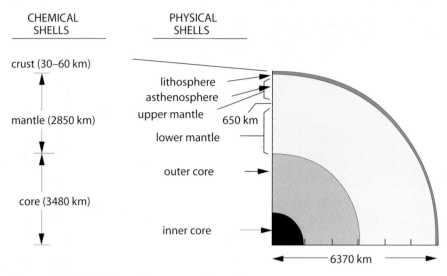

Figure 2.13. The layered structure of earth

hydrosphere, biosphere or atmosphere are not considered here). The crust
is the uppermost layer and is between some 5 to 7 and 30 km thick, depending on whether we deal with oceanic or continental crust. In regions of
deformed continental crust – for example underneath the Tibetan plateau
or the Alps – continental crust can get up to 80 km thick. Chemically, the
crust is highly-differentiated and very heterogeneous, but many of its mechanical and physical properties (e. g. density, conductivity or rheology) can
be well-approximated with those of quartz. The mantle is largely made up
of olivine and – at larger depths – its high-pressure breakdown products.
The seismically clearly visible contact between crust and mantle is called the
Mohorovičić-discontinuity (short: Moho). From the Moho the mantle reaches
down to a depth of about 2 900 km. The core consists mainly of iron and
nickel.

When considering the *physical properties*, the layered structure is quite
different. Then, the outermost layers of earth are the lithosphere and the
asthenosphere. The lithosphere is solid and acts like rock on geological time
scales. Therefore its name. It involves both a crustal and a mantle part.
The asthenosphere consists of the soft mantle that underlies the lithosphere.
Some authors call the entire upper mantle underneath the lithosphere the
asthenosphere. Others use the term only for the mantle section that lies above
the point where the adiabatic melting curve comes nearest to the temperature
profile (Fig. 3.9). According to Ringwood (1988) the mantle can be divided
into three zones:

1. The upper mantle, which reaches down to about 400 km and is characterized by a seismic p-wave velocity of about 8.1 $km\,s^{-1}$.
2. A transition zone from about 400 km to the 650 km discontinuity.

Figure 2.14. Nomenclature of different parts of the outer shells of the earth

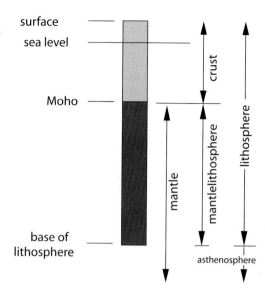

3. The lower mantle which reaches from the 650 km discontinuity to the core-mantle boundary at 2 900 km depth.

Below that is the core. The outer core is liquid and the inner core is solid.

2.4.1 Crust and Lithosphere

The lithosphere is the outer solid layer of the earth (s. sect. 3.4). As for the whole globe, the lithosphere can be divided according to its physical properties or according to its chemical (material) properties. Because there is overlap between layers distinguished on the basis of different properties it is crucial for the understanding of this book to be familiar with the nomenclature illustrated in Fig. 2.14. However, for many geodynamic questions it is not necessary to consider these subdivisions, as the lithosphere acts as a whole.

When considering its chemical properties, the lithosphere consists of a crust and a mantle part. The crust consists of highly-differentiated partial melts from the mantle. The mantle part of the lithosphere is largely made up of the similar material to that of the underlying asthenospheric mantle, but it acts like a solid, because of its lower temperature. However, we note that chemical differences between the mantle lithosphere and the underlying asthenosphere do exist and account for example for unusually thick, but apparently mechanically stable mantle lithosphere, underneath southern Africa. Modern research has been able to document much detail of the compositional variation within the uppermost mantle, both on chemical grounds (e.g. Mc-Donough and Rudnick 1998) and based on seismic velocities (Jordan 1981,

1989). Nevertheless we take in this book the simple-minded view that density variations between mantle lithosphere and asthenosphere may be largely attributed to differences in temperature (e.g. on p. 154).

A schematic but characteristic thermal and density profile of the lithosphere is shown in Fig. 2.15. A very large number of the geodynamic processes discussed in this book are a function of the fundamental shape of the curves on this figure.

Within the crust the temperature profile is curved, because of radioactive heat production. Within the mantle part of the lithosphere, the thermal profile is linear (in a steady state). The base of the thermal lithosphere is defined by the point where the temperature profile intersects the $1\,200°C$ or $1\,300°C$ isotherm (sect. 3.4). At higher temperatures, mantle material begins to flow rapidly on geological time scales and any temperature gradients will be eliminated by convection. Thus, temperature and density are constant below the depth z_l on the scale of Fig. 2.11. Both curves of Fig. 2.15 will be useful help throughout this book.

Definition of the Lithosphere. The term "lithosphere" comes from the Greek *lithos* = rock) and was first used by Barrell (1914). The term was later defined by Isacks et al. (1968) as a *near surface layer of strength* of earth. Even today it remains difficult to find a more precise definition than this. Most of the physical parameters, for example temperature or density, change continuously underneath the Moho and the transition from the rigid outer shell of the earth (the mechanical boundary layer) into the more viscous hot asthenosphere (from the Greek *asthenia* = soft) is also continuous. This transition zone is called *thermal boundary layer*) (Fig. 2.16; sect. 6.3.2; Parsons and McKenzie 1978, McKenzie and Bickle 1988). However, even on the definition of the term "thermal boundary layer" there is no clear consensus in the literature. Some authors refer with this term only to the transition zone between lithosphere and asthenosphere and others to the entire lithosphere as being a thermal boundary layer to earth (s. p. 81, p. 59 and Fig. 2.16).

One thing can be said with certainty. The definition of the lithosphere depends on the question that is being asked. For example, it can be shown that the thickness of the lithosphere is a function of the observed time scale. Seismic motion, isostatic uplift and ductile deformation occur on time scales of seconds, 10^4 y and $> 10^6$ y, respectively. The larger the time scale of the process, the smaller the thickness of the lithosphere. Seismically, the lithosphere is of the order of 200 km thick, while the elastic thickness of the lithosphere is only some tens of kilometers thick. Very generally the lithosphere may be defined *mechanically* as the outer part of the earth in which stresses can be transmitted on geological time scales (s. McKenzie 1967). According to a somewhat different mechanical definition the thickness of continental lithosphere may be defined as that thickness that is in isostatic equilibrium with the mid-oceanic ridges (Cochran 1982). This is meaningful, because the mid-oceanic ridges may be interpreted as manometers of the upper mantle (s. p. 156, Turcotte et al. 1977).

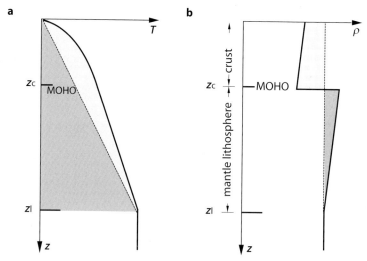

Figure 2.15. a Temperature and **b** density of the continental lithosphere as a function of depth. The depth of the Moho below surface is z_c, that of the whole lithosphere is z_l. **a** The curvature of the geotherm within the continental crust is caused by radioactive heat production. The light shaded area corresponds to the heat content of the lithosphere that can be attributed to radioactive decay in the crust. The dark shaded region corresponds to the heat content conducted into the lithosphere from the asthenosphere. The relative contributions of the radioactive and the mantle heat flow will be discussed in sect. 3.4.1 and sect. 6.2.1. However, it may be seen here, that the heat content of the crust consists to roughly equal part of mantle heat and of radioactive heat. **b** The slope of the density profile within the crust and within the mantle lithosphere is a function of the thermal expansion. A comparison of the shaded areas shows that the density deficiency in the crust (light shaded area) is comparable to the density excess in the mantle lithosphere (dark shaded area)- both relative to the asthenosphere. Within the asthenosphere convective flow equalizes all density and temperature heterogeneities. Both curve are therefore vertical

According to a *thermal* definition the lithosphere is the part of earth in which thermal energy is largely transferred by heat conduction, in contrast to the asthenosphere, where heat is transferred by convection (s. sect. 3.4 for more detail). In some ways the thermal definition encompasses the mechanical definition because many of the mechanical properties of rocks depend on the ratio of their temperature to their melting temperature. In stable continental lithosphere, thermal and mechanical definitions indicate thicknesses of 100–150 km. The thickness of the crust and its content in radioactive minerals is crucial to the thickness of the lithosphere, because they strongly influence the Moho-temperature (s. Fig. 3.18). Many studies that are concerned with the rheology of the lithosphere avoid to use a number for lithospheric thickness. Rather, the lithosphere is defined via the Moho-temperature (s. sect. 6.2.2).

Figure 2.16. Definition of the mechanical and thermal boundary layer as well as of the lithosphere and asthenosphere according to Parsons and McKenzie (1978) and McKenzie and Bickle (1988). The thick line is the geotherm. Its curvature in the upper crust was ignored. However, note that other definitions for *thermal boundary layer* are also in use in the literature s. p. 81, p. 59)

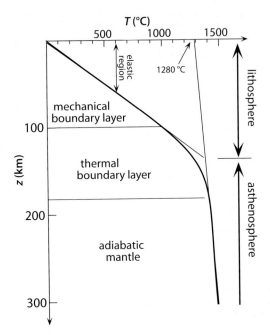

Types of Lithosphere. There is two fundamentally different types of lithosphere on earth: *oceanic* and *continental* lithosphere. Despite these names, the correlation of oceanic lithosphere with the geographic area of the ocean and vice versa is only very approximate (s. Fig. 2.1).

- *Oceanic lithosphere.* Oceanic lithosphere begins its life at the mid oceanic ridges. There, it consists only of an about 7 km thick oceanic crust, which is made up of crystallized partial melts from the uppermost mantle. The thickness of the mantle part of the oceanic lithosphere is zero near the mid-oceanic ridges. With increasing age – that is: with increasing distance from the ridge – the thickness of the mantle part of the oceanic lithosphere increases as the asthenosphere successively freezes to the base of the cold crust. In the oldest parts of known oceanic lithosphere the thickness of the oceanic mantle lithosphere is about as thick as continental mantle lithosphere. However, oceanic lithosphere is being produced and consumed at all times, so that there is hardly any oceanic lithosphere on earth that is much older than about 100 my (s. p. 2.1).

- *Continental lithosphere.* In contrast, the total area of continental lithosphere has remained largely constant in the entire Phanerozoic. Thus, the present day continents consist largely of Proterozoic continental lithosphere, which has been reworked in many places. Continental crust is chemically highly-differentiated, it has a high content in radioactive elements and in its stable state it is about 30–50 km thick. According to thermal and mechanical

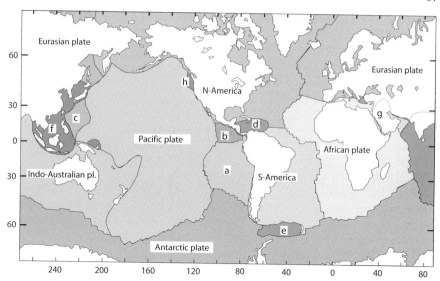

Figure 2.17. Plate tectonic division of the surface of earth. Continents are white and oceans are shaded. The difference between continental and oceanic lithosphere is not shown. For that, see Fig. 2.1. Note that the plate boundaries coincide only in a few places with the coast lines that delineate the continents (s. Table 2.2). The seven major plates are labeled with their names. The most important minor plates are labeled with letters. They are: *a* Nazca plate; *b* Cocos plate; *c* Philippine plate; *d* Caribbean plate; *e* Scotia plate; *f* Chinese subplate; *g* Arabic plate; *h* Juan-de-Fuca plate

definitions, the mantle part of the lithosphere is 70–100 km thick, so that the total thickness of stable continental lithosphere is of the order of 100–150 km. However, in old shield areas this thickness can be much more, probably due to a different chemical composition of the mantle lithosphere.

2.4.2 The Lithospheric Plates

The surface of the earth can be divided into seven major lithospheric plates plus a number of smaller plates (Fig. 2.17, Table 2.2). Not all major plates correspond to the seven continents and it is of some coincidence that the number of continents equals that of the major plates. Most major plates consist of both continental *and* oceanic lithosphere.

Among the seven major plates, the Antarctic plate and the African plate have a special position as they are surrounded on just about all sides by mid-oceanic ridges. Both plates increase therefore permanently in size and they form a good example where plates go from a compressive state into an extensional state purely as a function of their increasing age (Sandiford and Coblentz 1994; sect. 5.3.1).

Table 2.2. Approximate parts of oceanic and continental lithosphere of each of the major plates

plate	% oceanic lith.	% continental lith.
major plates		
Pacific plate	100	0
North-American plate	30	70
South-American plate	50	50
Eurasian plate	30	70
Antarctic plate	50	50
African plate	50	50
Indo-Australian plate	40	60
important minor plates		
Nazca plate	100	0
Cocos plate	100	0
Juan-de-Fuca plate	100	0
Scotia plate	100	0
Philippine plate	100	0
Caribbean plate	100	0
Arabic plate	10	90

Table 2.3. The twelve most important relative motions of plates (after DeMets et al. 1990). 1° corresponds to about 110 km

plate boundary	rotation pole longitude	latitude	angular velocity $\cdot 10^{-7\circ}/y$
Africa − Antarctica	5.6°N	39.2°W	1.3
Africa − Eurasia	21.0°N	20.6°W	1.3
Africa − North-America	78.8°N	38.3°E	2.5
Africa − South-America	62.5°N	39.4°W	3.2
Australia − Antarctica	13.2°N	38.2°E	6.8
Pacific − Antarctica	64.3°S	96.0°E	9.1
South-America − Antarctica	86.4°S	139.3°E	2.7
India − Eurasia	24.4°N	17.7°E	5.3
Eurasic − North-America	62.4°N	135.8°E	2.2
Eurasia − Pacific	61.1°N	85.8°W	9.0
Pacific − Australia	60.1°S	178.3°W	11.2
North-America − Pacific	48.7°N	78.2°W	7.8

2.4.3 The Plate Boundaries

Most geodynamically interesting processes occur along the plate boundaries. These boundaries can be divided according to:

- their kinematics, ✦
- the types of plates that are in contact.

Figure 2.18. Nomenclature of some of the most important parts of oceanic lithosphere

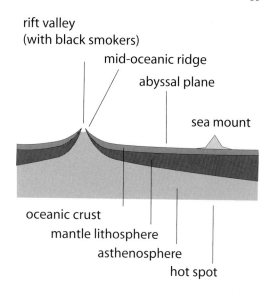

When choosing a division according to the types of bounding plates, we can discern:

– plate boundaries between two continental plates,
– plate boundaries between two oceanic plates,
– plate boundaries between a continental and an oceanic plate.

When choosing a division according to the kinematics, we can discern between *convergent, divergent* and *transform* plate boundaries (s. Tables 2.3; 2.4). Passive margins are formerly divergent plate boundaries between two continental plates which now consist of a passive contact between oceanic- and continental lithosphere. They are often listed as its own type of plate margin.

Divergent Plate Boundaries. Divergent plate boundaries are regions where two plates move into opposite directions or where one plate is splitting into two. Divergent plate margins exist only between two continental plates (e. g. central African rift system) or between two oceanic plates (e. g. mid-Atlantic ridge). The passive seams between continental and oceanic lithosphere are mechanically very strong and it would be a great coincidence if a divergent plate margin would form exactly along them. However, there are places on the globe where divergent plate margins *cross* passive margins. The Sheba-ridge in the Gulf of Aden and the Carlsberg-ridge in the Indian ocean are examples. Divergent plate boundaries on the continents are called rifts. The best know examples (in order of progressive rift development) are the Rheingraben, the Central African Rift system and the Red Sea. Divergent plate margins between two oceanic plates (mid-oceanic ridges) are – in most cases – the last stage of a rift (sect. 2.4.4, Fig. 2.18).

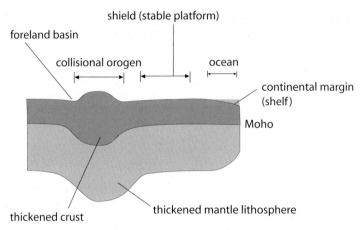

Figure 2.19. Nomenclature of important parts of continental plates

Convergent Plate Boundaries. Convergent pate boundaries may form between two continental plates (Fig. 2.15), between two oceanic plates or between a continental and an oceanic plate. In convergent plate boundaries between continental and oceanic plates, the oceanic plate dives beneath the continent, because of its higher density. This is called subduction and its surface expression is a trench (Fig. 2.20). The most famous example for subduction is the subduction of the Nazca-plate underneath the south-American continent along the Peru-Chile trench. Subduction leads to high pressure metamorphism of the subducted plate. This metamorphism is associated with dehydration and partial melting of the plate of the oceanic plate in the *Benioff zone*. Fluids that rise through the overlying mantle wedge react with the wedge material in endothermic reactions. However, additional heat input by convection in the wedge leads to partial melting (Hoke et al. 1994) and ultimately to the development of volcanic arcs on the surface.

The kinematics of subduction zones is complicated. The forces and velocities with which subducting plates sink into the asthenosphere are comparable to the forces exerted by mid-oceanic ridges onto the plate. Subduction zones can therefore move *backwards* (towards to mid-ocean ridge) if the downward velocity of the subducting plate is *larger* than the rifting rate at the ridge (e. g. South-Georgia, Scotia plate). They can move *forward* (towards the continent) if the rifting rate at the ridge is *larger* than the downward velocity (e. g. Pacific plate – Alaska). In other words, the distance between the trench and the continent in the far field hinterland increases, decreases or remains constant. Depending on details of the force and velocity field in subduction zone environments, forearc- or backarc basins may develop. In some cases of collision between oceanic and continental lithosphere, parts of the oceanic plate are welded onto the continental plate or even thrust over it. This is called *obduction*.

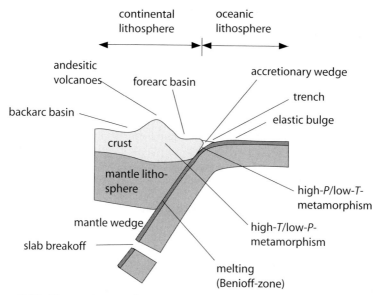

Figure 2.20. Nomenclature of important parts of subduction zones. Forearc- and backarc basins are even more clearly developed during subduction of oceanic lithosphere underneath another oceanic plate. Good descriptions of various phenomena on this figure can be found for slab break off by: Blanckenburg and Davies (1995); about the mantle wedge by: Spiegelman and McKenzie (1987) (sect. 3.5.2); about metamorphism by: Miyashiro (1973); about accretionary wedges in: sect. 6.2.3; about the elastic fore bulge: sect. 4.2.2

In contrast to subduction zones – where one of the two plates usually dives underneath the other – the convergence of two continental plates leads to a head-on collision and pervasive deformation of both plates. The reason for this is because continental lithosphere is (i) much thicker, (ii) less dense and (iii) much softer than oceanic lithosphere. This leads to the formation of the collisional mountain ranges that form most of the topographically high mountain belts of our globe (Fig. 1.1).

When two oceanic plates converge, no collision occurs and subduction zones form, similar to those that form when two plates of different kind collide. In contrast to the collision between two continental plates, *no* collision occurs between two oceanic plates because they are thinner, much stronger and because they are much denser and can therefore dive easier into the upper mantle (sect. 5.2.2). Because little internal deformation of the plates occurs, island arcs form that are clearly defined in space. Two beautiful examples for this are the subduction of the Pacific plate underneath the Philippine plate along the Mariana trench or the subduction of the Pacific plate underneath North-America along the Aleutes.

Transform Plate Boundaries. When two plates glide past each other without much convergence or divergence, their contact is called a trans-

Table 2.4. Different types of plate boundaries divided according to their kinematics and according to the plate type (O = oceanic lithosphere, C = continental lithosphere)

relative motion	type	plate tectonic feature	example
convergent	O-O	island arc	Philippines
	O-C	subduction zone	west coast of
		trench	South-America
	C-C	collisional mountain belt	Himalaya
divergent	O-O	mid-oceanic ridge	Atlantic
	C-C	continental rift	east African rift
passive	O-C	passive plate margin	eastern Australia

form plate boundary. No topographic features as significant as rift valleys or mountain ranges form. However, transform plate boundaries are well-known because they form some of the most important zones of seismicity on the globe. The best known examples are the San-Andreas-fault zone, (probably because it crosses one of the most densely populated parts of North-America) or the Alpine fault in New Zealand.

Triple Junctions. The spherical geometry of the earth requires that three or more plates touch each other in some places. This is called a triple junction. Triple junctions have a *stable* configuration if the relative motions of the bounding plates can maintain the geometry through time. They have an *unstable* configuration if their geometry is transient. Places where four or more plates touch are always unstable and will quickly dissolve into two or more triple junctions. Such an area occurs currently west of New Guinea, where the Philippine, Australian, Eurasian and Pacific plates touch. However, detailed mapping shows that this area may be divided into a series of micro plates and triple junctions and that the touching of more than three plates has only occurred transiently. Depending on the kinematics of plate boundaries, we can discern between a large number of different triple junctions. Using "R" (as in Ridge) for divergent plate boundaries, "T" (as in Trench) for convergent plate boundaries and "F" (as in Fault) for transform boundaries, we can describe RRR-, TTT-, FFF-, RTF- and a number of other triple junctions (McKenzie and Morgan 1969).

2.4.4 The Wilson Cycle

The Wilson-cycle is a model that brings the individual processes that we discussed in the last sections into an imaginary cycle (Fig. 2.21). This cy-

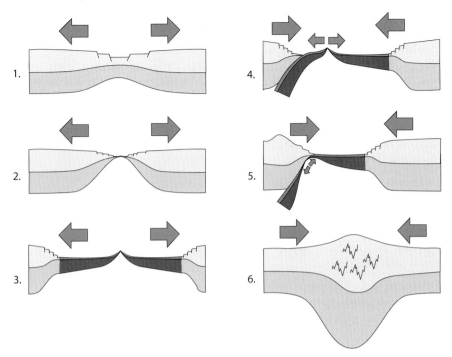

Figure 2.21. The Wilson cycle. The arrows indicate the relative plate motions

cle was first suggested by Wilson (1972) and supercedes the old terms of "geosynclines".

   The Wilson-cycle begins with extension of a continent. This stage of the Wilson-cycle can be currently observed in the Rheingraben. In the second stage, the continent breaks up and a spreading center forms in its middle. Oceanic lithosphere begins to form. This is the stage of the Red Sea. The third stage involves the development of a classic ocean with passive continental margins on both sides and a mid-oceanic ridge in its center. This is currently being observed in the Atlantic. The following stages describe processes that occur during opposite relative plate motion. These stages may therefore occur independent of the first three stages. The fourth stage describes the commencement of subduction of an oceanic plate underneath a continental plate as it is currently observes along the west coast of South-America. The fifth stage involves the subduction of a mid-oceanic ridge underneath the continental plate. A present day example for this is the subduction of the ridges bounding the Juan de Fuca plate underneath North-America. During the sixth and final stage the collision between two continents occurs. Clearly, the India-Asia collision is the most spectacular example for that.

2.5 Problems

Problem 2.1. *Small circles and subduction zones (p. 20):*
Make a very rough estimate of the small circle radius of the Aleute arc (from
Fig. 2.4 or any atlas) and estimate the subduction angle of the Pacific plate
underneath Alaska using the model of Fig. 2.5. This subduction angle is actually known. It is $\approx 45°$. Discuss possible reasons for the difference between
your estimate and this value. Compare your estimate with an estimate of the
subduction angle of the Indoaustralian plate underneath the Eurasian plate
along the Java trench.

Problem 2.2. *Understanding longitude and latitude (p. 21):*
Find a point on the surface of earth (outside the north pole!) where the
following experiment is possible: You walk 1 km south then 1 km west and
then 1 km north and you are where you started from. (It is said that this
question was asked by Sir Ernest Shackleton in 1908 to applicants for his
south pole expedition.)

Problem 2.3. *Understanding longitude and latitude (p. 21):*
The city of Vienna is located at 16° east and 48° north; the city of Munich is
located at 11° east and the same latitude as Vienna. How much earlier does
the sun rise in Vienna?

Problem 2.4. *Understanding arc-minutes in decimals (p. 21):*
Improve your result of Problem 2.3 using the following values: Vienna:
16°21' east, 48°11' north; Munich: 11°39' east and the same latitude as Vienna.

Problem 2.5. *Estimating the curvature of earth (p. 23):*
a) How far is it from Vienna to Munich going directly from east to west along
the surface of the globe? (use eq. 2.2). b) How far is it from Vienna to Munich
going the shortest distance along a great circle? (use the formula for angular
separation of 2 points in Table B.4). c) How far is it from Vienna to Munich if
you could go along a tunnel connecting the two cities in a straight line through
the earth? (use the result from b) and the logic illustrated in Fig. 2.5). Use
the longitudes and latitudes given in Problem 2.3 and a perfectly spherical
earth with radius $R = 6370$ km.

Problem 2.6. *Spherical kinematics (p. 23):*
What is the west-east velocity of a plate that has rotation pole that coincides
with the geographic north pole and rotates relative to another plate with an
angular velocity of $10^{-15} s^{-1}$; a) at the equator; b) at 48° north? (Use $R = 6370$ km).

Problem 2.7. *Understanding torque (p. 24):*
A north-south striking mid-oceanic ridge has a west-directed force of 10^{12} N.
At which latitude is the torque of this ridge around the axis of the earth
$4 \cdot 10^{18}$ Nm? (The radius of earth is: $R = 6370$ km)

Problem 2.8. *The Mercator projection (p. 28):*
Derive the function that is used in the Mercator projection to convert longitude ϕ and latitude λ into the Cartesian coordinates x and y. Detailed study of Fig. 2.9 is helpful to answer this problem.

Problem 2.9. *Physics of the lithosphere (p. 35):*
Label the axes of the diagrams on Fig. 2.15. (This problem requires the use of parameter values that have not been discussed up to this chapter yet. It is meant as a self-test for the interested reader.)

Problem 2.10. *Modern and ancient plate boundaries (p. 37):*
a) Draw relative velocity vectors between the plates onto Fig. 2.17. b) Draw three future new plate boundaries and three plate boundaries that will disappear within the next 50 my on Fig. 2.17. c) Draw three ancient plate boundaries onto Fig. 2.17.

Problem 2.11. *Understanding triple junctions (p. 42):*
How many different types of triple junctions exist, using the three possible relative motions R, T and F? Not all of the triple junctions you should come up with are stable. Draw an example of a stable and an unstable triple junction.

3. Heat and Temperature

In this chapter we discuss geodynamic processes that may be described with the units of energy or temperature. Energy has the unit Joule [J], which is equivalent to mass $\times$ velocity^2 (1 J $= 1$ kg m^{2}s^{-2}), or volume $\times$ pressure (1 J $= 1$ m^3 $\times$ Pa $= 1$ N m). These conversions give us a first indication that thermal and mechanical energy are often hard to separate. *Thermal energy may be converted into temperature using heat capacity and density* which are parameters that we shall discuss in some detail on the next pages. The general theme of energy and temperature is an obvious starting point in geodynamics, as so many properties of rocks that may be observed in the field are a strong function of temperature, for example the formation of metamorphic parageneses or the mechanisms with which rocks deform. These and many other examples are parameters relevant for orogenic processes and an understanding of the thermal structure of the lithosphere is therefore crucial for the understanding of most geodynamic processes.

The production and redistribution of heat in the lithosphere is done by three fundamentally different processes:

- heat conduction,
- heat advection (or convection) and
- heat production.

The relevance of these three processes for regional metamorphism is summarized at the start of sect. 3.4 in Table 3.4. However, in the first part of this chapter we will present some basic methods how to estimate if and how important each of these three processes may be for the thermal budget of the lithosphere. In this context we also discuss the basics of their mathematical description. The importance of such estimates is enormous: it will enable the reader to critically evaluate in his or her metamorphic study terrain to which degree heat conduction (e. g. because of burial), heat production (e. g. friction heat or radioactivity) or active transport of heat (e. g. by fluid or magma) may have played a role during metamorphism. Thus, the information on the following pages provides a powerful tool for the interpretation of heat sources of a metamorphic terrain.

3.1 Principles of Heat Conduction

3.1.1 The Heat Conduction Equation

The heat conduction equation - more commonly known as the diffusion equation - is fundamental for the understanding of the transport of heat in the lithosphere. We will also show in other chapters that the very same equation cannot only be applied to the transport of thermal energy, but also to the diffusion of mass. It finds therefore application in many other fields, for example geomorphology or metamorphic petrology (s. p. 172; p. 324). Thus, the diffusion equation is the first equation of this book that we will discuss in some detail. The fact that it is a second order partial differential equation should not scare us off. We will show that it is possible to understand it quite intuitively. There is also some explanations on how to read differential equations in section A.1.

Fourier's Law of Heat Conduction. Fourier's 1. law is the basic law underlying the diffusion equation. This law states that the flow of heat q is directly proportional to the temperature gradient (Fourier 1816). This statement can easily be formulated in an equation:

$$q = -k\frac{dT}{dz} \quad .$$ (3.1)

In this equation q is short for heat flow, T stands for temperature and z for a spatial coordinate, for example depth in the crust. The ratio dT/dz is the change of temperature in direction z. We call this ratio the temperature gradient. k is the proportionality constant between the gradient and the flow of heat. In order to understand this law better (and understand the units of k), let us consider a more familiar analogue: the flow of water in a river. The same law applies. In a river the flow of water can be described by the volume of water passing per unit of time and per area of cross section of the river (in SI-units: $m^3\,s^{-1}\,m^{-2} = m\,s^{-1}$). This is called the volumetric flow. When normalized only to the width of the river and not to the cross sectional area of the river, the volumetric flow has the units of $m^2\,s^{-1}$ (sect. 4.3). In contrast, the flow of mass has the units $kg\,s^{-1}\,m^{-2}$. Fourier's law – applied to our example of water flow – states that the flow of water is proportional to the topographic gradient of the river. This corresponds well to our observations in nature: The steeper a river bed, the faster the flow of water in the river (per square meter of cross sectional area). Fourier's law seems to be a good model description for this observation. This simple example also explains why there is a negative sign in eq. 3.1. The flow is against the gradient: it is *positive* in the *downwards* direction of the gradient.

In the theory of heat conduction, the flow of heat has obviously not the units of volume per time and area, but *energy* per time and area. (in SI units:

$J s^{-1} m^{-2} = W m^{-2}$). The thermal gradient now replaces the topographic gradient of the river. Because of historical reasons heat flow is not always expressed in $W m^{-2}$, but in heat flow units, or hfu. One hfu corresponds to $10^{-6} cal s^{-1} cm^{-2}$ (s. Problem 3.2, Table C.8). The units of the proportionality constant k, in eq. 3.1, follows now easily from the units of the other components of the equation: Because temperature has the units of K (or °C) and z has the unit m, k must have the units $J s^{-1} m^{-1} K^{-1}$ so that the equation is consistent in its units. The constant k is called *thermal conductivity*.

We can now try to read eq. 3.1. We can see that the flow of heat trends to zero if the conductivity is very low, regardless of the thermal gradient. Correspondingly, if the conductivity is very large, the flow of heat becomes large, even if the thermal gradient is very low. The equation may therefore be understood quite intuitively.

Would the thermal gradient be constant everywhere, we could write it as $\Delta T / \Delta z$. However, in geological problems this gradient is never constant. Thus, we use the derivative dT/dz, which states that we want to be careful and consider our thermal gradient only to be constant within each infinitely small section of the thermal profile. If the gradient changes along the z direction, then eq. 3.1 states that the heat flow must change correspondingly.

Energy Balance. The second part of the diffusion equation (often called Fourier's 2. law) describes an energy balance. This energy balance relates heat and temperature and the change of heat flow with change in temperature. This relationship may be established independently from eq. 3.1 and may be written as:

$$\frac{\partial T}{\partial t} \propto -\frac{\partial q}{\partial z} \ . \tag{3.2}$$

This equation states that the rate of temperature change of a rock must be proportional to the rate with which its heat content changes ($\propto$ is the symbol for "proportional to"). The rate with which the heat content of a rock changes ($\partial q / \partial z$) is given by the difference between the flow of heat *into* the rock and the flow of heat *out of* the rock (Fig. 3.1). If the heat flow into the cube of Fig. 3.1 is larger than the flow of heat out of it, then the heat content of the cube will rise and its temperature will increase. If the heat flow into the volume is just as large as that that flows out, the temperature will remain constant. If more heat flows out of the cube than into it, then its temperature will decrease.

In the last sentences we have begun mixing the terms "temperature" and "heat". However, we have to remain careful no to confuse them as the rate of temperature change is not the same as the rate of heat content change. They relate by

$$H = T \rho c_p \tag{3.3}$$

where H is the volumetric heat content in J m^{-3} (s. sect. 3.6.4). The rate, with which the temperature will change for a given change in heat content

Figure 3.1. The flow of heat in a unity volume of rock. The heat production inside this volume S, is not considered until we discuss eq. 3.24

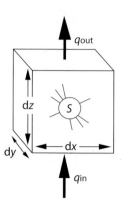

depends on another material specific proportionality constant. This is the *specific heat capacity* c_p. The specific heat capacity or short *"specific heat"* has the units of $J\,kg^{-1}\,K^{-1}$ and defines how many Joules are required to heat the mass of one kg of rock by one degree Kelvin. The most common abbreviation for specific heat is c. The subscript p symbolizes the condition that the specific heat is measured at constant pressure (s. sect. 3.2.2). If the specific heat of a rock is large, we need many Joules to heat the rock and even a rapid increase of its heat content will lead to slow temperature increase and vice versa. Specific heat is formulated in terms of the *mass* that is heated. Considering that the energy balance in eq. 3.2 is formulated in terms of the *spatial* coordinate z, and heat capacity is formulated in terms of *mass*, we need to multiply c_p with the density ρ, so that the relationship between the spatial change of *heat flow* and the temporal change of *temperature* is consistent with the units. We can write the proportionality of eq. 3.2 as:

$$\rho c_p \frac{\partial T}{\partial t} = -\frac{\partial q}{\partial z} \quad . \tag{3.4}$$

It should now be straight forward to understand eq. 3.4 intuitively using Fig. 3.1. The negative sign arises because the temperature *increase* when $\partial q = q_{out} - q_{in}$ is negative, that is, more heat flows into the rock volume than out of it. You may have noticed that the step from eq. 3.2 to eq. 3.4 was accompanied by the change from total- to partial differentials. This was necessary, because different parts of this equation are now differentiated with respect to different parameters (s. sect. A.1.1).

The Diffusion Equation. If we substitute Fouriers law of heat conduction (eq. 3.1) into the thermal energy balance of eq. 3.4, we arrive at eq. 3.5. This is the general form of the one-dimensional diffusion or heat conduction equation:

$$\rho c_p \frac{\partial T}{\partial t} = \frac{\partial \left(k \frac{\partial T}{\partial z} \right)}{\partial z} \quad . \tag{3.5}$$

If k is independent of z (e. g. if we consider heat conduction in an area without lithological contrasts), it is possible to simplify eq. 3.5 significantly. k can then be taken out of the differential and we can write:

$$\rho c_p \frac{\partial T}{\partial t} = k\frac{\partial \partial T}{\partial z \partial z} \qquad \text{or :} \qquad \frac{\partial T}{\partial t} = \kappa\frac{\partial^2 T}{\partial z^2} \quad . \tag{3.6}$$

The constants k, ρ and c_p are now summarized to $\kappa = k/(\rho c_p)$. κ is called thermal diffusivity. Eq. 3.6 can also be understood intuitively, without following the detailed derivation given above. Eq. 3.6 may be formulated in words as: *"The rate of temperature change is proportional to the spatial curvature of the temperature profile"* . If you do not understand the relationship between this sentence and eq. 3.6, then remember that the first differential of a function describes its slope (or: "gradient", or: "rate") and the second its curvature (s. sect. A.1, Fig. A.3, A.2).

Figure 3.2 illustrates this graphically. In our daily lives we encounter many examples that are described by this equation. Think for example that a piece of toast cools much quicker on its corners than along the edges or in its middle. This is because the *spatial curvature* of the isotherms in the toast is the largest at the corners! The same is true for the rapid cooling of the tip of a needle, the rapid erosion of ragged mountain tops and countless other examples in nature, all the way down to the rapid chemical equilibration of fine grained rocks in comparison with coarse grained rocks.

If we want to use eq. 3.6 we must solve it. For this we need boundary- and initial conditions. We also need some mathematical knowledge in order to integrate this equation. Various methods how to go about this are discussed in sect. A.1.1. A large part of this chapter will deal with various solutions of this equation. In this context we will often meet the terms *"boundary conditions"* and *"initial conditions"*. Make sure you understand what they mean (s. p. 354).

- *The magnitude of κ.* A quantitative application of eq. 3.6 requires the knowledge of κ and therefore the knowledge of k, ρ and c_p. The specific heat

Figure 3.2. The thermal equilibration of a random temperature profile. The temperature profile is drawn at two different time steps t_0 and t_1. Note that the largest change in temperature between the two time steps has occurred in those places of the profile where the curvature of the profile is the largest (s. eq. 3.4). Where the curvature of the profile is zero (at the inflection points) the temperature does not change at all

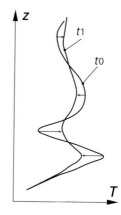

of rocks is about $c_p = 1\,000 - 1\,200$ J kg^{-1} K^{-1} (Oxburgh 1980). For most
rocks c_p does not vary by more than 20 % around this value. Thus, the nice
and even value of $c_p = 1\,000$ J kg^{-1} K^{-1} is a sound assumption that can be
used for many thermal problems. The density of many crustal rocks is of the
order of $2\,750$ kg m^{-3} and varies also not all that much around this value.
However, thermal conductivity, varies by the factor 2 or 3 between different
rocks types (Table 3.1). Fortunately, it is between 2 and 3 J s^{-1} m^{-1} K^{-1} for
many rock types. For $k = 2.75$ J s^{-1} m^{-1} K^{-1} and the values for specific heat
and density from above the diffusivity is: $\kappa = 10^{-6}$ m^2 s^{-1}. Because this value
is easy to remember it is commonly used in the literature. Note, however, that
κ may also be twice- or half as large if the thermal conductivity of rocks is
twice or half as large.

Heat Refraction. If rocks of different thermal conductivity are in contact,
the phenomenon of *heat refraction* may occur. What this is, is easily explained with eq. 3.1. In thermal equilibrium, the flow of heat in two adjacent
rocks must be equal. Following from eq. 3.1 we can formulate:

$$-q = k_1 \frac{\Delta T_1}{\Delta z} = k_2 \frac{\Delta T_2}{\Delta z} \quad , \tag{3.7}$$

where the subscripts $_1$ and $_2$ denote two different rocks as shown in Fig. 3.3.
We can see in this equation that, if the conductivities k_1 and k_2 are different,
the temperature gradient in the rock with the higher conductivity must be
lower and vice versa (Fig. 3.3). This is called heat refraction. Eq. 3.7 can also
be written in differential form. This means, the temperature gradient must
not change abruptly, but can also change continuously, if there are continuous
changes in thermal conductivity.

Let us illustrate the phenomenon with an example. A rock with extremely
high thermal conductivity, for example an iron ore body, will be practically
isothermal, even if it stretches over many vertical kilometers in the crust.
Its high conductivity will cause it to adapt some average temperature. Thus,
the upper part the body may have a significantly higher temperature than
its surroundings while its lowest part is colder than its surroundings. As a
consequence, it is conceivable that the process of heat refraction will even
leads to some kind of contact metamorphism (s. Problem 3.3).

Jaupart and Provost (1985) have noticed that there are some important
differences in thermal conductivity between the sediments of the Tethys zone

Table 3.1. Thermal conductivities of some rocks. The change of thermal conductivity as a function of pressure and temperature are negligible at geologically relevant temperatures in the crust (Cull 1976; Schatz and Simmons 1972)	rock type	k (J s^{-1} m^{-1} K^{-1})
	sandstone	1.5-4.2
	gneiss	2.1-4.2
	amphibolite	2.5-3.8
	granite	2.4-3.8
	ice	2.2
	salt	5.4-7.2

Figure 3.3. Illustration of the process of heat refraction. The flow of heat in the dark and the light shaded bodies is the same. However, the temperature gradient in the dark shaded body is larger, because its thermal conductivity k_1 is smaller. The subscripts $_1$ and $_2$ denote the dark and the light shaded body, respectively

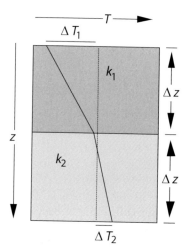

and the high Himalayan crystalline complex. They suggested that the process of heat refraction may have been of relevance in connection with the melting of the Himalayan leucogranites.

If we want to describe the process of heat refraction quantitatively, we can not assume the simplification that we have made in going from eq. 3.5 to eq. 3.6. We must stick with eq. 3.5 to describe conductive equilibration. If we form the derivative of the right side of eq. 3.5 using the rules of differentiation of products (Table B.1) we get:

$$\rho c_p \frac{\partial T}{\partial t} = \frac{\partial \left(k \frac{\partial T}{\partial z} \right)}{\partial z} = \frac{\partial k}{\partial z} \frac{\partial T}{\partial z} + k \frac{\partial^2 T}{\partial z^2} \quad . \tag{3.8}$$

In this form the heat conduction equation can be used for the description of many problems with variable conductivity.

Heat Conduction in Several Spatial Dimensions. Eq. 3.6 is a linear differential equation (s. p. 352). This means that heat conduction in two or three dimensions can be described by the sum of heat conduction in the individual directional components. In general, we can write:

$$\frac{\partial T}{\partial t} = \kappa \left(\frac{\partial^2 T}{\partial x^2} + \frac{\partial^2 T}{\partial y^2} + \frac{\partial^2 T}{\partial z^2} \right) \tag{3.9}$$

or, for the case that κ has different values in different spatial directions:

$$\frac{\partial T}{\partial t} = \kappa_x \frac{\partial^2 T}{\partial x^2} + \kappa_y \frac{\partial^2 T}{\partial y^2} + \kappa_z \frac{\partial^2 T}{\partial z^2} \quad . \tag{3.10}$$

In the literature eq. 3.9 is often written as:

$$\frac{\partial T}{\partial t} = \kappa \nabla^2 T \tag{3.11}$$

The symbol ∇ is called the "Nabla-" or: "Del"-operator and is defined in
eq. A.32. ∇^2 describes the same thing for the second partial derivative
(s. sect. A.3). Partial differentials (or derivatives) are discussed in detail in
section A.1.1. An important property of the diffusion equation is that it contains an energy balance. This means that no energy can be gained or lost by
diffusion processes. If a rock cools by conduction, then it is by heat loss at
the model boundaries, not because of the conduction itself.

• *Complicated geometries.* The diffusion equation we presented above is linear in temperature if all the material constants are independent of temperature and time. Therefore, it is possible to describe diffusion problems of
complicated geometries simply by a linear superposition of the solutions for
more simple problems. We will use this approach extensively when we describe the cooling history of intrusions in sect. 3.6. For the same reason is it
possible to describe time dependent problems by a superposition of a steady
state solution and a time dependent component.

Heat Conduction in Polar Coordinates. Many heat conduction problems in the earth sciences are much better described in cylindrical or spherical
coordinates. In *cylindrical* coordinates, eq. 3.6 adapts the form:

$$\frac{\partial T}{\partial t} = \kappa \left(\frac{\partial^2 T}{\partial r^2} + \frac{1}{r} \frac{\partial T}{\partial r} \right) \quad . \tag{3.12}$$

If we consider heat conduction in *spherical* coordinates (and restrict ourselves to heat flow in the radial direction), then the heat conduction equation
adapts the form:

$$\frac{\partial T}{\partial t} = \kappa \left(\frac{\partial^2 T}{\partial r^2} + \frac{2}{r} \frac{\partial T}{\partial r} \right) \quad . \tag{3.13}$$

In eqs. 3.12 and 3.13 r is the distance from the coordinate origin on the
cylinder axis or at the sphere center, respectively. Detailed derivations of these
equations will not be described here, but they can be found - among many
others – by Carslaw and Jaeger (1959); Crank (1975) or Smith (1985). Examples for spherical conduction problems that can be described with eq. 3.13 are
the Kelvin model for the cooling of the earth or the chemical diffusion of elements in garnet crystals. Note that eqs. 3.12 and 3.13 are one-dimensional and
their results can therefore be directly compared with eq. 3.6 (e. g. Fig. 3.29
and 3.35).

The Kelvin Model for the Cooling of Earth. The most famous example
for the application of the heat conduction equation is the estimate of the
age of the earth by Lord Kelvin (1864). However, we note that very similar
estimates were already performed by Fourier himself in 1820. Both physicists
realized that - in principle - one could estimate the age of the earth from
the present day thermal gradient at the surface (surface heat flow, s. Table
C.3) if the following assumptions are made (Fig. 3.4): 1. The whole earth was
at the time of its formation of constant temperature (which was assumed

by Kelvin to be $4\,000\,°C$); 2. the surface temperature has remained constant ever since and 3. heat conduction is the principle process of cooling. Using eq. 3.13, Kelvin concluded that the earth must be around 100 my old (he used a very similar solution to the one we introduce in sect. 3.6.1). Today we know that his calculation was wrong for two important reasons. The first (and most quoted) reason is that there are heat producing elements in the crust that prolong the cooling of earth. Radioactivity was unknown at Kelvin's time. The second reason is that convection in the upper mantle rises the isotherms. While this process actually leads to a faster cooling of earth, both *radioactivity* and *convection* cause the present day geothermal gradient in the lithosphere to be steeper than it would be without these two processes. This leads to a massive underestimate of the age of earth. Interestingly, the error that is caused in Kelvin's estimate by him neglecting convection is much larger than that caused by neglecting radioactivity. In 1895 John Perry improved Kelvin's estimates drastically, by assuming a "convective mantle conductivity" which he assumed to be ten times larger than the conductivity of the crust (s. p. V). With this assumption he arrived at an almost correct age.

Why Kelvin ignored convection in the mantle, although it had been known for some time, we do not know. However, in defense of his estimate it should be said that he *did* mention in his discussion that heat production, for example by chemical reaction, was not considered in his model but might change

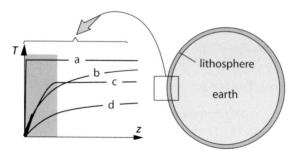

Figure 3.4. Schematic illustration of the model of Lord Kelvin for the estimation of the age of the earth. The small T-z-diagram is an enlarged section of outer 300 km of the globe. The shaded area is the lithosphere. a is the initial thermal profile of the earth at the time of its formation. b is the present day thermal profile that was guessed by Kelvin based on the measured surface heat flow q_s (thick drawn part on the thermal profile) and on a spherical heat conduction model. However, because of radioactivity (unbeknown to Kelvin), the lower lithosphere is much hotter than Kelvin thought possible. Moreover, in the asthenosphere convection destroys a conductive thermal gradient, which is very low in this part of earth. (convection was know at Kelvin's time but ignored in his estimates) d is the thermal profile that we should measure today, if the earth had only cooled by conduction since its formation at $\approx 4{,}5 \cdot 10^9$ y. The surface heat flow would be much lower than what we measure today

the results. The model of Kelvin remains a fantastic example for a conduction model that turned out to be completely wrong because other heat transport mechanisms were not considered. It should serve us as a reminder when interpreting the heat sources of a metamorphic terrain.

3.1.2 The Laplace Equation

Equations 3.6 and 3.11 describe the evolution of temperature as a function of time. Thus, we can use them to describe heating and cooling curves of rocks. However, for many geological questions we are not all this interested in the *temporal* changes of the temperature, but rather in the steady state shape of a temperature profile, for example the shape of a stable continental geotherm. For all those problems, where it is reasonable to assume that there is *no* temporal change of a temperature profile, we can write:

$$\frac{\partial T}{\partial t} = 0 \ . \tag{3.14}$$

Any geological situation where this equation applies is called the *steady state*. When we make this assumption, κ can be cancelled out of eq. 3.6 and eq. 3.11. We are left with:

$$\nabla^2 T = 0 \ . \tag{3.15}$$

Eq. 3.15 is called Laplace-equation. Just like the diffusion equation, it is an extremely important equation for many geological problems. In this chapter we will need it when we consider stable geotherms, but it also has applications in many other branches of the earth sciences.

3.1.3 The Error Function

If we want to use eq. 3.6 we must solve it for a given set of boundary- and initial conditions. If we try this, we would quickly realize that this is only possible for a very few boundary- and initial conditions. Periodic problems are some of those for which there are "real" solutions of this equation (sect. 3.7.1). For most problems there are simply no analytic solutions of eq. 3.6 possible. For example, for many geological problems we will see that it is useful to assume that the boundary conditions lie at infinity (at distances that are far away compared to the scale of the problem). In all such problems, the results of integrating eq. 3.6 will contain a term of the form:

$$\frac{2}{\sqrt{\pi}} \int_0^n e^{-n^2} dn = \mathrm{erf}\,(n) \tag{3.16}$$

This integral cannot be solved analytically, but can be determined numerically. However, because it occurs so often in solutions of the heat flow equation, it has its own name, the *error function*. The values of the error function

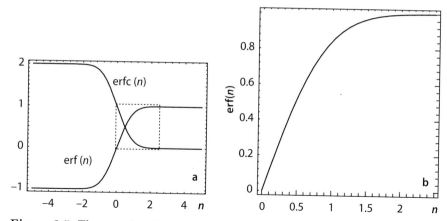

Figure 3.5. The error function and the complementary error function. The dashed frame in **a** shows the part of erf(n) that is shown enlarged in **b**

for different values of n have been determined numerically and can be looked up on tables, or it can be calculated with some numerical approximation (s. Table B.9). Fig. 3.5 shows the shape of the error function. In many solutions of eq. 3.6 the variable n from eq. 3.16 has the form $n = z/\sqrt{4\kappa t}$. There, time t, and distance z are inside the error function and they are in a quadratic relationship to each other. Most solutions that we will use for the description of contact metamorphism contain error functions of this form (sect. 3.6). We will need the term $n = z/\sqrt{4\kappa t}$ in sect. 3.1.4, 4.2.1 and others. The *complementary* error function erfc is defined as:

$$\mathrm{erfc}\,(n) = 1 - \mathrm{erf}\,(n) \quad . \tag{3.17}$$

3.1.4 Thermal Time Constants

The thermal time constant or *characteristic time scale of thermal equilibration* is a scaling factor for the duration of a thermal event. It is a fantastic aid for an enormous number of estimates, even when in the field. It can be used to estimate the width of contact aureoles, the duration of metamorphic events, the chemical zoning profile of crystals and much more (s. Problem 7.8). The thermal time constant τ is given by the relationship

$$\tau \propto \left(\frac{l^2}{\kappa}\right) \quad . \tag{3.18}$$

There, l is the spatial size, or: *characteristic length scale* of a thermal event (e. g. the diameter of an intrusion, of a hydrothermal vein or a metamorphic terrain). κ is the thermal diffusivity and $\propto$ means "proportional to". Eq. 3.18 tells us that the duration of diffusive equilibration is proportional to the square of the size of the equilibrating body. In other words, if for example

there is two granitic plutons with one being *twice* the size of the other, then
the duration of cooling will be *four* times longer for the larger body. The
proportional relationship from eq. 3.18 adapts various forms, depending on
the problem solved and on how exactly we want to define the thermal time
constant. Usually it is defined as:

$$\tau = \left(\frac{l^2}{2\kappa}\right) \quad \text{or}: \ \tau = \left(\frac{l^2}{(2.32)^2\kappa}\right) \ . \tag{3.19}$$

The different formulations are based on different assumptions about the
magnitude of the scaling factor for τ. In eq. 3.21 we will discuss this in
some more detail. Table 3.2 shows some numerical values for different sizes
of cooling bodies.

The value of τ can be used either way. It can be used to estimate the
duration of a diffusion process *or* it can be used to estimate the length scale
of a thermally influenced region after a given time scale. We should try to
remember that the typical duration of thermal equilibration of length scales of
10 m, 100 m, 1 km, 10 km and 100 km is of the order of 1 y, 100 y, 10 000 y,
1 my and 100 my, respectively. Table 3.2 gives a number of examples of
thermal time constants for different length scales.

The Meaning of the Thermal Time Constant. We now want to define
a bit more clearly what is meant by the terms "length scale" and "time scale"
which we have introduced in the last section. For this, we have to use some
information from sect. 3.6. There, we shall see that many solutions of the heat
conduction equation contain an error function of the form: erf $(l/\sqrt{4\kappa t})$. For
example, the solutions discussed on page p. 102 or in eq. 3.81 are all of the
form:

$$T(t) = a + b \left(\text{erf} \left(\frac{l}{\sqrt{4\kappa t}} \right) \right) \tag{3.20}$$

where a and b are constants. In eq. 3.20 the expression $(l/\sqrt{4\kappa t})$ corresponds to the variable n from eq. 3.16. The length scale l and time t are

Table 3.2. Different values of the thermal time constant τ for a series of geologically
relevant length scales l, and calculated for one of the two proportionalities given
in eq. 3.19 as well as for one used much later in this book (p. 264) for the same
purpose

l	$\tau = l^2/2\kappa$	$\tau = l^2/\pi^2\kappa$
10 m	$5 \cdot 10^7$ s ≈ 1.58 y	$1.01 \cdot 10^7$ s ≈ 16 weeks
100 m	$5 \cdot 10^9$ s ≈ 158 y	$1.01 \cdot 10^9$ s ≈ 32 y
1 km	$5 \cdot 10^{11}$ s $\approx 15\,000$ y	$1.01 \cdot 10^{11}$ s $\approx 3\,200$ y
10 km	$5 \cdot 10^{13}$ s ≈ 1.5 my	$1.01 \cdot 10^{13}$ s $\approx 320\,000$ y
100 km	$5 \cdot 10^{15}$ s ≈ 158 my	$1.01 \cdot 10^{15}$ s ≈ 32 my

both contained within the error function and do not appear elsewhere in the solution. The shape of the error function in Fig. 3.5 shows that it reaches asymptotically 1 as n get very large. Correspondingly, from eq. 3.20, the term inside the brackets will always remain 1 for very large l (regardless of t), or for very small t (regardless of l). At these conditions no temperature change will occur. Also, because time is in the denominator inside the error function, *complete* equilibrium is reached only after infinite time (when the term inside the brackets asymptotically approaches zero). In order to define a "duration of equilibration" it is therefore necessary to define an arbitrary point on this equilibration curve. This duration is what we call the "characteristic time scale of equilibration", τ. In eq. 3.19, there are two different formulations how to define this time scale. The first formulation for this time scale or "thermal time constant" was chosen so that the argument of the error function (for which we use n in eq. 3.16) is 1. This means that:

$$\left(\frac{l}{\sqrt{4\kappa t}}\right) = 1 \quad \text{or}: \quad t = \frac{l^2}{2\kappa} = \tau \ . \tag{3.21}$$

Figure 3.5 illustrates that for the argument to be 1 (where $t = l^2/(2\kappa)$), the thermal equilibration is 84.3 % complete. Just for simplicity, this arbitrary value is often chosen as a scaling factor for the equilibration history where it may be said that the diffusive equilibration is "largely complete". This time is referred to as the *thermal time constant*.

As the value of 84.3 % seems a bit arbitrary, thermal time constants are sometimes formulated using the reverse consideration: What is the thickness of a layer, which has equilibrated to 90 % within a given time? From Fig. 3.5 we can see that $\mathrm{erf}(n) = 0.9$, for $n \approx 1.16$ or $l \approx 2.32\sqrt{\kappa t}$. This expression for thickness is also used for the thermal definition of the lithosphere (sect. 3.4). It is what we call the *thermal boundary layer*. Both formulations were chosen in an arbitrary way, and either can be used depending whether one wants a slightly simpler formulation of τ, or the value of the percent of thermal equilibration. It is important to note that τ gives an approximation of the time it will take for most of the thermal equilibration to take place. In summary, it will be sufficient for most purposes to remember the basic message of eq. 3.18:

- During conductive processes the duration of thermal equilibration increases with the square of the length scale of the equilibration body.
- For geologically realistic thermal diffusivities, the characteristic time scale of equilibration of a 10^3 m length scale is of the order of 10^4 y. On the scale of the lithosphere (10^8 m), the time scale of equilibration is of the order of 10^8 y.

This means that regional metamorphism of nappe piles that are several tens of kilometers thick should last of the order of several tens of my. We shall discuss this in much more detail in sect. 6.2.1.

3.2 Principles of Heat Production

We discern three fundamentally different geological mechanisms that produce heat:

– radioactive heat production,
– chemical heat production,
– mechanical heat production.

In the next sections we derive the basic equations that are needed to describe these three mechanism and we discuss their respective geological relevance. In general, the rate of temperature change due to heat production may be described by

$$\frac{dT}{dt} = \frac{S}{\rho c_p} \quad . \tag{3.22}$$

There, T, t, ρ and c_p correspond to temperature, time, density and heat capacity as discussed on p. 49 and S is the volumetric rate of heat production in $J\,s^{-1}\,m^{-3} = W\,m^{-3}$. If S is positive, heat is produced, dT/dt is generally positive and rocks heat up. If S is negative, heat is consumed, dT/dt is negative and rocks cool. We can see in this equation that the heat production rate is divided by density and specific heat. This is so that we convert the volumetric heat production rate into a rate of temperature change, just as we have done with heat flow in section 3.1.1.

The heat production rate S can be of radioactive, chemical or mechanical origin:

$$S = S_{\text{rad}} + S_{\text{chem}} + S_{\text{mec}} \quad . \tag{3.23}$$

All three of these components may have a significant influence on the thermal evolution of rocks depending on the circumstances and all three have different characteristics that require different methods of description. Some heat production rates are very rapid (e.g. friction heat production during an earthquake), others are very slow (e.g. reaction heat production during retrograde metamorphism). The rate of heat production is particularly important in relationship to the rate of heat conduction.

From section 3.1 we remember that the rate of temperature change due to conduction is proportional to the difference between heat flow into and out of ($q_{\text{in}} - q_{\text{out}}$) a unit rock volume. However, the rate of temperature change depends also on the amount of heat that is produced *inside* this unit volume (Fig. 3.1). This value must be added to the difference $q_{\text{in}} - q_{\text{out}}$. Thus, if we adapt eq. 3.6 to formulate a thermal energy balance considering conduction *and* production, we can write:

$$\frac{\partial T}{\partial t} = \kappa \frac{\partial^2 T}{\partial x^2} + \left(\frac{S}{\rho c_p}\right) \quad . \tag{3.24}$$

Note that we now need to use partial derivatives. Whether or not we can use eq. 3.22 or need to consider eq. 3.24 depends on the relative rates of heat production and heat conduction. A comparison between the diffusion and the heat production term on the right hand side of eq. 3.24 can help us to judge if one of the two processes is much smaller than the other and can therefore be neglected in the problem that is being considered. Such a comparison could be made two ways:

1. *Comparison.* The characteristic time scale of diffusion τ of diffusion may be compared with the duration of heat production. If τ is much larger than the duration of heat production, then conduction can be neglected and the heat production will dominate the temperature change. However, if τ is much smaller than the duration of heat production, then all heat that is produced will be conducted away at a much faster rate than it is being produced. Even large amounts of heat production will have a comparably small influence on the temperature and it may be possible to neglect the production term. If heat production- and conduction occur on similar time scales, then the following comparison may be of value.
2. *Comparison.* Consider the case of stable temperature where $dT/dt = 0$. Then we can see from eq. 3.24 and eq. 3.6 that: $-S/k = d^2T/dx^2$. In words: if the ratio S/k corresponds to the curvature of the temperature profile, then the rate of heat production is balanced by the rate of heat conduction. An important example for such a balance is given by the steady state shape of isotherms in subduction zones (p. 99) and a corresponding example in the theory of mass transfer is discussed on p. 179.

However, for this chapter, let us remain with the case where heat production is much more important than heat conduction and we neglect any diffusion processes in the first instance.

3.2.1 Radioactive Heat Production

In order to get a feeling for the magnitude of the radioactive (or: radiogenic) heat production in the crust, Table 3.3 lists some average concentrations of the heat producing elements in the continental crust. The sum of the values listed in this table is about $4 \cdot 10^{-10}$ W kg^{-1}, which corresponds roughly to a heat production rate of 1 μW m^{-3}. However, it should be noted that

Table 3.3. Average concentrations of heat producing elements in the crust. In granites, the heat production is about 2–3 times higher than the values listed here. In oceanic crust they are only about half as large and in the mantle they are less than 10% of the values listed here

element	concentration (ppm)	heat production rate (10^{-10} W kg^{-1})
U	1.6	1.6
Th	5.8	1.6
K	20 000	0.7

there are a series of terrains where radiogenic heat production rates may be significantly higher (Fig. 3.6; Sandiford and Hand 1998b). However, on a global scale, it may be said that radiogenic heat production in the continental crust is responsible for about half of the heat flow that we can measure at the surface of the earth (Chapter 2, Fig. 2.15). Radioactive heat production is an important contributor to the heat budget of the crust as a whole (Chamberlain and Sonder 1990; MacLaren et al. 1999; Sandiford et al. 1998).

We know that the bulk of the heat producing elements of the earth is concentrated in the upper crust. This is partly because Potassium (one of the more important radioactive elements) is very light in weight and because Uranium and Thorium are incompatible and thus contained in granitic melts. Over time, such melts have transported those elements into the upper crust. Nevertheless, the vertical distribution of heat producing elements in the crust may be extremely variable (e. g. Haack 1983; Lachenbruch and Bunker 1971). In the section on the calculation of geotherms we will spend some time to discuss the influence of the vertical distribution of these elements on the temperature profile of the lithosphere (sect. 3.4.1). More information on the importance of radiogenic heat flow in the crust is also provided on page 273.

3.2.2 Mechanical Heat Production

The forces that deform rocks can be viewed as mechanical energy that is added to the rock. The work (being the product of force times distance) done

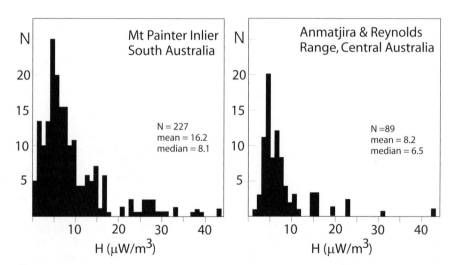

Figure 3.6. Radioactive heat production rates of two low-pressure high-temperature terrains from Australia (after Sandiford and Hand 1998b). While both terrains are unusually high, these examples show that some terrains may have significantly higher heat production than the global mean, N is the number of data points

on the system is the product of force applied to the system times distance over which it is deformed. This energy will be taken up by a variety of *mechanical energy sinks*. A part of this energy will be transformed into potential energy (s. sect. 5.3.1), some into dislocation energy in crystal lattices, some in noise and other forms of energy. However, most authors agree that the majority of this mechanically produced energy will be transformed into friction heat. We abbreviate this mechanical heat production with S_{mec}. The amount of friction heat (also called: *shear heat* or *viscous dissipation*) released in rocks is a direct function of the magnitude of deviatoric stresses that rocks can support. Stress has the units of Pascal. One Pascal is one Joule per cubic meter ($1 \, Pa = 1 \, J \, m^{-3}$). Thus, stress can be expressed as energy per volume and energy is stress *times* volume. These conversions between the different units should be straight forward, remembering the well-known relationships:

$$\text{force} = \text{mass} \times \text{acceleration} \quad \text{and} \quad \text{stress} = \frac{\text{force}}{\text{area}} \ .$$

The units of acceleration are $m \, s^{-2}$ and those of force are therefore: $kg \, m \, s^{-2}$. Stress and pressure therefore have the units of $kg \, m \, s^{-2} \, m^{-2}$ or $Pa = kg \, m^{-1} \, s^{-2}$ and energy has the units of $J = kg \, m^2 \, s^{-2}$. Accordingly, if high stresses are required to deform a rock a lot work is done on the system and the mechanical energy production rate is high. In contrast, if soft rocks deform under very small stresses, then the work done on the system is low and the friction heat production rate may be negligible. The rate of mechanical heat production S_{mec} is given by the product of deviatoric stress and strain rate. Both are tensors and the rate of mechanical heat production is therefore given by a tensor product. However if we consider a one-dimensional case (and only normal components i.e. we neglect shear stresses and strain rates), these tensors can be simplified and we can view the mechanical heat production rate as the scalar product:

$$S_{mec} = \tau \dot{\epsilon} \ , \tag{3.25}$$

where τ is the deviatoric stress and $\dot{\epsilon}$ is the strain rate (s. p. 200). Eq. 3.22 looks therefore like:

$$\frac{dT}{dt} = \frac{\tau \dot{\epsilon}}{\rho c_p} \ . \tag{3.26}$$

Some of the simplifications that have to be made to replace the general three-dimensional stress- and strain rate tensors by the scalars $\dot{\epsilon}$ and τ will be discussed in sect. 5.1.1 and 6.2.2. Note that eq. 3.26 is independent of the deformation mechanism. Both brittle and ductile deformation mechanisms will produce the same amount of friction heat if they support the same deviatoric stresses.

Both, stresses and strain rates on the scale of the crust are not very well constrained. Methods to measure geological strain rates show an upper limit of $\dot{\epsilon} = 10^{-12}$ to 10^{-14} s^{-1}. These numbers imply that deformation doubles the thickness of a rock package (strain of about 100%) within 1–10 my.

The magnitude of deviatoric stresses is much less constrained. Moreover, it is strongly temperature dependent. We know that the order of magnitude of plate tectonic driving forces is between 10^{12} and 10^{13} N m^{-1} (s. sect. 5.3). We will show on p. 285 that this implies a rock strength of 50–100 MPa – averaged over the thickness of the lithosphere. However, the vertical distribution of this strength in the lithosphere is largely unclear. In fact, in section 6.3.5 and 5.2.1 we will see that the middle crust may exceed those values dramatically.

Magnitude of Shear Heat Production. Let us now use eq. 3.26 to estimate the temperature increase a rock might experience for some realistic stresses and strain rates of lithospheric scale deformation. For this, we neglect heat conduction away from the site of mechanical heat production in the first instance. Our estimates are therefore an upper constraint but may be quite appropriate if the length scale of shear heat production is very large (e.g. deforming nappe piles of several tens of kilometres) and the time scale of heat production is short (e.g. less than a few millions of years).

Crust typically changes its thickness by a factor of two during orogenesis (it doubles in collisional orogens and halves its thickness in many extensional settings), so let us assume a stretch of 2. Remember that a stretch of 2 corresponds to a longitudinal strain (or elongation) of 1 ($\dot{\epsilon} \cdot t = \epsilon = 1$; s. eq. 5.1). We also neglect the temperature dependence of stresses in the first instance. Then, it is easy to integrate eq. 3.26 as all parameters are constants. The temperature at the end of deformation can be estimated simply with: $T = \tau/(\rho c_p)$. Using standard values for the density and specific heat: $\rho = 2\,700$ kg m^{-3} and $c_p = 1\,000$ J kg^{-1} K^{-1} we can see that a rock that has a shear strength of 100 MPa will be heated by about 37 °C. If rocks are twice as strong, then the temperature increase is twice as high. As the strength of rocks may be several hundreds of MPa under some circumstances, we must conclude that viscous dissipation may be of significant importance to the thermal energy budget of the lithosphere.

Examples where friction heat production has a significant influence on the temperature of rocks are well-known to us from pseudotachylites. There, friction heat was sufficient to even melt the rock. Pseudotachylites form during seismic events where extremely rapid deformation occurred on a very local scale. They are therefore not very appropriate to estimate the influence of friction heat on the thermal evolution of the entire crust where we have to deal with averaged strain rates and averaged stresses (e. g. Kincaid and Silver 1996; Stüwe 1998a). Regardless, even significant amounts of friction heat need not be reflected in significantly increased temperatures. Wether or not shear heating actually becomes geologically significant on a crustal scale depends largely on 2 factors:

– 1. It depends on the relationship between the *length scale* of heat production (which determines how rapidly heat may be conducted away from the site of production) and the *time scale* of heat production. The same considerations that we discussed on p. 61 apply. For example, if a 100 m thick shear zone is active for 1 my, then eq. 3.19 tells us that the characteristic time scale of diffusion of this shear zone is of the order of only 1 000 y. Thus, shear heat produced over a time interval of 1 my will be largely conducted away as it is produced. In contrast, if a 15 km nappe pile deforms under the same conditions, then its thermal time constant will be tens of my and all heat produced within 1 my will be largely retained in the pile.

– 2. It depends on the feedback between heating and softening of rocks. This will be dealt with in the next section

Temperature Dependent Strength. In chapter 5 (p. 216) we will show that the deviatoric stresses that rocks can support during viscous deformation are a strong function of temperature. Thus, eq. 3.26 should strictly have been written as:

$$\frac{dT}{dt} = \frac{\tau(T)\dot{\epsilon}}{\rho c_p} \quad . \tag{3.27}$$

We can see from this equation that the temperature increase that may occur due to shear heating is in itself a function of temperature. There is a negative feedback between shear heat production and thermal weakening so that any incremental amount of shear heating will instantaneously soften the rocks and subsequent heating becomes more and more difficult as temperature gets higher. Shear heating is said to be self limiting.

For some assumptions about the deformation mechanism it is possible to integrate eq. 3.27, even though temperature occurs both on the left and right hand sides of the equation. For example, for the constitutive relationship of eq. 5.41, it is possible to integrate eq. 3.27 (Stüwe 1998a). Eq. 5.41 describes a power law dependence of τ on strain rate and an exponential dependence of τ on temperature and is known to describe ductile deformation of rocks very well. Thus, temperature increase during shear heating and subject to the feedback between temperature increase and softening may be evaluated for quite realistic rheologies. While the analytical solution presented by Stüwe (1998a) is quite complicated and will not be repeated here, a graph of this solution is shown in Fig. 3.7.

There, effects of shear heating are evaluated for the material constants appropriate to viscous power law deformation of quartz and of olivine and for three different strains and strain rates. If rocks are very soft (e.g. at very high temperatures) there is no shear heating and temperatures before and after deformation are the same. However, at temperatures below about 600°C shear heating may be quite significant. Due to the self limiting nature of shear heating, curves become parallel to the vertical axis at even colder temperatures.

Note that Fig. 3.7 can only be used to evaluate the effects of shear heating if conduction of heat may be neglected. The thermal time constant of the lithosphere is of the order of hundreds of my, while the duration of continental deformation processes is only 1–10 my. Thus, on the scale of the lithosphere, the duration of mechanical heat production is at least one order of magnitude less than the duration of heat conduction. Thus, conduction of heat may be neglected if shear heating is considered on the scale of the lithosphere (s. comparison on p. 61). Also note that Fig. 3.7 was calculated assuming that the strain rate remains constant. While this describes some geological scenarios, there are many others where *stress* is a constant and the strain rate changes in order to balance it (s. p. 231).

Shear Zones of Finite Width. If we consider problems on a small scale, e. g. the influence of shear heating on the contact zone of a shear zone, then the influence of conduction may not be neglected. For the solution of such problems, eq. 3.26 must be enlarged by a term that describes diffusion. The equation we must solve is eq. 3.24, subject to initial and boundary conditions that describe the shear zone geometry. Fortunately, many shear zones have reasonably simple planar geometries that may be described with simple initial and boundary conditions. If we define the width of a shear zone to be $2l$ and assume a one-dimensional spatial coordinate z that extends normal to the

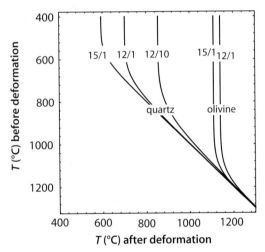

Figure 3.7. Temperature of rocks at the start of deformation, plotted against the temperature at the end of deformation. Mechanical heat production is assumed to be the only heating process and heat conduction is neglected. The diagram was calculated with the solution presented by Stüwe (1998a), which uses eq. 3.27 and eq. 5.41. The various curves are labeled according to the following syntax: before the dash: $_{10}$Log($\dot{\epsilon}$) between $\dot{\epsilon} = 10^{-12}$ and 10^{-15} s^{-1}; after the dash: elongation from 1–10. Obviously, the increase in temperature is larger for larger strains. However, during crustal shortening, a stretch of 2–3 is rarely exceeded. Thus, the effects of shear heating are likely to be much larger in strike slip regimes where much larger strains are possible than during crustal thickening. Rheological assumptions for quartz are: $Q = 1.9 \cdot 10^5$ J mol^{-1}, $A = 5 \cdot 10^{-6}$ MPa^{-3} s^{-1}, $n = 3$. Rheological assumptions for olivine are: $Q = 5.2 \cdot 10^5$ J mol^{-1}, $A = 7 \cdot 10^4$ MPa^{-3} s^{-1}, $n = 3$

shear zone and has its origin in the shear zone center, then the boundary conditions for a planar shear zone of constant width may be formulated as follows:

- $T = 0$ at the time $t = 0$ in the half space $z > 0$.
- For all $t > 0$ there is heat production in the region $0 < z < l$ at a constant rate S.
- For all $t > 0$ the thermal gradient at $z = 0$ is: $dT/dz = 0$.
- At $z = \pm\infty$ the temperature stays at $T = 0$ at all times.

You can visualize yourself these initial and boundary conditions by sketching them into a diagram where z is plotted against T. This one-dimensional description is relevant for shear zones in which the thermally influenced region is narrow compared to the extent of the shear zone. The boundary conditions describe only half of a shear zone of the width $2l$. It is sufficient to describe only half the zone, because the temperature profile will be symmetric about the center of the shear zone. The assumption $dT/dz = 0$ in the center is sufficient and simplifies the problem. Using these initial and boundary conditions it is possible to find two solutions of eq. 3.24 that describe the temperatures inside and outside the shear zone, respectively. These solutions are discussed by Carslaw and Jaeger (1959) who also summarize a large number of solutions for other geometries of heat producing zones and for a range of heat production rates as a function of time and temperature. For the conditions described above, temperature as a function of time inside the shear zone is given by:

$$T_{(0<z<l)} = \frac{\kappa St}{k} \left(1 - 2i^2 \text{erfc}\left(\frac{l-z}{\sqrt{4\kappa t}} \right) - 2i^2 \text{erfc}\left(\frac{l+z}{\sqrt{4\kappa t}} \right) \right) \quad . \tag{3.28}$$

For the region outside the shear zone the temperature is given by:

$$T_{(z>l)} = \frac{2\kappa St}{k} \left(i^2 \text{erfc}\left(\frac{z-l}{\sqrt{4\kappa t}} \right) - i^2 \text{erfc}\left(\frac{z+l}{\sqrt{4\kappa t}} \right) \right) \quad . \tag{3.29}$$

The function "i^2erfc" in the equation above is defined as follows:

$$i^2 \text{erfc}(a) = \frac{1}{4} \left((1 + 2a^2)\text{erfc}(a) - \left(\frac{2}{\sqrt{\pi}} \right) a e^{-a^2} \right) \quad . \tag{3.30}$$

In eq. 3.30, a is used for the expression inside the function "i^2erfc". We need not be concerned how those equations were derived or even what this function is. The values of i^2erfc for different a can be looked up in mathematical tables. Eq. 3.29 was taken directly from Carslaw and Jaeger (1959) and can be easily implemented for the illustration of temperature profiles across shear zones (Fig. 3.8). In Fig. 3.8 the largest error is introduced by the assumption that the mechanical energy production rate was assumed to be constant. In reality, there is a feed back mechanism between heating and softening in the shear zone. Carslaw and Jaeger (1959) present for this problem some solutions in which the heat production rate can be varied as a function of time. However, for most realistic problems, where shear heat production feeds back on the rheology of the rock (e. g. in Fig. 3.7) it is wiser to use a numerical solution of eq. 3.24.

Is Shear Heating Geologically Relevant?

A range of authors have discussed the importance of shear heating on a geologically significant scale (e. g. Brun and Cobbold 1980; Lachenbruch 1980; Scholz 1980; Barton and England 1979; Graham 1976). However, the influence of shear heating on many tectonic and metamorphic processes remains contiguous. In this debate, we should not forget that the relevance of shear heating is a direct function of the rock strength. With this respect, the discussion of shear heating is analogous to the discussion of tectonic over-pressures: every argument for or against shear heating should be reduced to an argument about one of the following three points:

- What is the strength of the rocks under consideration?
- What is the strain rate?
- What is the relationship between the *duration* of deformation and the *size* of the deforming rock body?

Rock mechanics experiments are performed at strain rates of $\dot{\epsilon} = 10^{-6}\ \mathrm{s}^{-1}$. For the application of such data to deformation at geological strain rates, these data must be extrapolated over six to eight orders of magnitude of strain rate. The relevance of such experimental results remains therefore debated. As an alternative, various authors have attempted to reverse the problem and use geological observations in the field to deduce the shear strength of rocks. However, the strength of rocks and the importance of shear heating

Figure 3.8. Temperature profile across a shear zone of the width $2l = 50$ m at three different time steps labeled in my. The heat production rate in the shear zone is $10^{-4}\ \mathrm{W\,m^{-3}}$; the thermal conductivity k is $2.7\ \mathrm{J\,s^{-1}\,m^{-1}\,K^{-1}}$. The figure was calculated with eq. 3.28 and eq. 3.29. For another example where the same model may apply see p. 181

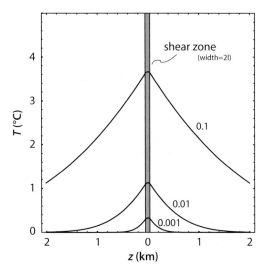

remains one of the most debates topics in the earth sciences. We can conclude from this active discussion in the literature, that the question is partly open and that it is unwise to take sides without some valued consideration of the problem.

Adiabatic Processes. "Adiabatic" means: "without change in heat content" or: "without change in enthalpy". If a rock is buried, the pressure rises because of the weight of the increasing overburden. The rock is compressed and there is work done to change its volume. The energy of the rock increases by the product of the applied lithostatic force and the distance of shortening during the volume change. It is a type of mechanically produced energy. If no change of the heat content of the system is allowed, then the rock must get warmer. The rock heats *adiabatically*. Correspondingly, an adiabatically heated rock will cool and expand when it is decompressed. We can observe adiabatic cooling on gas bottles that freeze on their surface when they are opened and adiabatic heating on bicycle tires that warm up when they are pumped up. Rocks are compressible enough so that adiabatic processes can become a geologically relevant process. For the consideration of adiabatic processes we define the *isothermal compressibility* β as the relative volume change, ∂V, per increment of pressure change, ∂P, at constant temperature. This has the units of Pa^{-1} (sect. 5.1.2):

$$\beta = -\frac{1}{V}\left(\frac{\partial V}{\partial P}\right)_T \quad . \tag{3.31}$$

We discuss the compressibility in sect. 5.1.2. More commonly we will encounter the thermal expansion coefficient α. Corresponding to the isothermal compressibility, α is given by the relative volume change per increment of temperature change at constant pressure This is given by:

$$\alpha = -\frac{1}{V}\left(\frac{\partial V}{\partial T}\right)_P \quad . \tag{3.32}$$

Rocks have typical values of $\alpha \approx 3 \cdot 10^{-5}$ K^{-1} and $\beta \approx 10^{-11}$ Pa^{-1}. When discussing adiabatic heating processes we need yet another parameter, which is the *adiabatic compressibility* (which is different from the compressibility at constant temperature). This is given by the change in density with change in pressure at constant entropy and should not be confused with the isothermal compressibility discussed in eq. 3.31. The adiabatic compressibility is smaller than the isothermal compressibility.

In the asthenosphere, convection is rapid enough so that conductively caused temperature gradients are rapidly equalized. Geotherms in the asthenosphere follow therefore the adiabatic gradient. Without going into the derivation in any detail we state here that this gradient is given by:

$$\left(\frac{dT}{dz}\right)_S = \frac{\alpha g T}{c_p} \quad . \tag{3.33}$$

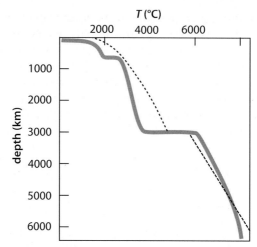

Figure 3.9. The temperature profile (thick line) and the melting curve (dashed line) between the earth's surface and the center of the earth in 6 300 km depth (after Jeanloz and Richter 1979; Jeanloz 1988). See Fig. 2.13 for scaling of this figure. In the mantle, the temperature increase is given by the adiabatic temperature gradient (eq. 3.33). The steps in the temperature profile arise from exothermic and endothermic reactions during phase transitions. Note that only the outer core is liquid according to this diagram. However, rapid exhumation of the upper mantle may lead to cross overs of the two curves above 500 km depth and therefore to partial melting. This is what happens underneath mid oceanic ridges or during continental extension

The subscript S indicates that this gradient is given at constant entropy. Do not confuse this subscript with the abbreviation S we have used for heat production elsewhere in this chapter. The constant g is the gravitational acceleration, T is temperature and c_p is the specific heat. Detailed derivations of eq. 3.33 are given by Turcotte and Schubert (1982) and in many other geophysical teaching texts. If we insert realistic numbers we see that the adiabatic temperature gradient in the mantle is about 0.3–0.5 °C km^{-1}. In other words, the adiabatic temperature change between the surface and the base of a 100 km thick lithosphere is about 50 °C. For most geological processes within the crust this is negligible. However, for geodynamic processes that require the consideration of large vertical length scales in the mantle, the process is of relevance. It is particularly important for melting processes in the upper mantle (Fig. 3.9).

3.2.3 Chemical Heat Production

Different rocks are characterized by different internal heat contents defined by the strength of bonding of the atoms in the crystal lattices in the rock-forming minerals. During chemical reaction, the difference in heat content between reactants and products is released or consumed as *latent heat of reaction.* By far the largest majority of chemical reactions are *endothermic* when the temperature increases. Because of this, temperature rise of rocks may be buffered by the phase transition. Correspondingly, most reactions are exothermic when crossed down temperature.   However, most chemical reactions have a positive slope in a pressure-temperature diagram. Thus exothermal reaction can not only be triggered by a *decrease* in temperature, but also by an *increase* in pressure (at constant temperature). In very general terms, we can summarize the importance of reaction heat in geological processes as follows:

- S_{chem} of reactions involving phase transitions is significant to the thermal budget and must be considered if phase transitions occur.
- In the solid state, dehydration reactions are the most important producer of reaction heat (Connolly and Thompson 1989; Peacock 1989). In the greenschist facies they produce of the order of $4 \cdot 10^6$ J per kg water. However, rocks contain only of the order of 4% H_2O and this water is being released over quite a large temperature interval. Thus, the heat of reaction is fairly insignificant during regional metamorphism. Connolly and Thompson (1989) estimated that metamorphic reaction produce of the order of 5–$10 \cdot 10^{-14}$ W cm^{-3}.
- S_{chem} of solid – solid reaction is negligible.

The geologically most important reactions that involve phase transitions are the melting reactions where the *latent heat of fusion* is released or the *latent heat of melting* is consumed. As a consequence, it is important to consider reaction heat when dealing with the thermal energy budget of migmatites and intrusions. A commonly used value for the latent heat of melting of rocks is $L = 320\,000$ J kg^{-1}. Evaporation and condensation reactions are also strongly exothermic and endothermic respectively, but they are not very important in the geodynamics of the lithosphere.

The rate of reaction heat production S_{chem} has the same units as any other heat production rate: W m^{-3}. It can be described by:

$$S_{chem} = L\rho\frac{dV}{dt} \quad . \tag{3.34}$$

In this equation L is the latent heat of reaction J kg^{-1}. L is typically listed in the units of energy per mass, rather than per volume and has therefore the units J kg^{-1}. Since we think of the chemical heat production rate as a *volumetric* heat production rate, it is necessary to multiply L by the density ρ to convert it into a volumetric heat content. The expression dV/dt is the volumetric proportion of the reaction that occurs per unit time (in s^{-1}). Note that V has the units of percent and not cubic meters. Thus, the equation determines the part of L that is freed in every time step of the reaction. Substituting eq. 3.34 into eq. 3.22 we can now formulate the temperature change during chemical heat production to be:

$$\frac{\partial T}{\partial t} = \frac{L}{c_p}\frac{\partial V}{\partial t} \quad . \tag{3.35}$$

If we consider diffusion of heat as well, but neglect other heat sources we can write:

$$\frac{\partial T}{\partial t} = \kappa\frac{\partial^2 T}{\partial x^2} + \frac{L}{c_p}\frac{\partial V}{\partial t} \quad , \tag{3.36}$$

where the first term on the right hand side is the diffusion term discussed at length in eq. 3.6. You will recall from there that $\kappa = k/\rho c_p$. In this form of the equation we also have neglected differences in density between the reactants and the products. It may help you to understand eq. 3.36 by formulating it in terms of energy rather than in terms of temperature. It then takes the form:

$$\frac{\partial H}{\partial t} = k\frac{\partial^2 T}{\partial x^2} + L\rho\frac{\partial V}{\partial t} \quad . \tag{3.37}$$

There, H is the heat content in $J\,m^{-3}$. The conversion between heat and temperature is illustrated in Fig. 3.10 and was briefly discussed in eq. 3.3. As the latent heat of melting reactions forms a significant part of the thermal energy budget of rocks, we will now discuss its influence.

Thermally Buffered Melting. Melting during prograde metamorphism in the upper amphibolite and granulite facies is a strongly endothermic process. Thus, the rate with which temperature increases during metamorphism at this grade will be buffered by the melting reactions. At univariant melting reactions, the temperature will remain constant until the phase transition from solid reactants to liquid products is complete. It is the very same reason why we have so much snow slush on our roads in spring: ice and water will both have a temperature of $0\,°C$, until all ice has melted, even if the air temperature has been above freezing for quite some time. For the same reason water will boil at a constant temperature of $100\,°C$, regardless of the heat added by the stove, until it all has evaporated. In the buffering interval, the amount of heat added to the rock from the outside is exactly balanced by the amount of heat consumed by the phase transition.

Most rocks consist of many chemical components in complicated chemical systems. As a consequence, they do not melt at a single temperature, but over a melting interval between their solidus (where the first melt appear during temperature increase) and liquidus (where the last remaining piece of rock melts). If we want to describe such rocks, we must find a formulation that allows us to release (or consume) the latent heat over a large temperature interval. For this it is useful to reformulate the rate of volume change from one phase to another from eq. 3.34 into:

$$\frac{\partial V}{\partial t} = \frac{\partial V}{\partial T}\frac{\partial T}{\partial t} \quad . \tag{3.38}$$

In this form it is easier to add the heat of reaction to the time derivative on the left side of eq. 3.36. This reads then as:

Figure 3.10. The relationship between heat content and temperature in the melting interval of a melting rock (shown between solidus temperature T_s and liquidus temperature T_l. The slope of the curves within the melting interval are for five different assumptions of the melting process according to eq. 3.41 (Fig. 3.11; after Stüwe 1995)

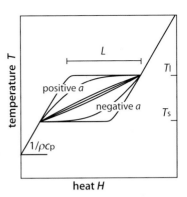

$$\left(c_p - L\frac{\partial V}{\partial T}\right)\frac{\partial T}{\partial t} = \frac{k}{\rho}\frac{\partial^2 T}{\partial x^2} \tag{3.39}$$

or:

$$\frac{\partial T}{\partial t} = \kappa_{mod}\frac{\partial^2 T}{\partial x^2} \ . \tag{3.40}$$

There, $\kappa_{mod} = k/(\rho c_{mod})$ is the modified diffusivity and c_{mod} is a modified heat capacity: $c_{mod} = c_p - L\rho(\partial V/\partial T)$. Chemical buffering of thermal processes can thus be described by modifying the diffusivity. The numerical problems that can arise when doing this were first described by Price and Slack (1954).

In principle, it is possible to implement the heat of reaction quantitatively and describe the evolution of PT paths under consideration of buffering processes. However, the estimates of Thompson and England (1984), Connolly and Thompson (1989), Peacock (1989), Barr and Dahlen (1989) and others shows that the magnitude of reaction heat does not justify to implement such a model in great detail.

How Do Rocks Melt? The modeling of problems where the latent heat of melting plays a role requires the knowledge of the relationship between melt volume and temperature. Only eutectic bulk compositions melt at a single temperature. Most realistic bulk compositions melt in a complicated sequence of reactions of high thermodynamic variance in an interval between a solidus and a liquidus temperature. Petrogenetic studies defining this relationship within the melting interval are still in their early stages (White et al. 2001). We are therefore confined to some simple model assumptions about this relationship. The relationship

$$V(T) = \frac{e^{aT} - e^{aT_s}}{e^{aT_l} - e^{aT_s}} \tag{3.41}$$

can be used to describe many aspects of such relationships. There, $V(T)$ describes the volumetric proportion of melt in the rock as a function of temperature between the solidus temperature T_s and the liquidus temperature

T_1. The constant a describes the slope of the function between the two and can be adjusted according to a chosen melt model (Fig. 3.11). If a has a large positive value, then most of the rock melts near the solidus. If a adapts large negative values, then most of the melting occurs near the liquidus. For values of $a \to$ zero, melting becomes linear in temperature. Many rocks contain hydrated phases at the onset of melting. As water is a great catalyst for melting processes, it is likely that more melting will occur near the solidus than near the liquidus. Realistic values for a are therefore likely to be positive and not too large. If we accept eq. 3.41 as a melting model, then the change of melt volume with temperature dV/dT can be derived from eq. 3.39 to be the following:

$$\frac{dV}{dT} = \left(\frac{a}{e^{aT_1} - e^{aT_s}} \right) e^{aT} \quad . \tag{3.42}$$

This relationship can be used directly to estimate the influence of melting on the thermal evolution of rocks. Stüwe (1995) showed that it may be large enough to account for the equilibration of parageneses in the low-pressure high-temperature metamorphic environment.

3.3 Principles of Heat Advection

Heat can be transported *actively* by the motion of warm rocks. We discern between *advection* and *convection* of heat. *Advection* is generally used if the active transport of heat is only in one-dimensional, for example the transport of heat by an intrusion that moves in the vertical direction. *Convection* is generally used when referring to material transport in a closed loop, for

Figure 3.11. A schematic model for the description of the relationship between melt volume and temperature. The curves are labeled for different values of the constant a in eq. 3.41. For $a = 0$, eq. 3.41 simplifies to the linear relationship $V(T) = T/(T_1 - T_s)$. The thick drawn line is probably the most realistic curve for the melting of hydrated metapelitic rocks

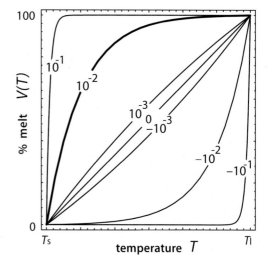

example the convection of mantle material in the asthenosphere, or that of fluids in a hydrothermal system. In this book, we only deal with advection. One-dimensional active transport of heat (for example in the vertical direction z), relative to the z direction may be described by:

$$\frac{\partial T}{\partial t} = u\frac{\partial T}{\partial z} \ .$$

(3.43)

In eq. 3.43, u is the transport velocity; the derivative $\partial T/\partial z$ describes the thermal gradient and $\partial T/\partial t$ is the change of temperature with time. For positive u, eq. 3.43 describes transport *against* the spatial coordinate z: transport is from high z towards lower z (Fig. A.8 illustrates the sign of u). In three dimensions, the heat advection equation is:

$$\frac{\partial T}{\partial t} = u_x\frac{\partial T}{\partial x} + u_y\frac{\partial T}{\partial y} + u_z\frac{\partial T}{\partial z} \ .$$

(3.44)

Although we shall discuss below some examples where eq. 3.43 may be solved analytically, the vast majority of geological problems require numerical solutions of eq. 3.43. In numerical solutions of advection problems we are often faced by the problem of *numerical diffusion*. (sect. A.2.4, A.2.4). One way to avoid problems with numerical diffusion is by converting the problem from an advection problem into one where there is no advection, but moving boundary conditions; i.e. we describe the problem in a Lagrangian, rather than a Eulerian reference frame (sect. 4.0.1). In Lagrangian descriptions the material is transported through the coordinate system. Eulerian reference frames move with the material.

Important examples of heat advection in the lithosphere are:

- advection of heat by magmas, e. g. magmatic intrusion;
- advection of heat by solid rock motion, e. g. during erosion or thrusting;
- advection of heat by fluids, e. g. during infiltration events.

There are fundamental differences between these three processes with respect to the way they need to be described. Thus, the three processes will be discussed separately.

3.3.1 Heat Advection by Intrusion

During intrusion of magma from deeper into shallower levels in the crust, the heat of the magma is transported to higher crustal levels by the motion of the magma itself. The process of magmatic intrusion is - in general - much faster than most other geological processes, for example the thermal equilibration during contact metamorphism. Thus, it is often unnecessary to describe the intrusion process itself by an advection equation. For questions related to the *thermal* evolution of contact metamorphism is usually suffices to assume that intrusion is infinitely rapid and can be described by instantaneous heating problems (s. sect. 3.6). Simple examples for good model assumptions to describe intrusions into the crust are given by Jaeger (1964) and for intrusion at the Moho (under plating) are given by Wells (1980) and Huppert and Sparks 1988). Thermal processes related to intrusion will be discussed at length in sect. 3.6. Nevertheless, we want to note that country rocks that heated during contact metamorphism are often referred to as being "advectively heated". Strictly speaking this is not correct as it is *conduction* from the intrusion into the country rocks, not *advection* itself that causes contact metamorphism (s. sect. 6.3.3).

3.3.2 Heat Advection by Erosion

During exhumation of rocks by erosion, the lithosphere (and its heat) are moved vertically upwards (s. sect. 4.1). The column is moved *through* a surface of constant temperature - the surface of earth. Erosion is therefore a heat advection process. In a similar way, any other motion of rocks, for example during thrusting or folding may be interpreted as an advective process, if viewed in an externally fixed Eulerian reference frame (s. p. 136). However, in the following we will only discuss one-dimensional, vertical advection of heat to and from the earth's surface. The time scale of continental denudation processes is comparable to the time scale of thermal equilibration on the scale of the crust and we can therefore not neglect to consider both processes at the same time. If we want to describe advection and diffusion of heat simultaneously, then we must expand eq. 3.43 by the diffusion term from eq. 3.6. The equation that must be solved becomes:

$$\frac{\partial T}{\partial t} = \kappa \frac{\partial^2 T}{\partial z^2} + u \frac{\partial T}{\partial z} \quad . \tag{3.45}$$

A schematic illustration how the two processes interact to shape a geotherm during erosion is shown in Fig. 3.12. You may also want to consider to expand this equation by yet another term describing heat production (e.g. the

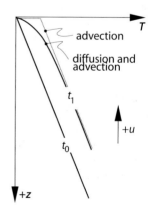

Figure 3.12. Schematic illustration of one-dimensional advection of heat by erosion. The coordinate system is fixed with $z = 0$ at the earth's surface. Temperature profiles through the crust are shown for two times: at the onset of erosion t_0, at which a linear geotherm is assumed and a later time t_1. The advection rate u is positive upwards. In the shown time interval the erosion process *advects* the geotherm by $u \times t_1$ meters upwards. Simultaneous *diffusion* causes the curvature of the temperature profile

term in eq. 3.22). Then, we would have a complete thermal energy balance to describe any thermal problem. However, it is strongly recommended to perform a careful evaluation of the relative importance of heat conduction, advection and production for a given problem to decide which terms must be considered and which not. Such a comparison may be done similar to that we have shown on p. 61. Here we consider only diffusion and advection.

Geotherms During Erosion and Sedimentation. Eq. 3.45 can be used to describe the evolution of a crustal geotherm that is advected through the earth's surface, for example during erosion (upwards advection) or sedimentation (downwards advection) (Benfield 1949). For complicated geotherms or variable erosion rates there is no analytical solution and the equation must be solved numerically. However, analytical solutions may be found if the following simple assumptions are made:

– The advection rate u is constant.
– The temperature profile through the crust is initially linear.
– The temperature at the earth's surface is constant.

These assumptions are a fairly good first order approximation for a large range of problems including the evolution of geotherms during erosion and sedimentation or – on a smaller scale – the evolution of temperature profiles in soil during accumulation or ablation of snow. In order to solve eq. 3.45 subject to these initial and boundary conditions, we also need to define a *lower* boundary condition. Depending on the lower boundary condition, solutions of eq. 3.45 do- or do not converge to a thermal steady state where the temperature profile does not change relative to the coordinate system. Examples for both steady state and time dependent evolution of geotherms will be discussed below:

• *Steady state geotherms during erosion.* If the basal boundary is fixed at a given temperature, solutions of eq. 3.45 will converge to a steady state where the upwards advection of heat is exactly balanced by conductive cooling from the surface. Here we present two different cases:

– 1. The lower boundary is fixed at infinity.
– 2. The lower boundary is fixed at a given depth.

For both cases eq. 3.45 may be solved to give time dependent descriptions of geotherms during erosion that converge to a steady state after some time. Here we restrict ourselves to the presentation of the steady state end member where $\partial T/\partial t = 0$. A general solution for this steady state scenario is presented in Spiegel (1968) and is well summarized by Mancktelow and Grasemann (1997) who also derived other solutions for a range of other problems including variably erosion rate and erosion of a heat producing crust (s. also Batt and Braun 1997).

1. If the upper boundary is fixed at $T_{(z=0)} = 0$ lower boundary is *fixed* with the temperature T_∞ as the depth goes to infinity ($T_{(z\to\infty)} = T_\infty$) then a steady state solution of eq. 3.45 is:

$$T = T_\infty \left(1 - e^{-(uz/\kappa)}\right) \quad . \tag{3.46}$$

2. If the upper boundary is fixed at $T_{(z=0)} = 0$ and the lower boundary is fixed at the temperature $T = T_L$ at a fixed depth $z = L$, a steady state the temperature is described by the solution:

$$T = T_L \left(\frac{1 - e^{-(uz/\kappa)}}{1 - e^{-(uL/\kappa)}}\right) \quad . \tag{3.47}$$

This latter case is quite a realistic scenario for the lithosphere where is may be fair to assume that the temperature at its base is constant (Stüwe et al. 1993b). Both cases are illustrated in Fig. 3.13.

• *Time dependent evolutions.* Time dependent solutions may also be found for the two problems discussed above. However, here we discuss a time dependent evolution for yet another scenario, namely one where there is no lower boundary condition, i.e. the advection of a semi-infinite half space through the surface. For that, and the assumption of an initially linear thermal gradient g, a solution of eq. 3.45 is:

$$T = gz - gut + \frac{g}{2}$$

$$\times \left((z + ut)e^{uz/\kappa}\mathrm{erfc}\left(\frac{z + ut}{\sqrt{4\kappa t}}\right) + (ut - z)\mathrm{erfc}\left(\frac{z - ut}{\sqrt{4\kappa t}}\right)\right) \quad . \tag{3.48}$$

(Benfield 1949; Carslaw and Jaeger 1959, chapter XV; Mancktelow and Grasemann 1997). Geotherms calculated with eq. 3.48 correspond to those qualitatively shown in Fig. 3.12.

Cooling Paths of Rocks. Eqs. 3.46, 3.47 and 3.48 describe temperatures as a function of time and depth. This is the point of view of a Eulerian observer who is fixed to the surface (s. sect. 4.0.1). For many questions is more interesting to follow the thermal history of a given rocks from the point of view of a Lagrangian observer. Within the reference frames of the geotherm descriptions, rocks move (are advected) through the coordinate system at the rate u. Thus, if they were at the initial depth z_i at the onset of erosion, they change their depth to:

$$z = z_i - ut \tag{3.49}$$

at time t. Thus, if we want to calculate cooling histories of a given rock that are due to its exhumation by erosion, we can simply use eqs. 3.46, 3.47 or 3.48 and substitute eq. 3.49 for z. Cooling curves that were calculated with this approach and eq. 3.48 are shown in Fig. 3.14.

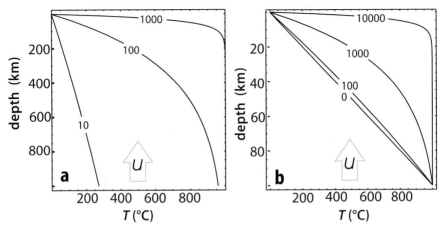

Figure 3.13. Examples of steady state geotherms during simultaneous upwards advection and diffusion. **a** Geotherms fixed at the surface with $T = 0$ and at infinite depth at temperature $T_\infty = 1000°C$ (calculated with eq. 3.46). **b** Geotherms fixed at the surface with $T = 0$ and at depth $L = 100\,000$ with $T_L = 1000°C$ (calculated with eq. 3.47). Contours are for different advection rates u labeled in meters per million years and $\kappa = 10^{-6}\,\mathrm{m^2 s^{-1}}$

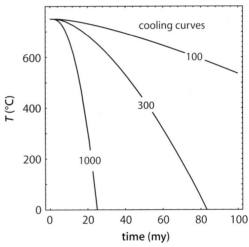

Figure 3.14. Cooling curves of rocks that cool as the consequence of exhumation only. The three curves are for three different erosion rates in $\mathrm{m\,my^{-1}}$. Calculated with eqs. 3.48 and 3.49. Other assumptions are: $g = 30°C\,\mathrm{km^{-1}}$ and $\kappa = 10^{-6}\mathrm{m^2 s^{-1}}$. An important feature of the cooling curves is that the cooling *rate* increases with decreasing temperature which is different from cooling curved following contact metamorphism

3.3.3 Heat Advection by Fluids

Heat may also be advected by fluids that circulate through rocks. Heating of rocks due to fluid advection is different from the previous examples, because only part of the rock volume is being advected, namely the fluids that fill the pore volume. Thus, when formulating an advective term in an advection-diffusion equation, we need to take care so that we describe only the advection of a fraction of the total rock volume. In a general one-dimensional form eq. 3.43 can be written as:

$$\frac{\partial T}{\partial t} = \phi v_{\mathrm{f}} \left(\frac{\rho_{\mathrm{f}} c_{pf}}{\rho c_p} \right) \frac{\partial T}{\partial z} \quad , \tag{3.50}$$

(McKenzie 1984). There, ϕ is the porosity of the rock (s. sect. 6.1.3) and v_{f} is the fluid flux in $\mathrm{m^3 m^{-2} s^{-1}}$. The product ϕv_{f} is the fluid volume that is transported per unit time and per unit area through the rock. This product has the units of $\mathrm{m\,s^{-1}}$, which corresponds to the standard definition of fluxes (sect. 4.3.4). It is called the *volumetric fluid flux*. ρ and ρ_{f} are the densities and c_p as well as c_{pf} are the specific heat capacities, both of the rock and the fluid, respectively. The term inside the brackets is the ratio of the volumetric heat capacities (volumetric heat capacity = heat capacity/density) of the fluid to that of the rock.

Eq. 3.50 may be used to describe the thermal effects of fluid advection. However, in many geological processes heat advection by fluids occurs on similar time scales as heat conduction. Thus, it is usually necessary to expand eq. 3.50 by a term describing diffusion as we did in eq. 3.50. The importance of the transport of heat by fluids for the thermal evolution of the crust was discussed by Bickle and McKenzie (1987), Connolly and Thompson (1989) as well as Peacock (1989). These authors agree that the fluid flux that may be caused by metamorphic dehydration reactions is less than about 1 kg fluid per square meter and per year. This is not enough to transport heat very efficiently by fluids. Peacock (1989) estimated that the thermal evolution of rocks can only be influenced by fluids if these are focused into narrow zones from regions as wide as 10 km.

3.3.4 The Peclet Number

In many geological processes the diffusion rates and advection rates are of the same order of magnitude. This is true for fluid infiltration processes, for erosion that occurs during thermal equilibration of the crust, thermal profiles around moving faults and many more (sect. 3.7.4). In such processes the Peclet number Pe is a useful parameter which can be used to estimate the relative influence of diffusive and advective processes. The Peclet number is defined as:

$$Pe = \frac{ul}{\kappa} \tag{3.51}$$

where u is again the rate of advection, κ the diffusivity and l the characteristic length scale of the advection process. If Pe is about 1, then diffusion and advection are of similar importance to a process. If Pe is much larger than 1, advection dominates the process. If Pe is much smaller than 1, then diffusion dominates the process. Aside from its importance for the description of thermal processes, the Peclet number finds many other application. For example, during fluid infiltration, Pe may be defined as:

$$Pe_f = \frac{v_f l \rho_f c_{pf}}{k} \quad . \tag{3.52}$$

There, v_f is the volumetric flux, ρ_f is density, c_{pf} is the specific heat capacity of the fluid, l the characteristic length scale and k is the conductivity. Bickle and McKenzie (1987) have used the Peclet number for some fundamental interpretations about the relative importance of diffusive and advective processes during fluid infiltration of rocks.

3.4 Heat in the Continental Lithosphere

In the past sections we have discussed the three fundamental processes that produce and redistribute heat in the lithosphere. In summary from there we can conclude that a full one-dimensional description of the thermal energy balance for the lithosphere has the form:

$$\frac{\partial T}{\partial t} = \left(\frac{k}{\rho c_p} \right) \frac{\partial^2 T}{\partial z^2} + u \frac{\partial T}{\partial z} + \left(\frac{S}{\rho c_p} \right) \quad , \tag{3.53}$$

where the diffusivity is the ratio of conductivity and density × heat capacity: $\kappa = k/\rho c_p$ and the heat production S may consist of three contributors (s. eq. 3.23). In this section we will apply our knowledge from the last sections and describe aspects of the thermal structure of the continental lithosphere using this equation. However, not all of the terms in eq. 3.53 may be relevant. Table 3.4 shows (as a summary from the last sections), which thermal energy parameters may be relevant to thermal estimates on the scale of the lithosphere. In order to set the scene, we begin with a brief explanation how the lithosphere may be defined thermally and what "geotherms" actually are.

Thermal Definition of the Lithosphere. The lithosphere may be defined thermally or mechanically (s. sect. 2.4.1). According to the thermal definition, the lithosphere is the outer shell of the earth, in which heat is transported primarily by conduction. In contrast, in the asthenosphere, heat is transported primarily by convection. Thus, the lithosphere itself is nothing but a *thermal boundary layer* of the earth. This boundary layer looses heat at all times through the earth's surface into the atmosphere and further – by radiation – into space. The average heat flow through the surface of the continents is 0.0565 W m^{-2}. The total surface area of the continents is about $A_c = 2 \cdot 10^8$ km^2. Thus, the total heat loss of earth from the continents is $1.13 \cdot 10^{13}$ J s^{-1}. This heat loss is balanced by radioactive heat production within the lithosphere and by heat flow into the lithosphere from the asthenosphere, so that this thermal boundary layer has a largely constant temperature profile, if it is not disturbed by orogenesis. Thermally stabilized lithosphere has a thickness between 100 and 200 km (Pollack and Chapman 1977).

Table 3.4. Summary of possible heat transfer mechanisms in the crust and their potential relevance for the heat budget of metamorphic terrains. The column "*relevance*" makes some very crude suggestions whether the processes listed in the 1st and 2nd column can be of relevance for terrain scale metamorphism. The relationships shown in the 4th column are the key relationships that should be considered when estimating the relevance of a given transfer mechanism in a given terrain. *Conduction* is only relevant if the time scale of conduction, τ, is comparable to the ratio of conduction length scale l and thermal diffusivity κ. *Advection* is only relevant if the Peclet number, Pe, is larger than 1, where u is the advection rate, l length scale of the advective process and κ the diffusivity. *Heat production* is only relevant if the heat production rate S times its duration t produces significant amounts of thermal energy (which may then be converted into temperature rise using density and heat capacity). For *radioactivity* this product is generally large on geological time scales, for *chemical* heat production this product is only large for melting reactions and for *mechanical* heat production it is given by the product of stress and strain rate. Both stress and strain rate are not well known and the relevance of mechanical heat production to metamorphism remains a much debated topic in earth sciences. Acronyms for selected key references are: ER79 = England and Richardson (1979); ET84 = England and Thompson (1984); HS88 = Huppert and Sparks (1988); BM87 = Bickle and McKenzie (1987); J64 = Jaeger (1964); L86 = Lux et al. (1986); L70 = Lachenbruch (1970); S98 = Sandiford et al. (1998); C90 = Chamberlain and Sonder (1990); ME90 = Molnar and England (1990a); S98 = Stüwe (1998); S95 = Stüwe (1995); P89 = Peacock (1989); CT89 = Connolly and Thompson (1989)

transfer mechanism	geological process		geol. relevance	relationship to consider	key ref.
conduction			large	$\tau = l^2/\kappa$	ER79; ET84
advection	by fluids		rare		BM87
	by magma	- intrusion	large		J64; L86
		- underplating	large	$Pe = ul/\kappa$	HS88
	by rock	- erosion	large		
		- deformation	large		
production	radioactive		large		L70; S98; C90
	mechanical		unknown		ME90; S98
	chemical	- melting	large	$S \times t$	S95
		- dehydration	small		P89; CT89
		- soldid-solid	negligible		P89

Definition of Geotherms. The function that describes temperature in the lithosphere as a function of depth is what we call a *geotherm*. We discern:

- stable or steady state geotherms,
- transient geotherms.

• *Stable geotherms.* Stable or *steady state* geotherms form by long term thermal equilibration of the lithosphere (sect. 3.4.1). In general, this is understood that the term "steady state" refers to a geotherm in a stationary lithosphere and we shall use it in this way in this section. However, in other reference frames, steady state geotherms may also occur in a moving lithosphere (for example a lithosphere that moves upwards relative to the surface during erosion, s. p. 76). In sect. 3.7.3 and 3.5.2 we discuss examples of steady state geotherms in Eulerian reference frames.

In most geological situations, the temperatures of steady state geotherms increase steadily with depth. Stable geotherms are only found in regions that have had at least about 100 my time for equilibration and have not changed in thickness during this time. The origin of this number is discussed in sect. 3.1.4. Thus, active orogens are *not* characterized by stable geotherms. Regardless, the calculation of steady state geotherms in orogens may help us to estimate the maximum or minimum temperatures that can be attained during an orogenic process at a given depth. This maximum or minimum possible temperature is often called *potential temperature* (e.g. Sandiford and Powell 1990).

• *Transient geotherms.* Transient geotherms are only valid for a particular point in time. In some geological situations, transient geotherms do *not* increase steadily with depth and the change of the geotherm with time can be different in different depths. For example, after rapid stacking of nappes, rocks may simultaneously heat above a major thrust, but cool below it (sect. 7.4.2). In principle it is possible to document such relationships of transient geotherms in space and time with careful observation in the field. If they can be documented, then such information is invaluable for the interpretation of the nature of a tectonometamorphic process (s. sect. 7.4.2). We discuss the transient evolution of geotherms during orogenic processes in several other chapters (e. g. sect. 3.6 and 6.2.1).

3.4.1 Stable Geotherms

In this section we calculate the quantitative shape of the geotherm shown in Fig. 2.15. We concentrate on *steady state* or *stable* geotherms and on *continental* lithosphere only. For the steady state case, the heat conduction equation (eq. 3.6) or the full thermal energy balance (eq. 3.53) can be simplified enough so that it is possible to find analytical solutions, even without a lot of mathematical knowledge. This is therefore a good example to familiarize ourselves with the involved thought process. The equation we must solve is the familiar heat conduction equation with a heat production term:

$$\frac{\partial T}{\partial t} = \left(\frac{k}{\rho c_p} \right) \frac{\partial^2 T}{\partial z^2} + \frac{S}{\rho c_p} \ . \tag{3.54}$$

We need the heat production term to account for radioactivity, which is of substantial importance to stable geotherms. However, we can neglect advective terms as we consider only steady state. For *steady state* geotherms, there is no change of the temperature with time (s. sect. 3.1.2). This means:

$$\frac{\partial T}{\partial t} = 0 \ .$$

Eq. 3.54 simplifies to:

$$\left(\frac{k}{\rho c_p}\right)\frac{\mathrm{d}^2 T}{\mathrm{d}z^2} + \frac{S}{\rho c_p} = 0 \ . \tag{3.55}$$

Note that eq. 3.55 is no *partial* differential equation anymore. By further canceling out of the constants we get:

$$k\frac{\mathrm{d}^2 T}{\mathrm{d}z^2} = -S \ . \tag{3.56}$$

The integration of this equation forms the basis for all calculations of stable geotherms.

• *Geotherms without radioactivity.* Equation 3.56 may be integrated the easiest, if we neglect radioactive heat production all together. In chap. 2 and sect. 3.2.1 we have shown that radioactive heat production is critical for the temperature profiles of the lithosphere (e. g. Fig. 2.15). However, in order to understand the integration of eq. 3.56, we begin by assuming the heat productiion is negligible: $S = 0$. Eq. 3.56 then simplifies to:

$$k\frac{\mathrm{d}^2 T}{\mathrm{d}z^2} = 0 \quad \text{or} \quad \text{even}: \quad \frac{\mathrm{d}^2 T}{\mathrm{d}z^2} = 0 \ . \tag{3.57}$$

As this still is a differential equation of the second order, me must integrate it twice to solve it. A first integration gives:

$$\frac{\mathrm{d}T}{\mathrm{d}z} = C_1 \tag{3.58}$$

and a second:

$$T = C_1 z + C_2 \ . \tag{3.59}$$

The two integration constants C_1 and C_2 must be determined by the geological boundary conditions (s. sect. A.1.1). For example, we can assume that the temperature at the surface of earth (at $z = 0$) is constant and has the value $T = 0$. Then, for eq. 3.59 to hold, C_2 must be zero so that the temperature is zero at $z = 0$. It we assume a thermal definition of the lithosphere, then we can determine the other constant with the assumption that $T = T_1$ at depth $z = z_1$. With this assumption C_1 must have the value $C_1 = T_1/z_1$. The temperature as a function of depth is therefore described by:

$$T = z\frac{T_1}{z_1} \ . \tag{3.60}$$

Equation 3.60 describes a linear temperature profile between the surface and the base of the lithosphere. This is not very surprising as we have assumed no heat production and no other reasons why the temperature profile should be curved. With a thermal conductivity of $k = 2\text{–}3$ W m^{-1} °C^{-1}, and $T_l = 1\,200$ °C as well as a lithospheric thickness of $z_l = 100$ km, our equation describes a surface heat flow of 0.024–0.036 W m^{-2}. This value is much lower than the average surface heat flow of the continents which is between 0.04 and 0.08 W m^{-2}. This is one of the proofs of the existence of radioactivity in the lithosphere. We can easily conclude that eq. 3.57 is not a very good model description and that it is wiser to integrate eq. 3.56 using a meaningful function that describes S as a function of depth. When we do so, we will always assume that $S = S_{rad}$, i.e. there is no other heat production sources but the radiogenic ones. In the steady state mechanical or chemical heat production sources are irrelevant.

The Contribution of Radioactivity. The radioactive heat production rate of rocks is of the order of some microwatts per cubic meter (s. Table 3.3). However, unusually high heat productions have been reported from a series of locations around the world, in particular from Australia (McLaren et al. 1999) (Fig. 3.6). A typical value measured from samples at the earth's surface is: $S = 5 \mu$W m$^{-3} \equiv 5 \cdot 10^{-6}$ W m^{-3}. If this value would be representative for the entire crust, then the radioactive contribution of a 30 km thick crust ($z_c = 30$ km) to the surface heat flow would be: $q = S \cdot z_c = 0.03$ W m^{-2}. From eq. 3.1 we know that the thermal gradient has the units of heat flow divided by the thermal conductivity. If the thermal conductivity is $k = 3$ W m^{-1} K^{-1}, then the assumptions from above indicate: $\mathrm{d}T/\mathrm{d}z = q/k = 0.05$ °C m$^{-1} = 50$ °C km^{-1}. This geothermal gradient of 50 °C per kilometer is only due to the contribution of radioactivity. The mantle heat flow would have to be added to this. Since the resulting thermal gradient would be much higher than just about all thermal gradients measured on earth, we can conclude that the radioactivity of rocks measured at the earth's surface must be higher than that of the rest of the crust (for more information on the radioactive heat flow contribution see p. 273).

The depth to which the crust produces radioactive heat has been elegantly determined using the relationship of two independent sets of data that can be measured at the surface: The surface heat flow and the heat production rate at the surface, S_0. Roy et al. (1968) explored this relationship in the eastern US and its significance was described by Lachenbruch (1968, 1970, 1971). They found a roughly linear relationship between these two parameters (Fig. 3.15). The straight line that fits these data has the form:

$$q_s = q_m + q_{rad} = q_m + z_{rad} S_0 \quad . \tag{3.61}$$

In this equation, q_s is the surface heat flow, q_m is the mantle heat flow, q_{rad} the radiogenically produced heat flow and z_{rad} is the thickness of a hypothetical layer in which radioactive heat is produced at the same rate as

Figure 3.15. Measured data of surface heat flow q_s and surface heat production S_0 in the eastern US. The best line that fits the data is described by the equation $q_s = 0.035 + 7\,413 S_0$. Accordingly, the thickness of the layer that produces heat at a rate S is $7\,413$ m thick and the contribution of mantle heat flow to the total heat flow is $0.035\ \mathrm{W\,m^{-2}}$ (after Roy et al. 1968)

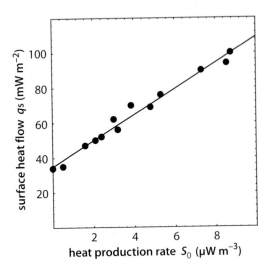

on the surface. q_m can be measured from the intersection of the line with the heat flow axis and the value of z_{rad} is given by its slope. The data of Roy et al. (1968) show that z_{rad} is about 7 km in the eastern US. Similar considerations in other areas indicate thicknesses of 10–15 km. Of course, the crust does not produce heat constantly in this layer and no heat at all below it, but the relationship is useful to estimate the total heat production in the crust. This is given by the product $z_{rad}S_0$. This product corresponds to the area underneath the different model curves in Fig. 3.16.

For the calculation of geotherms in the following sections we will use different model assumptions about the distribution of radioactivity with depth (Fig. 3.16). We will see that these different assumptions have important different implications for the temperatures in both the crust and the mantle part of the lithosphere. So we want to remember that both the *magnitude* and the *distribution* of radioactive heat production are important for the thermal budget of the crust (Sandiford et al. 1998, 2002; Chamberlain and Sonder 1990).

Geotherms With Heat Flow Boundary Condition. We begin our calculations of stable geotherms by integrating eq. 3.56 by assuming a somewhat different boundary condition at the base than what we have used to derive eq. 3.60. We will assume as boundary condition that the *heat flow* at the Moho is constant, instead of assuming the *temperature* at the base of the lithosphere. We do this for two reasons. Firstly because this boundary condition has been the assumption of England and Thompson (1984) for their classical calculations of continental geotherms. Secondly, we do this because the integration of eq. 3.56 is quite easy with this boundary condition.

• *Constant heat production.* If the heat production is independent of depth, S is a constant. Integration of eq. 3.56 results in:

Figure 3.16. Four simple models describing the distribution of heat production with depth in the crust (s. Haack 1983). The total heat production of the crust is given by the integral underneath the model curves. It is the same for all four models. For illustration, it is shaded for model c. a Constant concentration in the entire crust and no heat production in the mantle. b Constant concentration in the upper crust in a layer with the thickness z_{rad} and no heat production below that. c Exponential drop off of the heat production with depth. d Heat production peaking in the middle crust. Such a situation may occur if a crust with heat source distribution c is buried underneath a low heat producing sedimentary pile. Note that the surface heat production is different in the different models, while the total heat production is the same

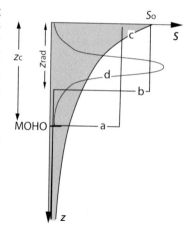

$$k\frac{\mathrm{d}T}{\mathrm{d}z} = -Sz + C_1 \quad . \tag{3.62}$$

The left side of this equation has the units of heat flow. The constant of integration C_1 must be derived from the boundary conditions. If we know the mantle heat flow at the Moho, and assume that it is constant through time, we can describe this condition as: q_m=constant at $z = z_c$. If no details are known about the changes of the heat flow at this depth, then this is the most simple assumption and corresponds therefore with the idea of a good model (sect. 1.1). With this assumption, the constant of integration must be:

$$C_1 = Sz_c + q_m \quad .$$

so that at the depth $z = z_c$ (z_c is the crustal thickness) we get: $k\left(\frac{\mathrm{d}T}{\mathrm{d}z}\right) = q_m$. After inserting C_1 into eq. 3.62 and a second integration we get:

$$kT = -\frac{Sz^2}{2} + Sz_c z + q_m z + C_2 \quad . \tag{3.63}$$

The most meaningful assumption for the second boundary is that the temperature at the surface at $z = 0$ has the value $T = 0$. This gives us that $C_2 = 0$. After some simplification we get the following expression:

$$T = \frac{Sz}{k}\left(z_c - \frac{z}{2}\right) + \frac{q_m z}{k} \quad . \tag{3.64}$$

Equation 3.64 describes the temperature as a function of depth in the crust. It is an *analytical solution* of the differential equation eq. 3.56. An analysis of the units can be used to confirm the internal consistency of this equation. A geotherm calculated with eq. 3.64 is shown in Fig. 3.17a. It is immediately obvious that the temperatures are much too high, although

Figure 3.17. Two calculated continental geotherm. It was assumed that the mantle heat flow has the value $q_m = 0.025 \ \mathrm{W \, m^{-2}}$. Curve a was calculated assuming that the entire crust has a constant heat production. It was calculated with eq. 3.64. Curve b was calculated with eq. 3.70 assuming that the heat production decreases exponentially with depth ($h_r = 10$ km). For a and b following other assumptions were made: $z_c = 35$ km; $S = 3 \cdot 10^{-6} \ \mathrm{W \, m^{-3}}$ ($= S_0$ in eq. 3.70; $k = 2 \ \mathrm{J \, s^{-1} \, m^{-1} \, K^{-1}}$

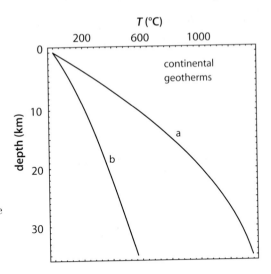

we made meaningful assumptions for mantle heat flow, crustal thickness and surface heat production. The reason is that we made the incorrect assumption that the heat production is constant in the whole crust and has everywhere the same value as on the surface. This assumption is in contrast to what we argued on p. 86, where we discussed the distribution of heat production with depth. The *real* distribution of the heat producing elements in the crust is not very well known, but we can help ourselves with one of the models from Fig. 3.16. If we make the assumption b from that figure (heat is only produced at a constant rate down to the depth z_{rad}), then there is a discontinuity in the heat production with depth. As a consequence, we can integrate eq. 3.56 only to the depth $z = z_{rad}$. We get a result that is very similar to eq. 3.64:

$$T = \frac{Sz}{k}\left(z_{rad} - \frac{z}{2}\right) + \frac{q_m z}{k} \quad \text{in the region:} \quad z < z_{rad} \ . \tag{3.65}$$

Below this depth the heat flow must be constant and the thermal gradient linear as there is no disturbing influence (s. eq. 3.60). The temperature at the base of the heat producing layer may be found by evaluating eq. 3.65 at $z = z_{rad}$. Below this depth the temperature $q_m(z - z_{rad})/k$ must be added to the temperature at the base of the assumed heat producing layer. It follows that:

$$T = \frac{Sz_{rad}^2}{2k} + \frac{q_m z}{k} \quad \text{in the region:} \quad z > z_{rad} \ . \tag{3.66}$$

Equations 3.65 and 3.66 are those that England and Thompson (1984) used for their classic model explaining regional metamorphism (s. sect. 6.2.1). If we insert reasonable values for the parameters in this equation, then we get temperatures at the base of the crust between 500 °C and 600 °C. We also get a depth for the base of the lithosphere (the depth of the 1 200 °C isotherm) around 100 to 150 km. Both corresponds to a large number of observations from the stable continental shields.

• *Exponential heat production.* A much more elegant model assumption for the distribution of the heat production in the crust is the assumption that there is a continuous exponential drop off in radioactive heat production with depth (model c in Fig. 3.16). This model has the great advantage that there is no discontinuity in the heat production in the crust. This also appears intuitively more appropriate. It is also much easier to formulate temperatures as a function of depth with a single equation. We assume that:

$$S_{(z)} = S_0 e^{\left(-\frac{z}{h_r}\right)} \ . \tag{3.67}$$

The variable h_r is called the *characteristic drop off* or *skin depth* of heat production (s. Problem 3.13). According to eq. 3.67, the heat production at depth $z = h_r$ is only the $1/e$ part of the heat production at the surface S_0. Our new starting equation is now:

$$k\frac{\mathrm{d}^2 T}{\mathrm{d}z^2} = -S_0 e^{\left(-\frac{z}{h_r}\right)} \ . \tag{3.68}$$

The integration of this equation is a bit harder than those of the previous section. A first integration gives:

$$k\frac{\mathrm{d}T}{\mathrm{d}z} = h_r S_0 e^{\left(-\frac{z}{h_r}\right)} + C_1 \ . \tag{3.69}$$

Note that we used the "product rule" for this integration (s. appendix B, Table B.1, B.2). With the same boundary condition as that used in the last section, the constant of integration is:

$$C_1 = -h_r S_0 e^{\left(-\frac{z_c}{h_r}\right)} + q_m \ .$$

After inserting this constant into eq. 3.69 and a second integration we get:

$$kT = -h_r^2 S_0 e^{\left(-\frac{z}{h_r}\right)} - z h_r S_0 e^{\left(-\frac{z_c}{h_r}\right)} + q_m z + C_2 \ . \tag{3.70}$$

The second constant must be $C_2 = h_r^2 S_0$ so that the condition $T = 0$ at the surface ($z = 0$) is fulfilled. Fig. 3.17b shows a geotherm calculated with eq. 3.70 after inserting this integration constant and using reasonable values for all parameters. We see that the geotherm has a realistic shape.

In the previous sections we have used a known mantle heat flow as the lower boundary condition. Our calculations were therefore confined to the description of the crust. The thickness of the mantle part of the lithosphere was implicitly determined by the boundary conditions (Fig. 3.18). With a thermal conductivity of $k = 3$, and a mantle heat flow at the Moho of 0.03 W m^{-2}, this gives a thermal gradient of 10 °C km^{-1} at the Moho. As we assumed no radioactive heat production in the mantle part of the lithosphere, this gradient

Figure 3.18. This illustration shows how the thickness of the mantle part of a thermally defined lithosphere is automatically determined by the condition of a constant mantle heat flow at the Moho. All three geotherms on this figure have the same thermal gradient at the Moho and therefore the same heat flow through the Moho. However, the mantle parts of the lithosphere (thickness shown by double arrows) is very different for the three examples

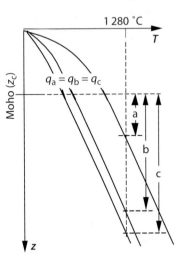

will be constant in the remainder of the mantle lithosphere. If the Moho-temperature is 500 °C then this gradient implies a thickness of the mantle part of the lithosphere of 70 km (if we define the lithospheric base thermally by the depth where the geotherm intersects the 1 200 °C isotherm).

This implicit determination of the thickness of the mantle lithosphere can be quite misleading if we want to relate our results of a geotherm calculation with thoughts about surface elevation, lateral buoyancy forces or other parameters that depend on the thickness of the entire lithosphere. Thus, the lower boundary condition we have used above is not always the best choice. In the next section we discuss alternative lower boundary conditions.

Geotherms with Fixed Basal Boundary Condition. If we define the lithosphere thermally, we implicitly state that we know the temperature at its base. An obvious choice for a lower boundary condition may therefore be: $T = T_1$ at the depth $z = z_1$. This choice allows us to describe temperatures in the entire lithosphere. However, we have to pay for this advantage by having to deal with various discontinuities of the parameters at the Moho, for example density. This complicates the integration of eq. 3.56.

• *Constant heat production.* In a model where we assume constant heat production rate in the crust and no heat production in the mantle part of the lithosphere, density *and* heat production are discontinuous at the Moho. This complicates the integration of eq. 3.56 dramatically. We will not present the equations here and refer the interested reader to the original works of Sandiford and Powell (1990) or Zhou and Sandiford (1992). However, for comparison with the thermal model of England and Thompson (1984) we show an example of a geotherm calculated with these assumptions as curve a in Fig. 3.19. We see that this model results in unrealistically high temperatures if we assume the surface heat production rate to be representative for the whole crust.

Figure 3.19. Examples of continental geotherms calculated with a lower boundary condition of a fixed temperature at the base of the lithosphere. Geotherm a was calculated assuming constant heat production in the crust and no heat production in the mantle lithosphere. Geotherm b was calculated for a continuous, exponentially decreasing heat production using eq. 3.74. The temperature T_l is assumed to be $1\,280\,°C$; all other parameters are the same as in Fig. 3.17

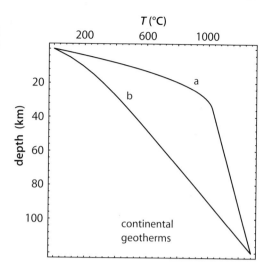

• *Exponential heat production.* If we assume a continuous heat production in the whole lithosphere that decreases exponentially with depth, then we can derive from eq. 3.68 an elegant and simple description of stable continental geotherms. After two integrations we get:

$$kT = -h_\mathrm{r}^2 S_0 e^{\left(-\frac{z}{h_\mathrm{r}}\right)} + C_1 z + C_2 \quad . \tag{3.71}$$

Both constants of integration can only be evaluated after this second integration. The lower boundary condition of $T = T_l$ at the depth $z = z_l$ can be evaluated by rearranging eq. 3.71:

$$C_1 = \frac{kT_l}{z_l} + \frac{h_\mathrm{r}^2 S_0 e^{\left(-\frac{z_l}{h_\mathrm{r}}\right)}}{z_l} - \frac{C_2}{z_l} \quad . \tag{3.72}$$

After inserting eq. 3.72 in eq. 3.71 we can evaluate C_2 from the second boundary condition for which we use again $T = 0$ at the surface where $z = 0$:

$$C_2 = h_\mathrm{r}^2 S_0 \quad . \tag{3.73}$$

After inserting both constants into eq. 3.71 we get:

$$T = \frac{zT_l}{z_l} + \frac{h_\mathrm{r}^2 S_0}{k} \left(\left(1 - e^{\left(-\frac{z}{h_\mathrm{r}}\right)}\right) - \left(1 - e^{\left(-\frac{z_l}{h_\mathrm{r}}\right)}\right) \frac{z}{z_l} \right) \quad . \tag{3.74}$$

Curve b in Fig. 3.19 is an example of a geotherm calculated with this relationship. Eq. 3.74 provides a realistic and useful description of stable continental geotherms and has been presented and used by a number of authors (Zhou and Sandiford 1992; Stüwe and Sandiford 1995; Mancktelow and Grasemann 1997: eq. 13).

• *More general formulations.* Using a lower boundary condition where we explicitly prescribe the thickness of the mantle part of the lithosphere (as we did in the previous section) is extremely useful to explore the influence of thickness variations of crust and mantle lithosphere on the temperatures in the crust. In order to do this more efficiently, it is useful to introduce two new parameters: the vertical thickening (or thinning) strain of the crust and that of the lithosphere. We call these parameters f_c and f_l and will discuss them in some detail in sect. 4.0.2 (s. a. sect. 5.1)(Sandiford and Powell 1990). A value of $f_c = 2$ means that the crust is twice as thick as in the reference state (Sandiford and Powell 1990; Zhou and Sandiford 1992). Using these parameters, eq. 3.74 can be generalized. All we need to do is multiply the reference crustal and lithospheric thicknesses z_c and z_l with their respective thickening strains. However, we also need to be careful in thickening the skin depth of the heat production as well. Eq. 3.67 becomes:

$$S_{(z)} = S_0 e^{\left(-\frac{z}{h_r f_c}\right)} \quad . \tag{3.75}$$

After substitution of these generalized formulations for the thicknesses of crust, lithosphere and skin depth into eq. 3.68 and integrate it subject to the same boundary conditions we used for eq. 3.74, we arrive at:

$$T = \frac{z T_l}{f_l z_l} + \frac{f_c^2 h_r^2 S_0}{k}\left(\left(1 - e^{\left(-\frac{z}{f_c h_r}\right)}\right) - \left(1 - e^{\left(-\frac{z_l f_l}{f_c h_r}\right)}\right)\frac{z}{f_l z_l}\right) \quad . \tag{3.76}$$

This equation is the most general and most elegant form of a geotherm equation that can be used to calculate temperature in the entire lithosphere. In the next section we will use this equation to calculate some important temperatures.

Figure 3.20.
Moho-temperatures of continental lithosphere for different crustal thickening strains (expressed by f_c) and for different total thickening strains of the lithosphere (expressed by f_l). The diagram was calculated with eq. 3.70. The assumption of the parameters are the same as in Fig. 3.19. The thick line in the bottom right hand corner of the diagram marks the limit of the allowed part of parameter space. It is explained in more detail in the context of Fig. 4.5

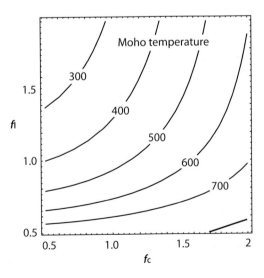

• *Examples of application of eq. 3.76.* For many thermal and mechanical considerations the Moho-temperature is a meaningful parameter that can be used to characterize the thermal structure of the entire lithosphere (sect. 6.2.2). If we want to calculate this temperature for a series of crustal and lithospheric thicknesses, we must evaluate eq. 3.76 at $z = z_c$ for a series of f_c and f_l. Figure 3.20 shows an f_c-f_l-diagram of Moho-temperature. It shows that the Moho-temperature does not change very much for homogeneous thickening of the entire lithosphere (which would appear on this figure as a diagonal path from $f_c = f_l = 1$ towards $f_c = f_l = 2$). This is because the heating effect of the increased thickness of the radioactive heat producing crust is almost balanced by the cooling effect of the thickened mantle part of the lithosphere. We have indicated this effect already in the discussion of Fig. 2.15a. However, the strong curvature of the contours on Fig. 3.20 shows that in very thin crusts ($f_c < 1$), the Moho-temperature depends largely on the thickness of the mantle part of the lithosphere. Also, in very *thick* crusts ($f_c \gg 1$), the Moho-temperature depends mostly on crustal thickness. This is because there the Moho-temperature is largely determined by the radioactivity in the crust. Small changes in crustal thickness will lead to quite significant temperature changes, while the thickness of the mantle part of the lithosphere is comparably insignificant.

• *Heat flow.* The surface heat flow q_s is one of the few important thermal parameters of the lithosphere that can be measured directly. It is therefore a much used parameter to define thermal features in the lithosphere (e. g. Zhou and Sandiford 1992). Within the models discussed above, heat flow is simply given by the derivative of the geotherm equations with respect to depth and divided by the thermal conductivity. The *surface* heat flow q_s can then be found by evaluating these derivatives at the depth $z = 0$. For a boundary condition at the base of the lithosphere we get from eq. 3.60:

$$q_s = k \frac{T_l}{z_l} \tag{3.77}$$

for $S = 0$ and from eq. 3.76:

$$q_s = k \frac{T_l}{f_l z_l} + S_0 f_c h_r \left(1 - \frac{f_c h_r}{f_l z_l} \left(1 - e^{\left(-\frac{f_l z_l}{f_c h_r} \right)} \right) \right) \tag{3.78}$$

for an exponential distribution of the radioactive heat production. Figure 3.21 shows an f_c-f_l-diagram, which is contoured for q_s. We can see that the surface heat flow depends much more on the crustal thickness than the Moho-temperature does.

3.5 Heat in the Oceanic Lithosphere

Oceanic lithosphere contains practically no radioactive elements. Thus, one could think that it is simple to describe stable oceanic geotherms. In analogy to continental geotherms we might want to formulate the geotherm equation as:

$$k\frac{d^2T}{dz^2} = 0 \quad . \tag{3.79}$$

After one integration we get:

$$k\frac{dT}{dz} = C_1 \quad .$$

Using the boundary conditions that are well-known to us from sect. 3.4.1 with: $q = q_m$ at the depth $z = z_c$, we get:

$$kT = q_m z + C_2 \quad .$$

If we also assume that $T = 0$ at the surface $z = 0$, then $C_2 = 0$ and we can write:

$$T = z\frac{q_m}{k} \quad , \tag{3.80}$$

... if the same boundary conditions are used as in sect. 3.4.1. However, eq. 3.80 is not a very good model to describe oceanic lithosphere. A simple consideration of the thermal time constant will show us why: oceanic lithosphere is produced at the mid-oceanic ridges and it gets its thickness only by its increasing age. The oldest oceanic lithosphere is about 100 my old. However, in sect. 3.1.4 we showed that the thermal time constant of thermally stabilized lithosphere is of the order of 150 my or more! We can conclude that oceanic lithosphere is *not* thermally stabilized. The assumption underlying eq. 3.79 is wrong. There is no thermally stabilized oceanic lithosphere! We can not assume that $dT / dt = 0$ and we must solve the time dependent diffusion equation (eq. 3.6).

Figure 3.21. The surface heat flow (contoured in 10^{-3}W m-2) for a range of crustal thicknesses (expressed by the vertical thickening strain f_c) and for a range of total thicknesses of the lithosphere (expressed by the vertical thickening strain f_l). The diagram was calculated with eq. 3.78. The assumptions for all parameters are the same as in Fig. 3.19, curve *b*

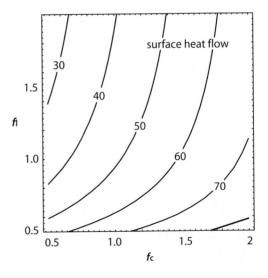

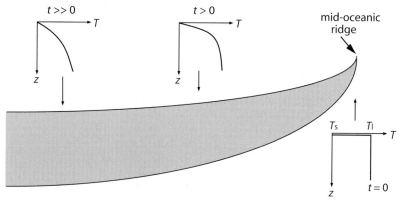

Figure 3.22. Thickness and thermal profile of oceanic lithosphere at a series of points. The oceanic crust is not drawn separately. On the scale of this sketch it would appear as a very narrow band of constant thickness near the upper boundary of the gray shaded region

3.5.1 Aging Oceanic Lithosphere

The oceanic crust, that is produced from partial mantle melts at the mid-oceanic ridges is of the order of 5–8 km thick. There, the thickness of the entire lithosphere is that of the crust (Fig. 3.22). The high potential energy of the ridges forces this crust to move away from the ridge. As the oceanic crust ages and moves further and further away from the mid-oceanic ridge, the asthenosphere cools and becomes oceanic mantle lithosphere. This process can be described with the diffusion equation using simple initial and boundary conditions. The description of oceanic lithosphere with these boundary conditions has become one of the most successful models of plate tectonic theory (s. a. Sclater et al. 1980). It is called the *half space cooling model* (s. sect. 3.6.1).

• *The half space cooling model.* As any other problem in the theory of heat conduction, the half space cooling model relies on the integration of eq. 3.6, using a set of boundary and initial conditions. The temperature at the surface of mid-oceanic ridges is that of the water temperature. For simplicity, we assume that it is $T_s = 0$. Below the ridge, the mantle temperature is almost constant – convection equalizes all temperature gradients. Thus, we can write a very simple initial condition describing the thermal profile below mid-oceanic ridges:

– $T = T_s$ at the depth $z = 0$ and:
– $T = T_l$ in all depths $z > 0$ at time $t = 0$.

This initial condition is illustrated in T-z-diagram on the bottom right corner of Fig. 3.22. For the upper boundary condition it is obvious to assume that the temperature at the ocean floor remains constant. As there is effectively no lower boundary, we assume that it lies at infinity and that the temperature there is $T = T_l$. We can write this as follows:

– $T = T_s$ at $z = 0$ for all $t > 0$ and:
– $T = T_l$ at $z = \infty$ for all $t > 0$.

(Fig. 3.22). The solution of the heat conduction equation for these boundary conditions is:

$$T = T_s + (T_l - T_s)\mathrm{erf}\left(\frac{z}{\sqrt{4\kappa t}}\right) . \tag{3.81}$$

We discuss this solution in some more detail in sect. 3.6.1 (s. a. sect. 3.1.3). Fig. 3.23a shows temperature profiles through oceanic lithosphere of different ages, that were calculated with eq. 3.81. The curves correspond to the two sketches of thermal profiles in the middle and on the left of Fig. 3.22. Fig. 3.23b shows the depth of a series of isotherms as a function of age.

Temperature profiles calculated with this model for the cooling oceanic lithosphere can not be tested directly, as we cannot drill deep enough into the oceanic lithosphere. Our observations are confined to parameters which we can measure near the surface. However, one of these parameters is very useful to infer the thermal profile: the surface heat flow. The surface heat flow is the product of thermal conductivity and the thermal gradient at $z = 0$. This can be calculated from eq. 3.81 and can be compared with measured data in the oceans. To obtain surface heat flow we must differentiate eq. 3.81 with respect to depth and evaluate it at $z = 0$. From eq. 3.81 this is:

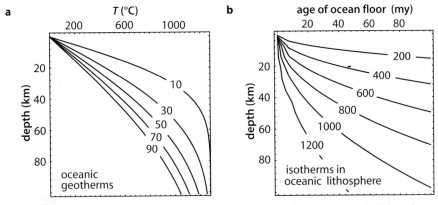

Figure 3.23. a Temperature T versus depth profiles through oceanic lithosphere at a number of different ages labeled in my. b The depth of isotherms (in °C) in oceanic lithosphere as a function of age between 0 and 100 my. The curves on both figures were calculated with eq. 3.81 assuming $T_l = 1\,280\,°C$. The age can be converted into distance from the mid-oceanic ridge by using $x = u/t$ where x is the distance from the ridge and u is the rifting rate. Compare the curves also with Fig. 3.37

$$q_s = k(T_l - T_s)\frac{d\left(\text{erf}\left(\frac{z}{\sqrt{4\kappa t}}\right)\right)}{dz}\Bigg|_{,z=0} . \tag{3.82}$$

As the error function itself is an integral (see eq. 3.16), it is easy to differentiate eq. 3.81 (sect. 3.1.3). We get:

$$q_s = k(T_l - T_s)\sqrt{\frac{1}{\pi\kappa t}} . \tag{3.83}$$

This equation can be generalized for the description of different oceanic plates with different rifting rates. For this, we express the rifting rate u as $u = x/t$. There, x is the distance from the mid-oceanic ridge and t is the age. If we replace t in eq. 3.83 by x/u, we get:

$$q_s = k(T_l - T_s)\sqrt{\frac{u}{\pi\kappa x}} . \tag{3.84}$$

Eq. 3.82 shows us that the surface heat flow as a function of distance from the mid-oceanic ridge is a square root function. Fig. 3.24 shows the surface heat flow in oceanic lithosphere as calculated with eq. 3.84. The heat flow data of Sclater et al. (1980) show that these curves correspond well with heat flow measured in the deep oceans. In sect. 4.2.1 we will show that the half space cooling model is not only a good description for the temperatures and heat flow in oceanic lithosphere, but can also be used to describe the water depth of the oceans. It can even be used to calculate the magnitude of the ridge push force (sect. 5.3.2). The relationship between all these parameters that are described with the half space cooling model is called the *age-depth-heat flow relationship* of oceanic lithosphere. This age-depth-heat flow relationship corresponds fantastically well with our observations up to an age of the oceanic lithosphere of 80 my. The age-depth-heat flow

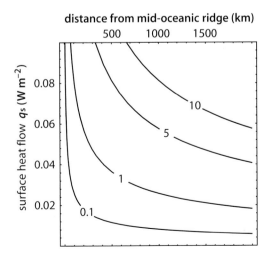

Figure 3.24. The surface heat flow of oceanic lithosphere as a function of age and therefore as a function of distance from the mid-oceanic ridge as calculated with eq. 3.84. Contours are for different rifting ate labeled in $\text{cm}\,\text{y}^{-1}$

relationship is generally accepted as one of the greatest successes of plate tectonic theory.

• *The CHABLIS-model.* The half space cooling model is the most famous model for the description of oceanic lithosphere, but there is two reasons why it may not be a perfect description:

- The half space cooling model assumes implicitly that the heat flow at the base of the lithosphere changes through time. It is not clear that this should be so.
- The model fails to describe the water depth and the surface heat flow of oceanic lithosphere beyond an age of about 80 my.

Alternatively we could assume that the mantle heat flow through the base of the lithosphere is kept constant by small scale convection (Doin and Fleitout 1996). This assumption leads to a description which corresponds much better with the long term evolution of oceanic lithosphere than the half space cooling model. This idea is know by the name of *CHABLIS-model.* This name is an acronym for: **C**onstant **H**eat flow **A**pplied to **B**ottom of **L**ithospheric **I**sotherm .

3.5.2 Subduction Zones

The description of the kinematics, the thermal evolution and the dynamics of subduction zones is a fundamentally two-dimensional problem. It is the first problem in this book for which we require more than one spatial coordinate to characterize the essence of the problem (Fig. 3.25). Many problems related to subduction zones concern the accretionary wedge that forms near the surface between the surface of the subducting slab and the upper plate. Dynamic and kinematic models for accretionary wedges will be discussed in sect. 6.2.3 and also in sect. 5.3.2. Here we discuss some general aspects about the deep thermal structure of subduction zone environments.

• *Isotherms in subduction zones.* Fig. 3.25 shows, schematically, the shape of isotherms in subduction zones. In the subducting slab, the isotherms will be bent and subducted with the slab. The further they are subducted, the more they merge to the center of the slab as both surfaces of the subducting slab equilibrate with the surrounding mantle temperatures. A thermal steady state will be reached when the curvature of the isotherms is large enough so that the rate of thermal equilibration is balanced by the subducting velocity (s. Fig. 3.2; s. a. Molnar and England 1995). In this stage, *diffusion* (which leads to the decay of the high curvature of the temperature profile in the tip of the subducting slab) will be balanced by *advection* (which moves the isotherms to larger depths). A steady state is reached that is very similar to the balance discussed in the second comparison on page 3.2 and also equivalent to the steady state that landforms between incising drainages may reach (discussed in Fig. 4.30). The time that is needed by subduction zones to reach this thermal steady state depends on the thickness of the plate and on the subduction rate. It can be estimated with the Peclet number (sect. 3.3.4).

At a temperature of about 1 600 °C, which is in about 400 km depth, olivine reacts to form spinel. The depth of this phase transition is called the *Clapeyron-Curve*. This reaction is *exothermic* with about $1.7 \cdot 10^5 \, \mathrm{J\,kg^{-1}}$. Thus, the isotherms in this depth have a kink. The positive slope of the Clapeyron curve in P-T-space causes that the Clapeyron curve is somewhat higher within the subducting slab than it is outside. At a depth of 650 km (about 1 700 °C) there is another kink in the geotherm, caused by the phase transition from spinel to oxide (not illustrated on Fig. 3.25). The qualitative direction of this kink is opposite to that of the Clapeyron curve as the spinel-oxide phase transition is *endothermic*.

Fig. 3.25 shows schematically that the isotherms within the upper plate are closer to the surface near the subduction zone than they are in the far field. This is because of the dehydration of the subducting plate and the consequential rise of partial melts and other hot fluids. This leads typically to high temperature metamorphism in the rocks of the upper plate and ultimately to the development of magmatic arcs (see next section). This is in contrast to the very *low* temperatures that occur in the subducting plate up to very large depths. This coupled occurrence of low-pressure - high-temperature and high-pressure - low-temperature metamorphism was recognized by Miyashiro (1973) as one of the characteristic features of metamorphic terrains in subduction zone environments. He called this a *paired metamorphic belt*.

• *Island arcs and subduction zones.* An interesting observation in the upper plate of subduction zones is that there is volcanic arcs that always form a narrow line that is exactly where the seismically active surface between the subducting slab and the upper plate is about 150 km deep (Isacks and Barazangi 1977). In subduction zones that have a dip of 45° this implies a horizontal distance of the arc from the trench of 100–150 km. If the sub-

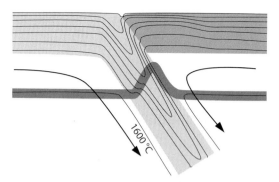

Figure 3.25. Schematic illustration of the temperature distribution in subduction zones. the subducted lithosphere is shaded light, the upper plate dark. The thick dark line that follows the 1 600 °C-isotherm outside the subduction zone is the Clapeyron-curve. It marks the olivine-spinel phase transition

duction angle is steeper, then this distance is shorter and vice versa. This observation is true for the distance of the Aleute-volcanoes to the Aleute trench, for the distance of the Indonesian volcanoes from the Java trench and many other volcanic arcs around the globe. This observation is not trivial to explain. The volcanics that erupt from these volcanoes are derived from partial melts in the mantle wedge that melted during fluid infiltration of fluids that were derived by dehydration of sediments on the surface of the subducting slab in the Benioff-zone. This zone stretches for several hundreds of kilometers along the surface of the subducting slab and is definitely much wider than the width of the volcanic arcs on the surface. Some authors have suggested that there are important pressure sensitive dehydration reactions that occur in exactly 150 km depth, but there is little petrological evidence for this.

An alternative explanation was suggested by Spiegelman and McKenzie (1987) (Fig. 3.26). Their model describes the motion of partial melts through the mantle wedge as the sum of two vector fields:

1. The motion of the asthenosphere in the mantle wedge. This motion follows the wedge and is illustrated in Fig. 3.26 by the dashed lines.
2. The motion of the partial melts. Partial melt is produced continuously along the surface of the subducting plate and moves vertically upwards.

The sum of the two velocity fields results in curved paths that converge at the tip of the mantle wedge (Fig. 3.26). This elegant model is a beautiful example for a successful model description of fluid flow in deforming rocks.

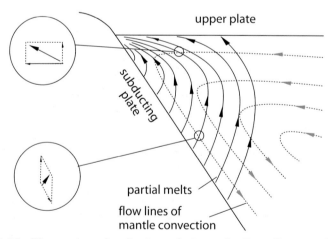

Figure 3.26. The motion of melts in and above the Benioff zone according to Spiegelman and McKenzie (1987). The dashed lines show the convective motion in the asthenosphere, the continuous lines are the motion of the partial melts. The enlarged sections show how the velocity field of the partial melts is given by the sum of the upward velocity of the melt and the motion in the mantle wedge

3.6 Thermal Evolution of Intrusions

Intrusion of magmatic rocks into higher levels in the crust is an important geodynamic process that can be responsible for a large range of thermal, chemical and mechanical changes in the crust. Intrusive rocks, as well as their contact aureoles, are familiar to us from field observations. Thus, their geodynamic interpretations can often be tested directly with structural and petrological data. The process of intrusion itself is a very efficient mechanism for the transport of heat. Thus, most intrusions do not cool very much on their intrusive path and their temperature may be used to infer the temperature of rocks at the depth of their origin. There are two important reasons, why the quantitative description of the thermal evolution of intrusions is quite simple:

- In comparison to the duration of a contact metamorphic event or the duration of an orogenic cycle, the rate of intrusion is very rapid. Thus, for the thermal modeling of intrusions it is possible to assume that their emplacement was *infinitely* rapid, compared to the time of the subsequent thermal equilibration. This is what is called an *instantaneous heating model*. The thermal equilibration of intrusions with their surroundings can therefore be described with the heat conduction equation (eq. 3.6), assuming the intrusion geometry as an "initial condition".
- Most intrusions are small if compared to the size of their surroundings, for example the distance to the surface or to the base of the lithosphere. Thus, the boundary conditions that are needed to solve the heat conduction equation can often be assumed to lie at infinity.

With these two assumptions, it is possible to solve the heat conduction equation. The solutions may then be used for the analytical description of the thermal evolution of intrusions. For simple intrusion geometries, for example dikes, the description is very similar to description of the thermal evolution of oceanic lithosphere (sect. 3.5.1). However, before we use such a model, we need formulate two more simplifying assumptions which we will make initially:

- The latent heat of fusion - which actually plays an important role for the thermal energy budget of intrusions - will be neglected. Thus we will analyses in the first instance none of the processes discussed in sect. 3.2.
- The thermal conductivity is assumed to the constant in space. Thus we will also not consider any of the problems discussed in sect. 3.1.1.

With these assumptions, it is possible to integrate eq. 3.6. We begin with some simple examples.

3.6.1 Step-Shaped Temperature Distributions

The most simple of all model examples describing the cooling of rocks in
the direct vicinity of intrusions is given by the thermal equilibration of step-
shaped temperature profiles in one dimension. This example is illustrated
in Fig. 3.27 and is one of the most useful examples for the understanding
of the cooling history of intrusions. We interpret the temperatures on both
sides of the step as the intrusion- and the host rock temperature; T_i and
T_b, respectively. The step itself is the intrusion contact. If we choose a one-
dimensional coordinate system in which the origin $z = 0$ is exactly at the
contact of the model intrusion, then the initial and boundary conditions of
this equilibration problem may be described by:

– Initial condition: $T = T_i$ for all $z > 0$ and $T = T_b$ for all $z < 0$ at $t = 0$.
– Boundary conditions: $T = T_b$ at $z = \infty$ and $T = T_i$ at $z = -\infty$ for all
  $t > 0$.

You may have noticed that there is *two* boundary conditions that are
located a very long distance from the contact *inside* the intrusion and far
away from the contact in the host rock (at $z = +\infty$ and $z = -\infty$). We need
two boundary conditions because the equation that is to be solved (eq. 3.6) is
a partial differential equation of the *second* order (s. sect. A.1.1). Integration
of eq. 3.6 using these boundary and initial conditions gives:

$$T = T_b + \frac{(T_i - T_b)}{2}\left(1 + \mathrm{erf}\left(\frac{z}{\sqrt{4\kappa t}}\right)\right) \quad .\tag{3.85}$$

We will not discuss how eq. 3.85 was derived (s. sect. 3.1.3). Because of
the choice of coordinate system we have made to formulate this problem
(Fig. 3.27), this solution looks very simple. In another coordinate system in
which the coordinate origin is located at a distance l from the temperature

Figure 3.27. Thermal
equilibration of an initially
step-shaped temperature
distribution. The curves were
calculated with eq. 3.85 for
$\kappa = 10^{-6}\ \mathrm{m}^2\,\mathrm{s}^{-1}$, $T_i = 700\,^\circ\mathrm{C}$
and $T_b = 200\,^\circ\mathrm{C}$. The different
curves are temperature profiles
at different times (in years) after
the intrusion event

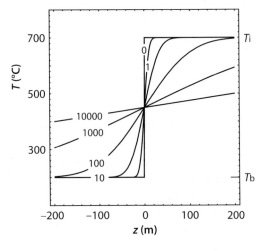

step, the initial condition must be reformulated to: $T = T_i$ for all $z > l$ and $T = T_b$ all $z < l$ at $t = 0$. The boundary conditions remain the same. The solution gets the form:

$$T = T_b + \frac{(T_i - T_b)}{2} \left(1 + \mathrm{erf} \left(\frac{z-l}{\sqrt{4\kappa t}} \right) \right) \quad . \tag{3.86}$$

On Fig. 3.27 it may be seen that the thermal evolution on both sides of the mean temperature between T_i and T_b develops symmetrically. This is to be expected as the initial and boundary conditions are also symmetric.

• *Cooling of half spaces.* In the last example, the temperature at $z = 0$ stays constant in time. It has the value $T_b + (T_i - T_b)/2$. Thus, eq. 3.85 is very similar to the description of the cooling of semi-infinite half spaces (sect. 3.5.1). This thermal problem is very important in the earth sciences and may be described by the following initial and boundary conditions:

– Initial condition: $T = T_b$ at all $z > 0$ and $T = T_i$ at $z = 0$ at time $t = 0$.
– Boundary conditions: $T = T_i$ at $z = 0$ and $T = T_b$ at $z = \infty$ for all $t > 0$.

These conditions are very similar to those of the last problem. Thus, the corresponding solution is also similar to eq. 3.85:

$$T = T_b + (T_i - T_b) \left(1 - \mathrm{erf} \left(\frac{z}{\sqrt{4\kappa t}} \right) \right) \quad . \tag{3.87}$$

We have met this equation in a slightly different form already in sect. 3.5.1. The result of eq. 3.86 and eq. 3.87 can be simplified if it is expressed as the dimensionless temperature $\theta = (T - T_b)/(T_i - T_b)$ (sect. 1.2). Then, using the complementary error function, eq. 3.87 simplifies to:

$$\theta = \mathrm{erfc} \left(\frac{z}{\sqrt{4\kappa t}} \right) \quad . \tag{3.88}$$

This simplification is shown here to illustrate that cooling curves of this problem have the shape of an error function. However, in the remainder of this section we will not use dimensionless temperatures. We begin with some geologically relevant examples that may be described with this solution.

3.6.2 One-dimensional Intrusions

One of the simplest but also most important applications of the equations introduced in the last section is the description of the cooling history of intrusions (Jaeger 1964). As the solutions shown above are one-dimensional, their application is particularly relevant to the description of the thermal evolution of dikes that are narrow compared with their lateral extent. When using the solutions described above to describe the cooling history of dikes, it is implied that the dike extends "infinitely" in the two spatial coordinates normal to the coordinate described in the cooling problem. For a coordinate system with its origin in the center of a dike with the thickness l, the initial conditions may be described by: $T = T_i$ for $-(l/2) < z < (l/2)$ and $T = T_b$ for $(l/2) < z < -(l/2)$ (Fig. 3.28). The boundary conditions remain the same as for the step problem. With these conditions, a solution of eq. 3.6 may be found to be:

$$T = T_b + \frac{(T_i - T_b)}{2} \left(\mathrm{erf} \left(\frac{0.5l - z}{\sqrt{4\kappa t}} \right) + \mathrm{erf} \left(\frac{0.5l + z}{\sqrt{4\kappa t}} \right) \right) . \qquad (3.89)$$

It may be easily seen, that the solution is made up of descriptions for two opposing step-shaped temperature profiles at $z = -l/2$ and $z = l/2$. Fig. 3.29 shows the thermal evolution described by eq. 3.89. As the diffusion equation is a linear differential equation, the diffusive equilibration of just about any one-dimensional geometry may be described by the summation of solutions for various initial conditions.

In contrast to Fig. 3.27, the temperature at the intrusion contact departs from the temperature $(T_i + T_b)/2$ after some time in Fig. 3.29. The contact of the dike begins to cool. This is because the dike contact at $z = +l/2$ begins to follow the thermal effects of the temperature step at $z = -l/2$. Correspondingly, the other dike contact at $z = -l/2$ cools, because it "feels" the cooling at $z = +l/2$.

Cooling History of Simple Intrusions. In the following paragraphs we will use eq. 3.89 to infer some characteristic features of contact metamorphism. Firstly, eq. 3.89 shows us that – in the absence of other thermal processes – the maximum temperature that may be reached by contact metamorphism is much lower than the intrusion temperature. Only at the very contact of the intrusion the temperature reaches the mean temperature between host rock and intrusion. We can conclude that field observations of contact metamorphic haloes documenting haloes of considerable width and temperature imply that thermal processes other than conductive equilibration have played a role in their formation (s. sect. 3.6.4).

• *Cooling curves.* In order to interpret heating and cooling curves of rocks in the contact metamorphic environment, it is useful to plot eq. 3.89 in a temperature-time diagram (Fig. 3.30a). This figure illustrates that rocks located at different distances from the intrusion may experience very different

Figure 3.28. Schematic illustration of the initial condition and the parameters of the dike cooling problem of eq. 3.89

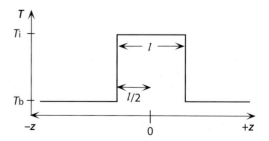

Figure 3.29. Thermal
equilibration of one-dimensional
intrusions, for example
magmatic dikes of large lateral
extent (calculated with eq. 3.89
and labeled in years after initial
intrusion). All parameters are
the same as in Fig. 3.27. Cooling
curves of rocks from a range of
distance from the dike center are
shown in Fig. 3.30a

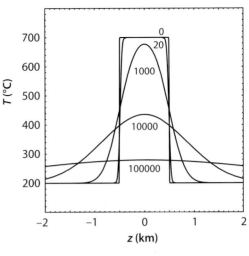

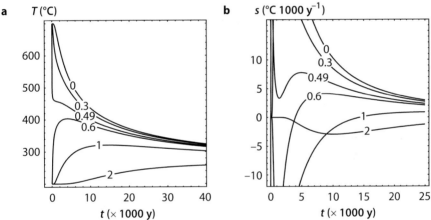

Figure 3.30. a Temperature-time paths (cooling curves) for a series of rocks within
and within the contact aureole of a 1 km thick magmatic dike ($l = 1$ km) shown
up to 40 000 years after intrusion (all constants are the same as in Fig. 3.29). The
figure was calculated with eq. 3.89. The curves are labeled for distance from the
center of the intrusion in km. As the thickness is $l = 1$ km, the first three curves are
within the magma and the others in the country rock. **b** Cooling rates for the same
points as shown in **a** plotted against time. Curves are calculated with eq. 3.90

cooling curves. For example, it may be seen that some rocks cool, while others
heat up, or that rocks cool with different rates. In fact, near the contact of the
intrusion, cooling curves have extremely complicated shapes including more
than one maximum in the cooling rate (e.g. the 490-m-curve in Fig. 3.30a).
As many metamorphic processes involving grain growth and diffusion are
strongly dependent on cooling rate (e.g. Dodson 1973), the interpretation of
these curves is extremely important (s. sect. 7.2).

• *Cooling rates.* The cooling rate s for a chosen point near the intrusion is described by the time derivative of eq. 3.89. As the error function itself is an integral, it is not too difficult to find this derivative, although we will not explain it in detail here. It is:

$$s = \frac{dT}{dt} = \frac{(T_i - T_0)}{4t\sqrt{\pi \kappa t}} \left(\frac{z - 0.5l}{e^{((0.5l-z)^2/4\kappa t)}} - \frac{z + 0.5l}{e^{((0.5l+z)^2/4\kappa t)}} \right) \qquad (3.90)$$

(s. sect. 3.1.3, 3.5.1). Cooling rates as a function of time as calculated with this equation are shown in Fig. 3.30b. They correspond directly to the curves in Fig. 3.30a. For many petrological questions the cooling rate at a given *temperature* is much more important than the cooling rate at a given *time*. For this, a *parametric* plot of temperature against cooling rate is useful. *Parametric* plots are diagrams in which two independent functions of the same variable are plotted against each other. Thus, the parametric plot of T against s shown in Fig. 3.31 is a combination of Figs. 3.30a and 3.30b. However, the time dependence of temperature or cooling rate can not be illustrated on this figure. Thus, Fig. 3.31 may appear a bit confusing at first view. Nevertheless, such diagrams are crucial for a meaningful interpretation of petrological and geochronological data.

• *Other important information.* Aside from cooling curves or cooling rates, there is even more important information on the thermal evolution of intrusions that may be extracted from eq. 3.89. For example, the *time* of the contact metamorphic thermal peak $t_{T_{max}}$ for the model of eq. 3.89 may be found analytically. At the thermal peak the rate of temperature change is zero: $s|_{t=T_{max}} = 0$ (read: s at $t = T_{max}$). Thus, this time is given by setting eq. 3.90 to zero and solving for time. We get:

Figure 3.31. Parametric plot of cooling rate against temperature. The diagram was constructed from Figs. 3.30a and 3.30b. Such a diagram is extremely useful for the interpretation of the equilibration of mineral parageneses (sect. 7.2)

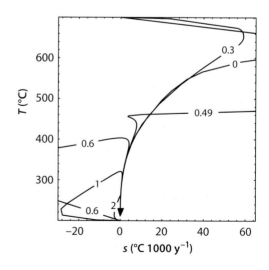

$$t_{T_{max}} = -\frac{zl}{2\kappa \ln \left(\frac{z-0.5l}{z+0.5l}\right)} \quad . \tag{3.91}$$

From this, we can also get the thermal peak temperature as a function of distance from the intrusion. We get this by substituting the time obtained from eq. 3.91 into eq. 3.89 and solving that for temperature (Fig. 3.31; Fig. 3.32). With this diagram we can make some fundamental predictions about the nature of contact metamorphism:

1. The contact metamorphic peak temperature drops rapidly with *increasing* distance from the heat source.
2. The time of peak contact metamorphism increases rapidly with *increasing* distance from the heat source and with *decreasing* metamorphic grade. This predicts that – if contact metamorphism occurred – low grade metamorphic rocks should experience their metamorphic peak later than high grade metamorphic rocks (Den Tex 1963).

These predictions may be extremely helpful when interpreting the heating mechanisms of metamorphic terrains, in particular since this relationship is reversed, for example during regional metamorphism (sect. 6.3.3 and sect. 7.4.1). In the chapter on *P-T-t*-paths we will discuss further implications of the timing relationships of various cooling curves.

As a final useful parameter, we can derive from eq. 3.89 the time of the maximum cooling rate $t_{s_{max}}$. The maximum cooling rate occurs at the inflection point of the cooling curves in Fig. 3.30. There, the curvature of the cooling curve is zero. Thus, this point may be found by differentiating eq. 3.90 a second time with respect to time and setting it to zero. This gives:

$$e^{\left(\frac{lz}{2\kappa t}\right)} = \left(\frac{6(0.5l-z)\kappa t - (0.5l-z)^3}{6(0.5l+z)\kappa t - (0.5l+z)^3}\right) \quad . \tag{3.92}$$

Figure 3.32. Contact metamorphic peak temperature and the time of the contact metamorphic temperature peak of the simple one-dimensional intrusion from Figs. 3.28 and Fig. 3.29. Calculated by substituting eq. 3.91, into eq. 3.89. The thickness of the intrusion l is 1 km, $T_i = 700\,^\circ\mathrm{C}$, $T_b = 200\,^\circ\mathrm{C}$ and $\kappa = 10^{-6}\,\mathrm{m^2\,s^{-1}}$

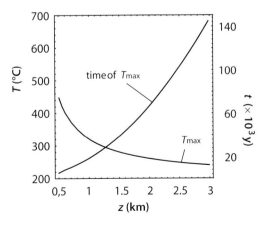

This equation may be solved iteratively for time. For the determination of the maximum cooling rate this time must be substituted into eq. 3.90. The maximum cooling rate is an important parameter for the characterization of cooling curves.

Other One-dimensional Geometries. One of the most severe problems of the model discussed in the last section (eq. 3.89 and its derived relationships) is that it predicts much narrower contact metamorphic aureoles than are generally observed around intrusions. One of the reasons for this may lie in the geometry of the heat source. Other causes will be discussed in sect. 3.6.4.

• *Dike swarms.* In terrains that are penetrated by many intrusions, the mean contact metamorphic temperature may be much higher than that would be observed around a single intrusion. This may be described schematically by the combination of a series of temperature steps. For example, if there is N intrusions of the thicknesses l_n, that intrude at the depths $z = z_n$, then

Figure 3.33. Contact metamorphism between two neighboring intrusions. The figure was calculated with eq. 3.93 using $z_1 = -1\,000$, $z_2 = 1\,000$ and $l_1 = l_2 = 1\,000$. Curves are shown in thousands of years (e. g. the line labeled for 0.2 corresponds to 200 years). Note that the contact metamorphic aureole *between* the intrusions is much wider than on their outside

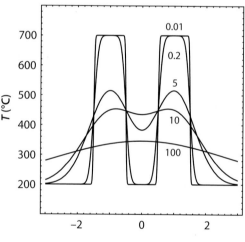

Figure 3.34. Schematic illustration of the diffusive equilibration of a step-shaped temperature profile (for example a one-dimensional intrusion like a sill) superimposed on a linear geotherm. Temperature profiles are shown at two different time steps t_0 and t_1. As the diffusion equation is a linear differential equation, the temperatures of eq. 3.89 can simply be added to the geotherm equation (in this case a straight line)

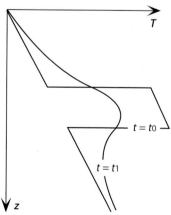

this can be described by the summation of $2N$ steps in the temperature distribution. The solution is:

$$\theta = \frac{1}{2} \sum_{n=1}^{N} \left(\text{erf} \left(\frac{z - (z_n - 0.5l_n)}{\sqrt{4\kappa t}} \right) + \text{erf} \left(\frac{(z_n + 0.5l_n) - z}{\sqrt{4\kappa t}} \right) + ... \right) \quad (3.93)$$

Fig. 3.33 shows an example of two intrusions calculated with this equation. It may be seen that the contact metamorphic halo between the intrusions is much wider and of a higher temperature than on their outside. Because of this observation, Barton and Hanson (1989) suggested that multiple intrusion may be the principle heat source for low-pressure high-temperature metamorphism (s. sect. 6.3.3). Fig. 3.34 shows another schematic example for a one-dimensional problem that may be solved by the summation of a series of one-dimensional model geometries (s. p. 54).

• *Spherical intrusions.* The thermal evolution of spherical intrusions is also a one-dimensional problem in polar coordinates. The equation we need to solve is eq. 3.13. If we want to compare the thermal evolution of spherical intrusions with that described by eq. 3.89, then we formulate the following boundary conditions: $T = T_b$ at $r = \infty$ and $r = -\infty$, and the initial condition $T = T_i$ in the region $-R < z < R$ and $T = T_b$ in the region $-R > z > R$ at $t = 0$, where R is the radius of the intrusion and is analogous to $l/2$ in the Cartesian example discussed in eq. 3.89 and r is the distance from the origin. The solution of eq. 3.13 under these conditions is:

$$T = T_b + \frac{(T_i - T_b)}{2} \left(\text{erf} \left(\frac{R - r}{\sqrt{4\kappa t}} \right) + \text{erf} \left(\frac{R + r}{\sqrt{4\kappa t}} \right) \right)$$

$$- \frac{(T_i - T_b)}{r} \sqrt{\kappa t / \pi} \left(e^{-\left(\frac{(R-r)^2}{4\kappa t} \right)} - e^{-\left(\frac{(R+r)^2}{4\kappa t} \right)} \right) . \quad (3.94)$$

Temperature profiles across cooling spheres at different times are shown in Fig. 3.35. They may be directly compared with those in Fig. 3.29. It may be seen that spheres cool much faster than dikes, which is intuitively clear as they have a much larger ratio of surface to volume.

3.6.3 Two-dimensional Intrusions

The description of two-dimensional thermal problems is – for most boundary conditions – not much harder than one-dimensional problems because heat conduction is several spatial dimensions can be described as the sum of conduction in the individual directional components (s. eqs. 3.9 to 3.11). In general it can be said that two-dimensional heat conduction problems may be described by the product of the one-dimensional solutions if 1. the initial conditions may be expressed as the product of two functions $f(z)$ and $f(y)$

Figure 3.35. Cooling history of a spherical intrusion. For easy comparison, all chosen parameters and the shown time steps (labeled in years) are the same as those in Fig. 3.29

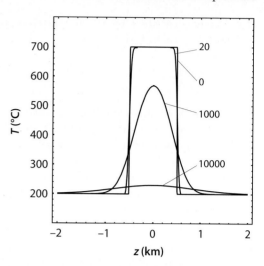

and 2. the boundary conditions are given by a constant temperature or constant heat flow. Solutions of the two-dimensional heat conduction equation can therefore be expressed as:

$$T = \text{solution in } z\text{-direction} \times \text{solution in } y\text{-direction}$$

For example, the cooling history of a *corner* is described by the product of the cooling of two initial step-shaped temperature profiles as we discussed them in eq. 3.87:

$$T = T_b + (T_i - T_b)\text{erf}\left(\frac{z}{\sqrt{4\kappa t}}\right)\text{erf}\left(\frac{y}{\sqrt{4\kappa t}}\right) \quad . \tag{3.95}$$

(Fig. 3.36). The diffusive equilibration of *squares* finds a series of important applications in the fields of geomorphology (sect. 4.3) and petrology (sect. 7.2). The solution for this can be derived directly from eq. 3.89:

$$T = T_b + \frac{(T_i - T_b)}{2}\left(\text{erf}\left(\frac{0.5l - z}{\sqrt{4\kappa t}}\right) + \text{erf}\left(\frac{0.5l + z}{\sqrt{4\kappa t}}\right)\right)$$
$$\times \left(\text{erf}\left(\frac{0.5l - y}{\sqrt{4\kappa t}}\right) + \text{erf}\left(\frac{0.5l + y}{\sqrt{4\kappa t}}\right)\right) \quad . \tag{3.96}$$

This solution may be used to describe the cooling history of a rectangular hot region. In Chapter 4.3 we will see that weathering of granites can be described with the same solution.

3.6.4 Other Examples of Boundary Conditions

In the previous sections we have described the thermal evolution of intrusions always under the assumptions that they are small, if compared to the size of

Figure 3.36. Temperatures in a cooling corner, 10 000 y after its intrusion with $T_i = 700\,°C$ into rocks of $T_b = 200\,°C$

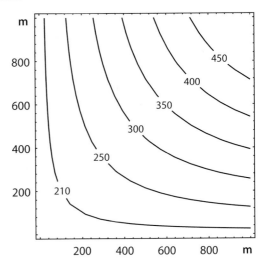

the their surroundings, for example to the distance from the surface. Thus, we have used boundary conditions at infinity. In the following we discuss some examples, where this simplification is not justified.

Fixed Boundary Conditions. Some thermal problems of intrusions may be described with boundary conditions of fixed temperature; for example, intrusion near the surface, where the temperature is constant, or intrusions through which magma flows to keep them at constant temperature. If only one of the boundaries is fixed in temperature, then such problems may be described as a *half space problem* (sect. 3.6.1). However, if both boundaries of a one-dimensional problem are fixed in temperature, then none of the solutions we discussed so far can be used. Under fixed boundaries the integration of eq. 3.6 is quite difficult and the solutions are not as simple as those we have discussed in sect. 3.6.2 and 3.6.3. Eq. 3.6 may only be solved by a Fourier transform (s. sect. A.4). Such solutions contain infinite summations. In their most general form, boundary and initial conditions of diffusion problems with fixed boundaries are given by:

- Initial condition: In the region $0 < z < l$ the temperature is given by $T = f_1(z)$ at time $t = 0$.
- Boundary conditions: At times $t > 0$ the temperature at $z = 0$ is given by $T = f_2(t)$ and at $z = l$ it is given by $T = f_3(t)$.

The function $f(z)$ can be used to describe the shape of a cooling body and the functions f_2 and f_3 can describe a large number of thermal processes at the model boundaries. The fact that f_2 and f_3 are functions and not fixed values is also a sort of "fixed boundary" (s. sect. A.2.1). In other words, fixed boundary conditions need not be a fixed value of temperature, but they need to be defined at all times by external parameters. Eq. 3.6 can only be solved using Fourier series and such solutions therefore contain infinite summations and trigonometric functions. However, for many special cases of the space and time dependent functions $f_1(z)$, $f_2(t)$ and $f_3(t)$ Fourier series solutions are relatively easy to derive and can be looked up in the literature (e. g. Carslaw and Jaeger 1959).

The Stefan Problem. A series of observations in contact metamorphic aureoles of intrusions show that these are much wider and of a higher temperature than those we have predicted in the last section. There is two important reasons for this:

– All problems we have discussed so far are "instantaneous cooling" problems. This means, we have assumed that the cooling history commences at the time of intrusion. This need not be the case. In fact, in dikes through which magma is fed into a pluton, this would be highly unlikely. In this case it would be more appropriate to assume that the temperature inside the dike stays constant while magma flows through it and cooling commences only once the flow stops. For the first stage of this process, the temperature evolution of the host rocks can be described with eq. 3.87 (Fig. 3.37). For the description of the cooling history in the second part of the process, we must use the temperature distribution at the time of cessation of the magma flow as a new initial condition, and track the conductive equilibration thereafter. This temperature distribution is given by a complicated curve and the subsequent cooling history is easiest to calculate numerically.

– So far we have neglected the latent heat of fusion as part of the cooling history. This latent heat of fusion amounts to about 320 kJ per kg of rock or roughly $8.64 \cdot 10^8$ $\mathrm{J\,m^{-3}}$. During crystallization of intrusions this heat is added to the thermal energy budget available to cause temperature change and causes buffering of the cooling history. The description of this process was first performed on the example of ice formation in the polar oceans (Stefan 1891), but the problem is identical to that of crystallizing intrusions. In the following we discuss this problem.

In sect. 3.2.3 we have mentioned, that the latent heat of reaction is an important part of the heat budget of high grade metamorphic terrains. As we need about 1 000 Joules to heat one kg of rock by one degree ($c_p = 1\,000$ $\mathrm{J\,kg^{-1}\,K^{-1}}$), the latent heat of fusion in enough to heat a rock by about 320 °C (because $L = 320$ kJ $\mathrm{kg^{-1}}$). An intrusion of $T_i = 700$ °C, that intrudes host rocks of $T_b = 200$ °C, is $\Delta T = 500$ °C hotter than its surroundings. This corresponds to $\Delta T c_p = 500\,000$ $\mathrm{J\,kg^{-1}}$ additional energy that is brought into the rock by this temperature difference. However, the total *heat content* of the intrusion (including its latent heat) that is brought into the rock is $\Delta T c_p + L = 820\,000$ $\mathrm{J\,kg^{-1}}$. The excess energy is therefore about 1.64 times as large as the excess temperature. This means that we have underestimated the cooling history in the previous sections substantially. For the description of the total heat budget of an intrusion process it would be more appropriate to describe a 1 000 m thick intrusion with a $1\,000 \times 1.64$ wide step-shaped temperature distribution, if we want to refrain from considering the influence of latent heat properly (s. p. 116).

However, this rough approximation only describes the total heat budget better, but not the thermal evolution. In order to describe the thermal evolution of a cooling crystallizing intrusion it is useful to imagine the processes involved in the freezing of a lake. During cooling of the air below the freezing point, at first the lakes surface will freeze. The crystallization heat that is freed during this process buffers the further freezing process. During further cooling of the surface, the subsequent thickening of the ice layer will slow down, as the frozen layer insulates the water to the outside and the latent heat of crystallization freed at the ice-water interface remains contained in the water. For this reason it is rare to find pack ice on the polar oceans that is thicker than about 2 m.

For the one-dimensional case and if the magma of a cooling intrusion crystallizes at a single eutectic temperature, there is an analytical solution that can be used to describe its thermal evolution under consideration of the latent heat of fusion. This is the solution found be Stefan (1891) for the description of pack ice formation on the polar oceans and it therefore bears his name. However, most rocks crystallize in divariant reaction over a large temperature interval between a solidus and a liquidus temperature, rather than at a single temperature. Then, numerical solutions of the heat flow equation must be used to consider the effects of latent heat (s. sect. 3.2.3). Regardless, the solution of Stefan (1891) gives important insights into the thermal processes involved. It solves the heat conduction equation under consideration of the latent heat crystallization at the rock- magma interface for the geometry described by:

Figure 3.37. Temperature profiles around a dike which is kept at constant temperature by the flow of magma. The curves were calculated with eq. 3.87 (time separation of the different curves in my). Compare this figure with Fig. 3.23 and Fig. 3.27

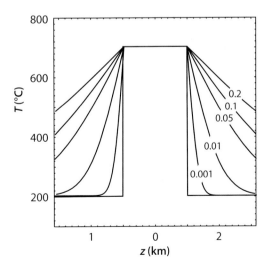

Figure 3.38. Temperatures in the contact aureole of a crystallizing interface between magma and host rock after 1 000 and 5 000 years, as calculated with eq. 3.97. Note that the contact metamorphic aureole is wider and of higher temperature than when only heat conduction is considered (Fig. 3.29). All parameters assumed here correspond to those used for Fig. 3.29

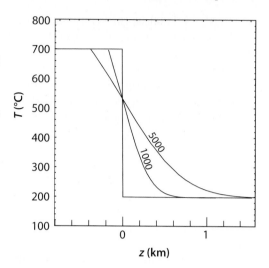

– $T = T_b$ at all $z > 0$ and $T = T_i$ at $z < 0$ at time $t = 0$

and boundary conditions at infinity. It is also assumed that the magma crystallizes at temperature $T = T_i$ The solution is given by:

$$T = T_b + (T_i - T_b) \left(\frac{\text{erfc}\left(\frac{z}{\sqrt{4\kappa t}} \right)}{1 + \text{erf}(\lambda))} \right) \quad . \tag{3.97}$$

In this equation, λ is given by the unsolvable function:

$$\frac{L\sqrt{\pi}}{c_p(T_i - T_b)} = \frac{e^{-\lambda^2}}{\lambda(1 + \text{erf}(\lambda))} \tag{3.98}$$

The derivation of this equation is complicated and of no relevance here. Eq. 3.98 may be solved iteratively or its solutions can be looked up in tables. Values for λ can then be inserted into eq. 3.97. This equation describes the temperature profile around a crystallizing dike margin. Fig. 3.38 shows some curves calculated with eq. 3.97. The thickness of the dike is no part of this solution, so that it is implicitly assumed that the dike is thick, if compared to the width of its contact metamorphic halo. The position $z = 0$ is the location of the dike margin at the onset of crystallization. The solution may be used up to the time where the crystallization interfaces of the opposite walls of the dike meet at its center. Thereafter numerical solutions must be used to describe the subsequent cooling history.

Heat Content. All thermal problems can be described either via temperature or via heat. The conversion factors between heat and temperature are the density and the specific heat capacity:

$$H = T\rho c_p \quad . \tag{3.99}$$

or, if formulated in terms of rate of heat loss/gain and cooling/heating rate:

$$\frac{\partial H}{\partial t} = \frac{\partial T}{\partial t} \rho c_p \quad . \tag{3.100}$$

Eq. 3.99 states that the heat content H (in $J\,m^{-3}$) is given by the product of temperature T (in K or °C), density ρ (in $kg\,m^{-3}$) and specific heat capacity c_p (in $J\,kg^{-1}\,K^{-1}$) (s. eq. 3.3). Most problems we have discussed in this section were formulated in terms of temperature. However, for some problems it is wiser to use the thermal energy balance in terms of heat. We only need to be careful with the units. The term ρc_p has the units of $J\,K^{-1}\,m^{-3}$, which is the energy that is necessary to heat one cubic meter of rock by one degree. If we multiply this term with the intrusion temperature, we get the heat content of the intrusion per cubic meter of magma. If we are interested in the heat content of the intrusion as a whole, then we must multiply this value with the volume of the intrusion.

The volume of a dike is given by the product of its thickness and its surface area. However, if we consider a one-dimensional dike cooling problem, then we can confine ourselves to consideration of the heat content per square meter of dike surface. Thus, the heat content of a dike of thickness l, per unit area of dike surface (as illustrated in eq. 3.89 and Fig. 3.29), may be read from the area underneath a one-dimensional $T - z$ curve. In other words, the heat content of a dike in $J\,m^{-2}$ is given by: $(T_i - T_b)l\rho c_p$. Using typical values of $\rho = 2\,700$ $kg\,m^{-3}$ and $c_p = 1\,000$ $J\,K^{-1}\,kg^{-1}$, the intrusion of Fig. 3.29 has a heat content of $H = 1{,}89 \cdot 10^{12}$ $J\,m^{-2}$. This is the heat content per square meter of dike surface.

• *Using heat content as a boundary condition.* The model of eq. 3.89 was based on the assumption of boundary conditions at infinity. The same boundary conditions may also be formulated in terms of the heat content of the intrusion. As no heat can be lost from the system if it cools by conduction only, the area underneath the curves in Fig. 3.29 must stay constant between $z = +\infty$ and $z = -\infty$. This is schematically shown on Fig. 3.39. During cooling of the intrusion, the area A must always be as large as the sum of $2B + 2C$. When solving the heat flow equation numerically on grids of finite extent, problems are often easier to handle this way than they are if the temperature is assumed to be constant at infinity. In this case the heat flow at the model boundary (given by the angle α) could be adjusted so that the areas C have the correct size. From Fig. 3.39 it may be seen that $A = 2B + 2C$ and $\tan(\alpha) = \Delta z / \Delta T$. From this, the thermal gradient at the model boundary is given by:

$$\alpha = \text{tg}^{-1}\left((A - 2B)/\Delta T^2\right) \quad . \tag{3.101}$$

This relationship is a useful in many numerical descriptions of diffusion problems with boundary conditions at infinity.

Figure 3.39. Schematic illustration of the heat content of a dike. The sum of the areas of the two dark-shaded regions B, plus the white triangles C must be the same as the light-shaded area A. Then, the heat content of the total system remains constant

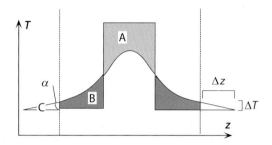

• *Heat content of metamorphic terrains.* In high grade metamorphic terrains containing syn-metamorphic intrusives it is often discussed if the volume of the intrusions is sufficient to contribute significantly to the metamorphism. In other words, it is discussed if the metamorphism is *contact metamorphism* in the widest sense (e. g. Problem 3.19). In order to do this estimate properly, the heat content of the intrusives must be compared with the heat content of the terrain. If the specific heat capacities of the intrusives and the metamorphic host rocks are the same, then the comparison of energy contents can also be made as a comparison of the temperatures. However, we must consider the latent heat of fusion that is part of the heat of the intrusives, but not of the host rocks. As simple calculation shows that:

$$T_{max} = T_b + \left(T_i - T_b + \frac{L}{c_p}\right) \frac{V_{intrusives}}{V_{terrain}} . \tag{3.102}$$

There, T_b and T_i are the temperature of the host rock before metamorphism and that of the intrusion, respectively. $V_{intrusives}$ is the volume of the intrusives and $V_{terrain}$ is the volume of the entire metamorphic terrain. T_{max} is the maximum temperature that can be reached by contact metamorphism.

If the *areal* proportion of intrusives to host rock are representative for the *volumetric* proportion of intrusives in the terrain, then the volumes of eq. 3.102 may be replaced by areas. Using $T_i = 700\,°C$, $T_b = 300\,°C$, $L = 320\,000$ J kg^{-1} and $c_p = 1\,000$ J kg^{-1} K^{-1}, eq. 3.102 shows the following: only about 55 % of the terrain must consist of syn-metamorphic granites in order to heat the entire terrain to 700 °C, even if the intrusion temperature itself was only 700 °C. If the intrusives are 1 200 °C hot mafic magmas, then only 30 % of the terrain must be intrusives in order to heat the terrain to 700 °C.

3.7 Selected Heat Transfer Problems

In this section we discuss a selection of geologically important heat transfer problems that do not belong directly to any of the past sections. Aside from the problems of the temperature distribution around moving faults (sect. 3.7.4) we discuss mainly problems which are subject to spatially or temporally changing temperatures that may be described with periodic boundary conditions. Such problems have the great advantage that analytical solutions of the heat conduction equation can be found for them. Thus, they differ from all other solutions discussed so far, which all contained error functions or infinite summations (we discussed in sect. 3.1.3 that the error function is an unsolvable integral and solutions containing them are – strictly speaking – no an analytical solutions). In the following sections we have selected three geodynamically relevant sets of problems which may be described with periodic boundary conditions. Analytical solutions for many others can be found in the literature (e. g., Carslaw and Jaeger 1959)

3.7.1 Temporally Periodic Temperature Fluctuations

The temperatures at the surface of earth are subject to the daily or annually periodically changing temperatures of the atmosphere. Problems where this is relevant, range from understanding the thickness of permafrost soils, to the regulation of temperatures in tunnels and insulation of walls of buildings. Many of these problems can be described with a one-dimensional coordinate system and with boundary conditions that describe a periodic fluctuation of the temperature at the surface. This may be written as:

- Initial condition: $T = T_0$ at all z at time $t = 0$.
- Boundary condition: $T = T_0 + \Delta T \cos(ft)$ at $z = 0$ for all $t > 0$, and $T = T_0$ at $z = \infty$ for all $t > 0$.

There, ΔT is half the amplitude of the annual fluctuation, t is time and f is the frequency (Fig. 3.40 a). T_0 is the mean annual temperature. The time dependent diffusion equation (eq. 3.6) can be integrated using these assumptions. We will not go through this integration here in any detail. However, the result is amazingly simple. It is given by:

$$T = T_0 + \Delta T e^{\left(-z\sqrt{\frac{f}{2\kappa}}\right)} \cos\left(ft - z\sqrt{\frac{f}{2\kappa}}\right) \ . \tag{3.103}$$

This equation may be used to describe temperature fluctuations at depth as a function of a periodic temperature variation at the surface. It may be seen that this equation contains a trigonometric function and an exponential function. At each time t it describes a cosine function of temperature which decays in amplitude exponentially with depth. At the depth $z = (f/2\kappa)^{-0.5}$ the amplitude of the temperature oscillation is only that of the surface divided by e. This depth is often called the *characteristic depth of equilibration* or *skin depth* (s. sect. 3.4.1). Eq. 3.103 is extremely important for many near surface problems, for example temperature profiles in snow, insulation problems of building walls, air temperatures in caves and many more.

3.7.2 Folded Isotherms

A beautiful problem illustrating the comparison of length and time scales
(s. sect. 3.1.4) concerns the shape of isotherms during folding. Before we
discuss this problem it should be said that – strictly speaking – isotherms
can not be folded, as they are not material lines. However, a number of
model descriptions use the term "folding" even when talking of isotherms
and we use it here as well. Fig. 3.41a shows how isotherms may be folded
during deformation of rocks. This process may occur if the axial plane of the
folds is not parallel to the isotherms. Whether or not folding of isotherms
has a thermal influence on rocks depends on the relationship between the
wavelength and amplitude of the fold, on the folding rate and of course on
the diffusivity κ. If the wavelength, the amplitude and the folding rate are
large, then isotherms will be folded together with the stratigraphy. If the
folding rate is small, then the rate of thermal equilibration is rapid compared
with the folding and the isotherms will remain flat. Field observations like
the updoming of the Tauern Window in the eastern Alps, together with the
updoming of the Alpine metamorphic isograds show that it may be important
to estimate the magnitude of this effect. If the estimates show that folding
is rapid enough to fold the isotherms as well, then we should observe that
antiforms cool and synforms heat. With the methods of modern petrology
it is conceivable that such details of the thermal evolution can be tested
(sect. 7.2).

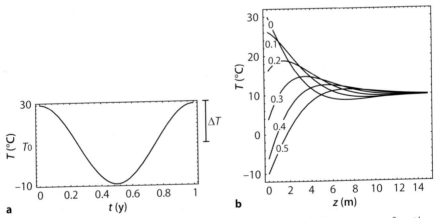

Figure 3.40. The temperature in the upper few meters of the crust as a function
of annually changing surface temperature. **a** shows the upper boundary condition:
the temperature T as a function of time t at $z = 0$. The annual mean temperature
was assumed to be $T_0 = 10\,°C$. The variation about this mean was assumed to be
$\Delta T = 20\,°C$. **b** shows six thermal profiles in the ground at six different times through
the course of half a year. The curves are labeled for time in years. It may be seen
that below depths of about 2 m temperatures never sink below the freezing point

A first estimate of the potential magnitude of an isotherm folding process can be made using the Peclet number (sect. 3.3.4). For a geologically reasonable shortening rate of 1 cm y^{-1} an amplitude of a single antiform of 5 km forms in about 0.5 my. The rate u, with which the material lines are deformed is therefore 10^4 m my^{-1}. Using l as the amplitude and $\kappa = 10^{-6}$ m^2 s^{-1} we get from eq. 3.51, that $Pe \approx 1.5$. While this estimate is only one-dimensional, it shows that diffusion and advection processes are both relevant for the assumed parameters and that a more detailed investigation of the problem is justified.

A quantitative and very simple model that can be used for the description of this process was developed by Sleep (1979). This model is based on the following assumptions:

1. Before folding, isotherms are parallel to each other and parallel to the surface of earth. The isotherms have a constant distance to each other and the initial temperature distribution as a function of depth can be described by:

$$T_{(t=0)} = gz \tag{3.104}$$

There, g is the geothermal gradient and z is depth.

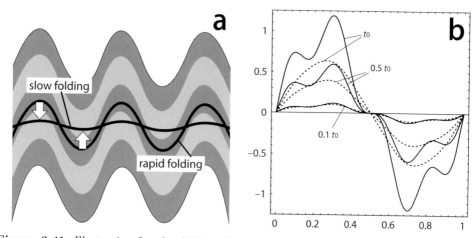

Figure 3.41. Illustration for the folding of isotherms. **a** Schematic illustration. The light and dark shaded regions are folded strata, the thick lines are isotherms. During slow folding, equilibration is faster than folding and isotherms will not be folded with the rock. During rapid folding of the rock conductive equilibratuion can not keep up with the material advection and the isotherms will be folded with the rock. **b** Folding of stratigraphy and isotherms as calculated with the model of Sleep (1979). The continuous lines are the shape of a given layer at three different points in time. The folding has a wave length of 1 and is overprinted by a parasitic fold of the wavelength 0.25. This means, in eq. 3.108, $A_1 = \lambda_1 = 1$ and $A_2 = \lambda_2 = 0.25$. The dashed lines are isotherms. It may be seen that the parasitic fold influences the isotherms only up to time $0.1t_0$. At time t_0 only the largest fold is visible in the shape of the isotherm. The diagram was calculated with eq. 3.108 and 3.109

2. Folding occurs by shear folding along axial planes that are perpendicular to the isotherms and a sine-shaped fold is produced that has the form:

$$v = v_0 \sin\left(\frac{2\pi x}{\lambda}\right) \quad . \tag{3.105}$$

In this equation, v is the displacement rate at each point of the fold, v_0 is the displacement rate of the fold hinge and x is the horizontal coordinate. This assumption describes shear folding at constant volume and does not consider shortening perpendicular to the axial plane of the folds. λ is the wavelength of folding. Using these assumptions, eq. 3.45 may be integrated. According to Sleep (1979) this gives:

$$T = g\left(z - z_0\left(1 - e^{-t/t_0}\right)\sin\left(\frac{2\pi x}{\lambda}\right)\right) \quad . \tag{3.106}$$

There, the constants z_0 and t_0 are:

$$z_0 = \frac{v_0 \lambda^2}{4\pi^2 \kappa} \quad \text{and} \quad t_0 = \frac{\lambda^2}{4\pi^2 \kappa} \quad . \tag{3.107}$$

Folding events that last shorter than t_0 will cause a significant deformation of the isotherms. If they last longer, then the additional deformation is negligible. For most field examples of folding, the geometry assumed for the folding process in this model is too simple. However, Sleep (1979) notes that *all* fold geometries that have formed by shear folding can be represented by a summation of sine-functions (using Fourier transforms). Thus, eq. 3.106 could be applied to all sine-shaped components of a random folding geometry and the results could then be summed up. For example, if a principle fold with the amplitude A_1 and the wavelength λ_1, is overprinted by a parasitic fold with the amplitude A_2 and the wavelength λ_2, then the shape of this fold at time t can be described by:

$$z_{\text{layer}} = t v_0 \left(A_1 \sin\left(\frac{2\pi x}{\lambda_1}\right) + A_2 \sin\left(\frac{2\pi x}{\lambda_2}\right)\right) \quad . \tag{3.108}$$

This equation corresponds to eq. 3.105, with the only difference being, that more than one wave with different wavelengths are added. Also, we multiplied the velocity here with time, in order to get the depth of any point of the folded layer (z_{layer}). Correspondingly, the position of the isotherm for more than one overprinting wave can be described by a reformulation and expansion of eq. 3.106:

$$z_{\text{isotherm}} = \frac{T}{g} + z_{01}\left(1 - e^{-t/t_{01}}\right) A_1 \sin\left(\frac{2\pi x}{\lambda_1}\right)$$

$$+ z_{02}\left(1 - e^{-t/t_{02}}\right) A_2 \sin\left(\frac{2\pi x}{\lambda_2}\right) \quad . \tag{3.109}$$

Fig. 3.41b shows an example in which the amplitude and wavelength of the overprinting parasitic fold is 1/4 of that of the principle fold.

3.7.3 Isotherms and Surface Topography

An important example of a heat conduction - advection problem concerns the influence of the surface topography on isotherms at depth. In a landscape with large vertical elevation differences, rocks that lie inside mountains are thermally insulated, while rocks nearer the surface of an incising valley are cooled by the surface (Fig. 3.42a). The surface temperature is largely constant because of the convection in the atmosphere and the isotherms underneath the surface follow it in a damped form. The variation of the distance of a given isotherm from the surface is of large importance for the interpretation of low temperature geochronological data, for the design of ventilation systems in tunnels and more. In this section we discuss some models that can be used to estimate the shape of isotherms underneath topography.

Description of the Topography. For our analysis we assume that the surface topography may be described by a sine-function with the wavelength λ and the amplitude $h_0/2$ where we interpret the wavelength λ as the distance between two parallel valleys and h_0 as the maximum elevation of the peaks above the valleys. Using the coordinate system illustrated in Fig. 3.42a,b (where z is the downwards axis with its origin at the valley floors and x is lateral distance across the topography), elevation h at any point of the topography is described by:

$$h = h_0 \frac{1}{2}\left(1 + \cos\left(\frac{2\pi x}{\lambda}\right)\right) \tag{3.110}$$

A large advantage of the description of topography by a such a function is that it allows us to evaluate the magnitude of the thermal effects as a function

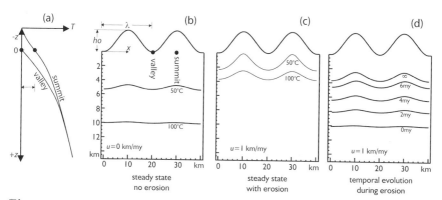

Figure 3.42. Isotherms underneath regions of high topographic relief. **b,c** and **d** show two-dimensional profiles through a mountainous topography. The two isotherms in **b** are for a non-eroding topography and in **c** for the thermal steady state in an eroding topography. In **d** the time dependent evolution underneath an eroding topography is shown for a single isotherm, in this case 100 °C. The two black dots in **a** and **b** are at equal depth, but they have a different temperature. Note that in the chosen coordinate system the surface is located at negative values for z, but (according to eq. 3.110) at positive values for h (s. Fig. 4.1)

of two simple parameters of topography: the wavelength and the amplitude. Any other topography may be described by the summation of a series of sine waves, so that eq. 3.110 is a very general form of a description.

• *Fudging the boundary condition.* In order to describe isotherms underneath topography, we have to assume a boundary condition along the surface as described by eq. 3.110, for example $T = T_s$. If we wanted to be really sophisticated we could also assume an atmospheric temperature gradient, but we will neglect this here. However, even if we assume a constant surface temperature, integration of the two-dimensional diffusion- or diffusion-advection equation (eq. 3.6 or 3.45) under this boundary condition is very difficult (s. sect. 3.6.3). A common way to surround this problem is by substituting this boundary condition of *constant* temperature at a *variable* spatial position (the topography), by a boundary condition of *variable* temperature at a *constant* elevation, for example at $z = 0$. This would be described by:

$$T = \Delta T \tfrac{1}{2} \left(1 + \cos\left(\tfrac{2\pi x}{\lambda}\right)\right) \text{ at } z = 0$$

where $\Delta T = h_0 g$ and g is the geothermal gradient in the absence of topography. This assumption implies that it is assumed that the thermal gradient inside the mountains is linear. ΔT corresponds to h_0 in the proper formulation (s. Fig. 3.42). This approximation is good if the wavelength of the topography is large compared to the amplitude because then the lateral cooling through the sides of the valley can be neglected. Note that this assumption is only an approximation if we were to describe isotherms underneath topography, but that it is quite a correct description if the surface temperature on a flat shield *does* vary periodically, for example because of the presence of lakes.

The Shape of Isotherms. The shape of isotherms underneath topography depends dramatically on the rate of erosion. In summary of the following three paragraphs, Brown (1991) has shown that isotherms are *not* strongly enough curved to influence apatite fission track results if the topography does not erode and Stüwe et al. (1994) have shown that the shape of isotherms *must* be considered if the erosion rate exceeds some hundreds of meters per my. Stüwe and Hintermüller (2000) expanded the problem to investigate the

Figure 3.43. The drop off of the amplitude of a periodic temperature fluctuation that has the wavelength $\lambda = 10$ m at the surface $z = 0$, with depth. Calculated with eq. 3.111, as well as $T_s = 0\,°C$ and $\Delta T = 20\,°C$. The curves are shown for $z = 0$, $z = 1$ and $z = 5 = \lambda/2$

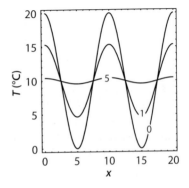

thermal influence of laterally migrating drainage divides. All these authors explored the thermal steady state where the rate of upwards advection of heat is balanced by cooling of the surface, so that there is no change in the position of the isotherms with respect to the surface. Mancktelow and Grasemann (1997) investigated the time dependent effects how this steady state is reached (Fig. 3.42). However, as the Peclet number of most topographies is small, it is usually not necessary to investigate the time dependent evolution of isotherms. The steady state solutions discussed below are generally well sufficient to study isotherms. In the following paragraphs we mention three special cases of isotherms underneath topography.

• *Non-eroding topography.* If topography is approximated with the variable temperature boundary conditions described in eq. 3.7.3, and there is no advection of heat by erosion, then there is a solution of the two-dimensional diffusion equation. It describes the temperature variation as a function of depth z and lateral distance x with:

$$T_{(x,z)} = T_s + \Delta T \frac{1}{2} \left(1 + \cos\left(\frac{2\pi x}{\lambda}\right) \right) e^{-2\pi z/\lambda} \quad . \tag{3.111}$$

Fig. 3.43 illustrates that the spatial temperature variation at the surface causes negligible temperature variations at a depth that are larger than $z = \lambda/2$. For example, if the distance between two mountain peaks is about 10 km, then the amplitude of the isotherms is negligible at depths larger than 5 km from the surface. This caused Brown (1991) to conclude that the curvature of isotherms underneath a topography that is not eroded has negligible influence on the interpretation of apatite fission tack data. Fig. 3.43 shows this result in form of temperatures at a constant depth as a function of x. It does *not* show the depth of a given isotherm. If solved for z, eq. 3.111 gives us the vertical distance of a given isotherm from its mean depth. If we want to know the depth of an isotherm, we must add a linear thermal gradient to the temperature. Thus, for a linear thermal gradient, the temperatures at depth as a function of a lateral temperature variation at the surface are given by:

$$T_{(x,z)} = T_s + gz + \Delta T \cos\left(\frac{2\pi x}{\lambda}\right) e^{-2\pi z/\lambda} \quad . \tag{3.112}$$

An example for this is shown in Fig. 3.42b.

• *Downwards eroding topography.* During erosion (i. e. advection of material towards the surface) isotherms may be partly exhumed and compressed into the topography. Depending on the erosion rate, this can dramatically increase the amplitude of the isotherms. How much the amplitude of isotherms may be increased by the effects of erosion depends on the duration and rate of erosion as well as the amplitude and wavelength of the topography. Stüwe et al. (1994) found a semi-analytical solution of the two-dimensional diffusion-advection equation to describe this problem. Fig. 3.42c and 3.44 shows some results from their study. They concluded that at erosion rates above 500 m my^{-1} it

becomes important to consider the topographic effects on the interpretation
of apatite fission track results.

• *Asymmetrically eroding topography.* If erosion rates vary spatially, for example due to differential rainfall on two different sides of a mountain range,
then it is possible that the incision direction of drainages has a lateral component. In a topographic steady state, a mountain range may incise downwards
and migrate laterally. In a reference frame fixed to the topography, this process may be illustrated by rock trajectories that are oblique to the topography (Fig. 3.45a; Stüwe and Hintermüller 2000). Fig. 3.45b shows the cooling
paths of four characteristic points on the topography. It may be seen that
these cooling paths have very different shapes and that they cool through any
one isotherm at different times. In particular, it may be seen that the cooling
rates *decrease* with decreasing temperature on the slow eroding side of the

Figure 3.44. The depth and shape of the
100 °C-isotherm in a mountainous region with a
relief of three kilometers and a distance between
mountains of 20 km. The isotherm is shown for
the steady state during erosion at four different
rates in km my^{-1}. The spatial coordinate system was fixed at the surface (Eulerian reference
frame). In such a coordinate system the surface does not move downward in the crust, but
the material is advected upwards, towards the
surface. For details of the assumptions s. Stüwe
et al. (1994)

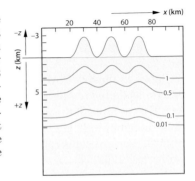

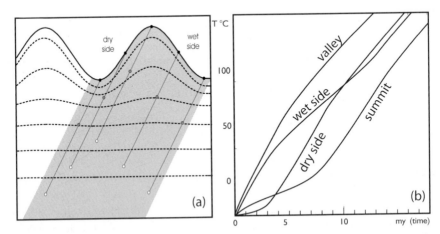

Figure 3.45. Cooling paths underneath an asymmetrically eroding topography
(from Stüwe and Hintermüller 2000). Cooling paths in **b** are drawn for a topography
with wavelength λ= 20 km, elevation of peaks h_0= 3 km, a vertical component of
erosion of 500m y^{-1} and a lateral migration rate of the topography which is twice
as large as the vertical component. "Dry side" and "wet side" stand symbolically
for the slow and rapidly denuding side of the range, respectively

range, while the cooling rates *increase* with decreasing temperature on the
rapidly denuding side of the range. With the advent of radiometric dating
methods that can be used to date temperatures substantially below 100 °C
(e.g. House et al. 1998, 2000) such differences in cooling curves will soon be
of use to document details of erosion processes.

3.7.4 Temperature Distribution Around Faults

The thermal evolution of rocks around faults is a typical two-dimensional
problem in which all three heat transfer mechanisms may play a role: *diffusion* in the foot and hanging wall of the fault, *advection* of heat by the motion
of the fault and *production* of friction heat in the fault itself (s. Fowler and
Nisbet 1982). In the following analysis we only discuss diffusion and advection. For the importance of mechanical heat production, the interested reader
is referred to the discussions by Molnar and England (1990a), Graham and
England (1976) and Pavlis (1986), as well as sect. 3.2.2.

Fig. 3.46 shows a vertical cross section through a package of rock that is
transected by a fault (in the shown orientation it is a *normal* fault). The
normal fault is inclined with 60° to the left and the initial isotherms were
inclined with 20° to the right. The displacement along the fault causes that
the hot rocks in the foot wall are opposed to relatively cold rocks in the
hanging wall. Near the fault the isotherms are stretched. Thus, the lateral
temperature gradient in the vicinity of the fault is *decreased* by the fault

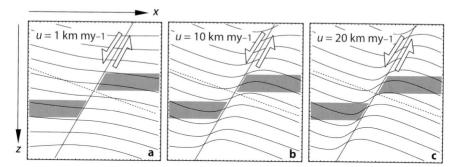

Figure 3.46. Isotherms in a crustal cross section around a normal fault that is
inclined with 60 degrees to the left. The shown area is 50 km by 50 km in size.
The figures are labeled for the rate of normal displacement u. **a**, **b** and **c** show the
temperatures after 10 my, 1 my and 0.5 my. Thus, the displacement is the same in
all three figures. The gray shaded region indicates schematically a lithological layer.
The isotherms are shown every 50 °C. The isotherms were inclined with 20 degrees
to the right before the fault became active (dashed line). The boundary conditions
on all four sides are given by constant heat flow. This implies that the crustal
section shown is surrounded on all sides by matter and that the diagram can be
rotated arbitrarily. The diagram was calculated with a numerical solution of a two-
dimensional form of eq. 3.44. For an analytical solution of related problems see e. g.
Voorhoeve and Houseman (1988)

displacement (Fowler and Nisbet 1982). If the fault were a reverse fault of
the same angle, then the situation would be reversed: the foot wall would
be cooled, the hanging wall would be heated and the isotherms would be
compressed. The heating of rocks by reverse faulting is one of several processes
that have been made responsible for the formation of inverted metamorphic
gradients (s. discussion by England and Molnar 1993).

• *Estimating the thermal influence.* In order to estimate the thermal influence
of a fault zone on its surroundings, we can employ two parameters we have
discussed in previous sections: The thermal time constant (eq. 3.18) and the
Peclet number (eq. 3.51). In Fig. 3.46 the displacement of the fault is 10 km.
In eq. 3.51 we can therefore assume $l = 10$ km. With $\kappa = 10^{-6}$ m^2 s^{-1},
Figs. 3.46a,b and c give Peclet numbers of $Pe \approx 0.3$, ≈ 3 and ≈ 6. For
Fig. 3.46a this implies that diffusion outweighs the influence of advection. A
displacement of the isotherms is therefore hardly visible. In Fig. 3.46c the
Peclet number is much larger than 1, implying that advection outweighs the
influence of diffusion. Thus, the displacement of the isotherms is large and the
thermal effects are confined to a narrow region around the fault. The duration
of displacement in Figs. 3.46a,b and c is 10, 1, and 0.5 my, respectively. Using
eq. 3.18 this indicates a length scale of the thermally influenced region of 20,
5 and 3 km. These estimates correspond well with the shape of the curves on
Fig. 3.46.

• *Temperature-time evolution during exhumation in the vicinity of faults.* In
many regions of active and ancient mountain belts it is observed that displacement along fundamental structures leads to different rates of exhumation of
the foot wall and the hanging wall. This may lead to a complicated thermal
evolution of rocks as there are two competing thermal processes:

1. The respective heating and cooling of the two sides of the fault.
2. The cooling of rocks with proximity to the surface.

Fig. 3.47 illustrates this with a vertical cross section through the crust.
This figure differs from Fig. 3.46 in that the upper boundary (the earth's
surface) is now defined by a boundary condition of constant temperature.
Thus, in contrast to Fig. 3.46, no isotherms intersect the surface. The crustal
section shown is divided into two blocks separated by a vertical fault. The
two blocks exhume with two different rates, u_1 and u_2.

In order to understand the thermal evolution of rocks near the fault, consider rocks in the slower exhuming block (left block on Fig. 3.47) near the
fault. There, the *cooling* influence of the surface is opposed by the *heating*
influence of the block on the right hand side. Because of this, it may happen,
that rocks that cool during exhumation experience a late stage heating event
caused by the other side of the fault. An example of such a thermal evolution
was documented by Grasemann and Mancktelow (1993) at the Simplon-line
in the central Alps.

Figure 3.47. Isotherms in a vertical cross section through the crust with a vertical fault. Both sides of the fault exhume with different rates u_1 and u_2. The shown section is 50 km by 50 km. The boundary condition at the top boundary is given by a constant temperature. The boundary conditions on all other sides are given by constant flow of heat. In contrast to Fig. 3.46 all isotherms are preserved, because of the fixed top boundary

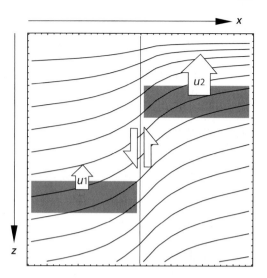

3.8 Problems

Problem 3.1. *Converting different units of energy (p. 47):*
In nuclear reactions, mass is converted to energy. How long can a 60-W-light globe be lit with 1 g of mass assuming that this mass may converted completely into energy? Remember: energy = mass × speed of light^2. The speed of light is $\approx 300\,000$ km s^{-1}. A related problem is found in Problem 5.3.

Problem 3.2. *Converting different units of energy (p. 49):*
What is the conversion factor between hfu (1 hfu $= 10^{-6}$ cal cm^{-2} s^{-1}) and the SI-units W m^{-2}.

Problem 3.3. *Understanding heat refraction qualitatively (p. 52):*
A spherical iron ore body of 10 km diameter (with extremely high thermal conductivity) lies in the middle of a 30 km thick crust. Draw a cross section through the crust and sketch schematically some isotherms. Discuss the thermal consequences that such a geometry may have for the surrounding rocks.

Problem 3.4. *Understanding heat refraction quantitatively (p. 52):*
Calculate the temperature at 10 km depth in the crust using the following assumptions. The crust is made up of schists with a conductivity of $k = 2$ J s^{-1} m^{-1} K^{-1}. These schists are intruded by a 2 km thick sill that has a conductivity of $k = 4$ J s^{-1} m^{-1} K^{-1}. The upper contact of the sill is in 5 km depth. The thermal gradient at the surface is 20 °C km^{-1} and the temperature at the surface is 0 °C. The temperature profile is in steady state (there is no exchange of heat between the layers) and there is no heat production in the crust (i.e. the different sections of the geotherm are linear). You can use eq. 3.7.

Problem 3.5. *Using thermal time constant (p. 57)*:
Two continental plates collide and deform a 35 km thick crust. The deforma-
tion rate $\dot\epsilon$ is a) 10^{-12} s^{-1} and b) 10^{-16} s^{-1}. During and after deformation,
the thermal structure of the crust re-equilibrates by diffusion. For the given
deformation rates, estimate whether the equilibration of the crust is faster,
slower or of similar rate as the deformation? The time scale of the thermal
equilibration may be estimated with the thermal time constant (eq. 3.18).
What consequences may your results of a) and b) have for structures and
parageneses that may be observed in a thin section?

Problem 3.6. *The importance of radiogenic heat production (p. 61)*:
A roughly spherical radioactive ore body of 10 km diameter produces about
$S_{rad} \approx 100$ micro Watts of heat per cubic meter. Estimate how hot the center
of the body will get after about $t= 10^5$ years. Use eq. 3.22 but before you
do, make an argument why it may be realistic to neglect heat conduction.
Use eq. 3.19 and estimates for the necessary parameter values as discussed in
the text, for example $\kappa = 10^{-6}$m^{2}s^{-1}, $c_p=1\,000$ J kg^{-1}K^{-1} and $\rho = 2\,700$kg
m^{-3}.

Problem 3.7. *The importance of mechanical heat production (p. 62)*:
a) What is the mechanical heat production rate of a 5 km wide shear zone
which deforms at a rate of i) $\dot\epsilon = 10^{-13}$ s^{-1} or: ii) $\dot\epsilon = 10^{-15}$ s^{-1}? Assume that the deviatoric stress that the shear zone material supports is be-
tween 30–300 MPa. Give minimum and maximum values using both strain
rates and both shear strengths using eq. 3.25. b) How warm could the shear
zone possibly get if deformation lasts for 1 my? ($c_p = 1\,000$ J kg^{-1} K^{-1};
$\rho = 2\,700$ kg m^{-3}). c) Which parameters control if (and how much) the shear
zone heats up during this process? (Another problem related to mechanical
heat production is Problem 5.3)

Problem 3.8. *The importance of adiabatic processes (p. 69)*:
a) Demonstrate that eq. 3.33 is dimensionally consistent. b) Use eq. 3.33 to
estimate the adiabatic temperature rise between the surface and the center of
earth using $c_p = 1\,000$ J kg^{-1} K^{-1} and $\alpha = 3 \times 10^{-5}$K^{-1} as well as estimates
for the remaining parameters. c) The adiabatic compressibility is smaller than
the isothermal compressibility. Why?

Problem 3.9. *Understanding heat of reaction (p. 70)*:
How much mass is converted to energy when burning 5 kg of wood?

Problem 3.10. *The importance of latent heat of fusion (p. 72)*:
A high grade metamorphic rock contains 30 % partial melts. All of this partial
melt was formed by a single melting reaction at a fixed temperature. The
rock cools conductively from its metamorphic peak (which was much higher
than this melting temperature) with a constant cooling rate of 100 °C my^{-1}.
Estimate how long the rock will be buffered to a constant temperature when

this melt crystallizes? Give the result in my. The latent heat of fusion is $320\,000$ J kg^{-1}, the rock has a density of $\rho = 2\,700$ kg m^{-3} and a specific heat capacity of $c_p = 1\,000$ J kg^{-1} K^{-1}. Discuss the potential effect of your result on geothermometry that may be planned for this rock.

Problem 3.11. *Understanding the Peclet number (p. 80):*
A regional metamorphic event occurred over the whole of a 30 km thick crust. A mountain belt at the surface eroded at the same time and exhumed the metamorphic rocks during this process. Estimate whether the regional thermal evolution can be described by only considering heat conduction (that caused the regional metamorphic event) or if heat advection (due to erosion) must also be considered. Use erosion rates of a) 100 m my^{-1}; b) $1\,000$ m my^{-1}; c) $5\,000$ m my^{-1} and eq. 3.51.

Problem 3.12. *Contribution of radioactivity to heat flow (p. 85):*
Use eq. 3.61 to estimate how the surface heat flow changes if the radiogenically caused heat flow doubles and the mantle heat flow is decreased by 50%. Note that *both* will occur when the lithosphere is doubled in thickness.

Problem 3.13. *Distribution of radioactivity (p. 85 – 90):*
If we plot surface heat flow data against data for the surface heat production S_0, then the slope of this data may be interpreted as the thickness of a hypothetical radioactive layer z_{rad} in the crust in which the concentration of the radioactive elements is the same as that on the surface (Fig. 3.15). However, in reality, not all radioactive elements in the crust will occur in this layer. Rather, they are distributed through the crust. An elegant description for a more realistic distribution is given by eq. 3.67. Both models are illustrated in Fig. 3.16 as curves b and c, respectively. Estimate the depth in the crust z_w above which the two models describe the same *total* crustal heat production (area underneath the curves on Fig. 3.16, assuming both models describe the same surface heat production S_0 and the characteristic depth in eq. 3.67 is $h_r = 2z_{rad}$.

Problem 3.14. *Calculation of stable geotherms (sect. 3.4.1):*
Integrate eq. 3.56 to calculate a stable geotherm for the crust assuming the following. The distribution of heat production in the crust is given by $S = S_0(1 - \cos(2\pi z / z_c))/2$. z_c is the thickness of the crust. As boundary conditions, assume that both the temperature at the surface $T_s = 0$ and the heat flow at the Moho q_m are constant. The assumption of a cosine function of the heat production with depth is illustrated as model d in Fig. 3.16 and may describe what scenario?

Problem 3.15. *The geotherm equation (sect. 3.4.1):*
a) Derive an equation that can be used to calculate the heat flow through the Moho for a generalized description of the thickness ratio of crust and mantle lithosphere. Eq. 3.76 describes temperatures for the assumptions that should

be used here. It can be used as a start for your derivation. Remember: the heat flow through the Moho is $q_m = k dT/dz$ at $z = z_c$. b) Check how the mantle heat flow changes if the crust or the lithosphere is thickened. Draw your results into an f_c-f_l-diagram for $k = 2\ \mathrm{J\,s^{-1}\,m^{-1}\,K^{-1}}$, $z_c = 35$ km, $z_l = 100$ km, $T_l = 1\,200\,^\circ\mathrm{C}$, $h_r = 10$ km and $S_0 = 5 \cdot 10^{-6}\ \mathrm{W\,m^{-3}}$. c) Discuss the implications of your result for the heat sources of regional metamorphism in collisional orogens.

Problem 3.16. *Cooling of oceanic lithosphere (p. 93):*
The temperature distribution in oceanic lithosphere may be described with the *half space cooling model* (eq. 3.81). a) Calculate the depth of the $1\,000\,^\circ\mathrm{C}$-isotherm for an 80 my old oceanic lithosphere. Assume the temperature of the asthenosphere is $T_l = 1\,200\,^\circ\mathrm{C}$ and the temperature at the surface is $T_s = 0\,^\circ\mathrm{C}$. The diffusivity κ is $10^{-6}\ \mathrm{m^2\,s^{-1}}$. Use eq. 3.81 and Fig. 3.5 or the approximation in Table B.9 to solve the error function. b) Draw a temperature profile through 10 my old oceanic lithosphere with the same assumptions.

Problem 3.17. *Qualitative thermal evolution of intrusions (p. 103 – 108):*
a) Estimate the total duration of cooling of a 50 m wide dike with $\kappa = 10^{-6}\ \mathrm{m^2\,s^{-1}}$. Use eq. 3.19. b) Assume that the dike has intruded with $700\,^\circ\mathrm{C}$ into host rocks that are $300\,^\circ\mathrm{C}$ hot. What is the maximum contact metamorphic temperature? c) How much additional heat does the dike bring into the rock it intrudes? Use eq. 3.99 and eq. 3.100 and $\rho = 2\,700\ \mathrm{kg\,m^{-3}}$, $c_p = 1\,000\ \mathrm{J\,kg^{-1}\,K^{-1}}$. d) Draw a qualitative temperature profile across the dike at 40 years after intrusion. Help yourself for this with the result from a).

Problem 3.18. *Quantitative thermal evolution of intrusions (p. 103 – 108):*
The thermal evolution of magmatic dikes can be well-described with the model of eq. 3.89 (Fig. 3.28). For the dike we discussed in the last problem, following questions are asked: a) What is the maximum contact metamorphic temperature that rocks can reach that are 10 m away from its contact. (For this problem you can find an elegant and a not so elegant way of solving it). b) Draw the thermal evolution of three rocks onto a temperature-time diagram: 1. A rock from the center of the dike, 2. A rock that lies 10 m from the contact and 3. A rock inside the dike, which lies 1 m near the contact. c) Calculate a temperature profile across the dike, 40 years after the intrusion. Compare this profile with your result from Problem 3.17d.

Problem 3.19. *Inferring metamorphic heat sources (p. 114):*
Estimate if the following metamorphic terrain could have been heated by contact metamorphism. The terrain has reached a peak temperature of $600\,^\circ\mathrm{C}$. About 10% of the area of the terrain are syn-metamorphic mafic intrusions that had an intrusion temperature of $1\,100\,^\circ\mathrm{C}$. Another 30% of the area of the terrain are syn-metamorphic granitoids that had an intrusion temperature of $700\,^\circ\mathrm{C}$. The peak metamorphic pressure was 5 kbar, which corresponds to a depth of 18.5 km. Before metamorphism the terrain was at the same depth

Figure 3.48. Illustration for Problem 3.20

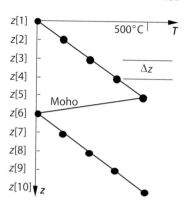

and had a temperature corresponding to a stable geothermal gradient of $16.2\,°C\,km^{-1}$. The density of the rocks is $\rho=2\,700\ kg\,m^{-3}$ and the specific heat capacity is $c_p=1\,000\ J\,kg^{-1}\,K^{-1}$. Answer the question graphically and algebraically with and without consideration of the latent heat of fusion of the magmas ($L=320\,000\ J\,kg^{-1}$).

Problem 3.20. *Inverted isograds and numerical solutions of the heat flow equation (sect. 3.1.1, A.2):*
Calculate the thermal evolution above and below a thrust that has doubled the entire crust of $z_c=40$ km. Assume that the thrusting rate was much faster than the rate of thermal equilibration. Thus, we can assume a "saw tooth" geotherm as our initial condition (Fig. 3.48). The initial geotherm is defined as follows: In the region $0 < z < z_c$ it is given by $T = T_{Moho}(z/z_c)$; in the region $z_c < z < 2z_c$ it is given by $T = T_{Moho}((z - z_c - \Delta z)/z_c)$. ($T_{Moho} = 500\,°C$ and $z_c = 40$ km). The distance between discrete points for your calculation is $\Delta z = 10$ km. Use eq. A.16 and eq. A.18 in order to approximate eq. 3.6. For the mathematical stability of your solution you have to make sure that the constant $R = (\kappa \Delta t)/(\Delta z^2)$ is smaller than 0.25. This condition gives you the maximum time step Δt that you can use. κ is $10^{-6}\ m^2\,s^{-1}$.

4. Elevation and Shape

In this chapter we discuss the position, shape and the motion of rocks; in short: geodynamic processes measured in *meters*. This includes discussions of the elevation of mountain ranges, and the depth of the oceans, as well as the *change* of such parameters: kinematics. We begin with a consideration of different coordinate systems and reference frames that we will use.

4.0.1 Reference Levels

All geological motion, for example the motion along faults, the uplift of a mountain range or the approaching of two lithospheric plates may only be observed *relatively* (s. Fig. 2.6). Thus, for any kinematic problem it is necessary to define a coordinate system to which the motion may be related. For some examples we do this quite intuitively. For example, when talking of a *dextral fault*, we all understand that we mean the motion of *one* side of the fault as seen relative to a reference frame fixed to the *other* side of the same fault. For other examples, in particular those revolving around vertical motions or those around plate motions on a large scale there has been much confusion with reference frames in the literature. For example, in all previous sections (when discussing geotherms), we have fixed our coordinate system to the surface of the crust. This reference level is useful when describing geotherms as it is irrelevant to a thermal evolution whether the entire lithospheric column is being uplifted or remains stationary. However this reference level is not very meaningful when considering vertical motion of this surface itself: the elevation of a mountain range relative to its own surface is always zero. It cannot be described as long as the reference frame is fixed to the surface. Thus, for problems dealing with surface elevation, we need to change our reference frame and fix it to sea level or some other externally fixed reference frame (Fig. 4.1). In sect. 4.1.1 we will show in some detail how careless handling of reference levels may lead to grave misinterpretations. However, we begin by discussing some important reference levels for geodynamic problems.

Sphere, Spheroid and Geoid. When considering geological problems for which it is *not* possible to make the two-dimensional approximation that the earth is flat (s. sect. 2.2.1), it is necessary to make another approximation: we must find a good model description of the shape of our three-dimensional

Figure 4.1. Schematic sketch of two different vertical axes that are used by earth scientists to describe vertical motions in the lithosphere. The surface elevation H is usually described relative to sea level or the reference lithosphere and is generally assumed to extend positively upwards. The depth of rocks in the crust is generally measured relative to the surface and is measured positively downwards. Note that both *direction* and *origin* of the axes are different for the two reference levels, which has led to much confusion in the literature

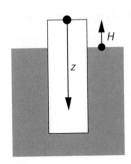

globe. The best first order approximation is that the earth is a *sphere*. For a long time one meter was defined as the 40 000 000th fraction of the circumference U of the globe. According to this definition, the radius R of a spherical earth would be: $R = U/(2\pi) \approx 6366$ km. This approximation of the earth's radius is sufficient for most plate tectonic problems, for example those that consider the curvature of mountain belts or subduction zones, or those that consider the torques exerted by mid-oceanic ridges (s. sect. 2.2.2). However, the "real" shape of the earth deviates from a sphere by being flattened in the direction of the rotation axis: the centrifugal forces that arise because of the rotation of the globe cause the earth to have more the shape of an ellipsoid. This flattening of the globe at the poles is much more pronounced in the atmosphere than it is in the solid part of earth. In fact in the atmosphere it is pronounced enough so that Mt Everest could not be climbed if it were at any higher latitudes (Fig. 4.2). The polar radius of earth is $R_P = 6356.75$ km.

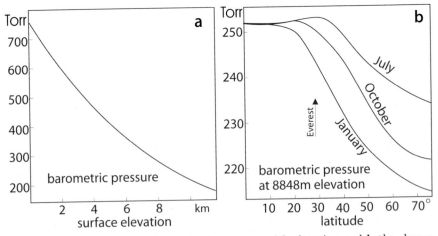

Figure 4.2. a The drop of atmospheric pressure with elevation and **b** the change of atmospheric pressure at the elevation of the top of Mt Everest with latitude. The pressure is given in Torr which corresponds to millimeters Hg. The conversion to SI units is 1 Torr = 133.32 Pa (s. Table C.8)

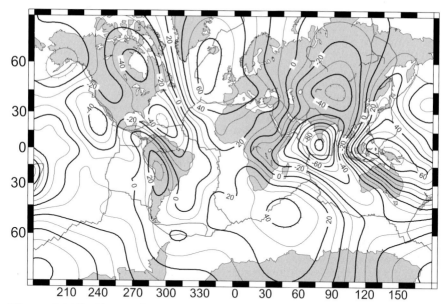

Figure 4.3. Geoid map of the world. The largest geoid anomaly (labelled in meters) is in the Indian ocean where the geoid has an 80 m deep hole relative to the spheroid described by eq. 4.2

The equatorial radius is $R_A = 6\,378.139$ km. Because the difference between polar and equatorial radius is very small (only about 20 km which is $\approx 0.3\,\%$ of the radius), the ellipsoid of earth is also often referred to as a *spheroid*. The fractional difference between the two radii is given by:

$$f = \frac{R_A - R_P}{R_A} = 0.0034 \tag{4.1}$$

and is called the *ellipticity*. The radius of earth at any latitude λ may then be calculated with the relationship:

$$R \approx R_A \left(1 - f\sin^2\lambda\right) \quad . \tag{4.2}$$

However, strictly speaking, eq. 4.2 describes no ellipse. The spheroid defined by eq. 4.2 is the reference level used for the interpretation of gravity anomalies. The *geoid* is the surface of constant gavitational potential energy (Fig. 4.3. In areas of high density, the gravitational acceleration is relatively high and the surface of the geoid lies low and vice versa (as to compensate the high acceleration with a lower mass of the column: energy = mass × acceleration). If the earth would not rotate (so that there is no centrifugal force), then the *spheroid* surface would be identical with the *geoid*. The differences between the spheroid as defined by eq. 4.2 and the measured geoid surface are called *geoid anomalies*.

The Undeformed Reference Lithosphere. For most geological problems of the field geologist reference levels like sea level, the geoid, or the spheroid are not very instructive. A much more useful reference frame is usually the *undeformed reference lithosphere* (Le Pichon et al. 1982). The undeformed reference lithosphere is a hypothetical lithospheric column which remains unchanged relative to the orogen under observation. If the undeformed state of the lithosphere under observation is unknown, then it is most simple to assume that the surface of the undeformed reference lithosphere corresponds to the mean surface of the continents. This is 840 m above sea level and the mean depth of the oceans is 3700 m. However, about 80 % of the land surface is only between 100 and 200 m above sea level (Fig. 1.2). All heights and elevations that are given in the following sections *without* a detailed mention of the reference level, are understood to be relative to the surface of the reference lithosphere. That is, they are relative to the elevation of the respective lithospheric level (e. g. the surface, the Moho etc.) *prior* to onset of thickening or thinning.

Lagrangian and Eulerian Reference Frames. Most problems we have discussed in the previous chapters of this book were discussed in fixed reference frames. That is, temperatures change and rocks move relative to the coordinate system. Such a coordinate system is called a *Eulerian reference frame*. The description of processes in Eulerian reference frames is referred to as *Eulerian* or *spatial* description. For many problems in the earth sciences, it is more useful to choose a coordinate system that moves with the rock. Such coordinate systems are called *Lagrangian reference frames*. Lagrangian descriptions are also referred to as *material description*. Fig. 4.4 illustrates both reference frames. Fig. 4.4a illustrates the evolution of geotherms during

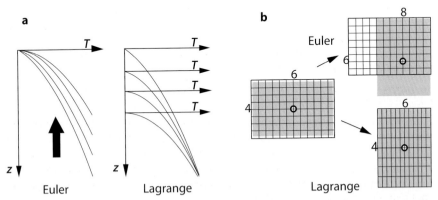

Figure 4.4. Differences between reference frames according to Lagrange and to Euler. **a** One-dimensional model of the evolution of geotherms during erosion. **b** Two-dimensional deformation of a square block. In both cases, the Eulerian description is within a coordinate system that is fixed externally (material is transported relative to the coordinate system) and the Lagrangian description is in a coordinate system that moves with the material

erosion. From the point of view of a Eulerian observer (who stands on the eroding surface) rocks (and the geotherm) are moved *through* the surface. From the point of view of a Lagrangian observer, the coordinate system is fixed to the rock and the location of the surface boundary condition changes its location downwards in the coordinate system.

Both reference frames have advantages and disadvantages and it depends on the nature of a problem in terms of which reference frame it should be dealt with. The *Eulerian* description has the advantage that it is easy to describe the motion of material relative to an unmoved area, for example relative to the reference lithosphere. One advantage of *Lagrangian* descriptions is it that the history of a given rock is much easier to track. For example, in numerical calculations using a Lagrangian reference frame, the time dependent changes of a variable at a given grid node will always describe the evolution of a rock at this location, while in Eulerian reference frames, rock trajectories will go *through* the grid.

As the description of a problem changes with the choice of reference frame, so does the relevant equation. Consider the example of Fig. 4.4a. In a Eulerian description, erosion may simply be described as advection of the material through the grid (e.g. the advection rate u in eq. 3.45). The boundary condition at the surface remains fixed at $z = 0$. For example, it could be $T = 0$ at $z = 0$ as it was in many examples in chapter 3. The thermal evolution of individual material points is tracked by changing the observation depth according to: $z(t) = z_i - ut$ (eq. 3.49), where z_i is the initial depth of a material point, t is time and $z(t)$ is the depth of the material point through time. Within a Lagrangian description, the same problem may be described without advection. The process of erosion is described by moving the boundary condition with the velocity ut, for example $T = 0$ at $z = ut$. Within this reference frame the boundary condition is moving towards to observer that is fixed to the rock. We will encounter the description of a very similar problem in two different reference frames in the context of Fig. 4.30.

Figure 4.4b illustrates another example of Lagrangian and Eulerian descriptions of the same problem, using a two-dimensional deformation model. The gray area represents a body of rock and the grid is a two-dimensional coordinate grid. The rock is deformed towards the walls on the right and at the top by forces acting from the left. Within a Eulerian reference frame, this deformation appears like material is being transported through the coordinate system. The coordinates of individual rocks change with time. For a Lagrangian observer the deformation appears to be caused by an approaching side wall and a retreating upper wall. Individual rocks remain at the same coordinates and the grid is distorted together with the rock (Fig. 4.4). In this example, the principle advantage of the Lagrangian description is that a given coordinate remains fixed to a rock. In order to calculate – for example – a *P-T*-path of a rock, a single coordinate must be tracked through time. The Eulerian description has the advantage that it is easier to keep track where the system is located with respect to its environs. The disadvantage of the Eulerian description is that rock trajectories are much more difficult to follow (e. g. eq. 3.49 and sect. 4.3.2).

4.0.2 The f_c-f_l-Plane

The elevation of the earth's surface in isostatic equilibrium is a direct function of:

- 1. the thickness of the crust,
- 2. the thickness of the mantle part of the lithosphere,

(s. Fig. 2.15). During orogenesis, these two parts of the lithosphere may change their thickness at different rates and by different amounts. It is therefore instructive to explore the influence of thickening of the two parts of the lithosphere explicitly but simultaneously. This may be done by plotting the vertical strain of the crust (which we call f_c), against the vertical strain of the lithosphere (which we call f_l). We can write:

$$f_c = \frac{z_{\text{defc}}}{z_c} \quad \text{and} \quad f_l = \frac{z_{\text{defl}}}{z_l} \quad , \tag{4.3}$$

where z_{defc} and z_{defl} are the thicknesses of crust and lithosphere at a given time during orogenesis and z_c and z_l are the thicknesses of undeformed reference crust and lithosphere, respectively. The f_c-f_l-plane – in which f_c is plotted against f_l – was first introduced by Sandiford and Powell (1990, 1991) and has proven extremely useful (e.g. Zhou and Stüwe 1994, Hawkesworth et al. 1995, Turner et al. 1995). Orogenic evolutions may be plotted in this diagram as paths. Similarily to PT paths in PT diagrams such paths are parametric in time (s. p. 322). Note that, according to the original definition of Sandiford and Powell (1990), f_l is the thickening strain of the *entire* lithosphere and not only that of the mantle lithosphere. While other definitions are of course possible (s. Problem 4.2), we retain this original nomenclature here. Note also that f_l is the inverse of the stretching factor β ($f_l = 1/\beta$) that is commonly used in the description of continental extension (sect. 6.1.4).

Fig. 4.5a shows the f_c-f_l-plane with some schematic lithospheric profiles of different deformation geometries of crust and mantle part of the lithosphere. As it is impossible that the whole lithosphere is *thinner* than the crust, the f_c-f_l-plane is not defined in the region $f_l z_l < f_c z_c$ (shaded region in Fig. 4.5). The slope of the limiting line of this region is given by the initial thickness ratio of crust and whole lithosphere in the undeformed reference state: $\phi = z_c/z_l$. Fig. 4.5b shows a range of deformation paths that are end members of different orogenic evolutions.

The principal value of the f_c-f_l-plane is that it may be contoured for a range of important geodynamic parameters, for example surface elevation, Moho temperature (Fig. 3.20), potential energy (Fig. 5.32), strain rate and others. The influence of the deformation geometry of the lithosphere on these parameters may then be explored in this diagram. As such, the f_c-f_l-plane is in contrast to Fig. 3.18, where the thickness of the mantle part of the lithosphere is only *implicitly* determined by thermal considerations (s. Fig. 6.17 and 6.18). One of the disadvantages of the f_c-f_l-plane is that both axes of the horizontal plane are used to describe the thickening strains. Thus, the use of the f_c-f_l-plane and the two dimensions of this page limit us to the illustration of a single variable for which the plane may be contoured. For example on Fig. 3.20 the f_c-f_l-plane is contoured for Moho-temperature as a characteristic temperature for the entire lithosphere.

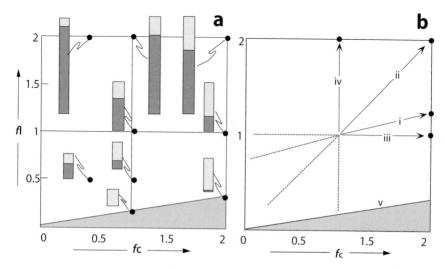

Figure 4.5. The f_c-f_l-plane. **a** The little schematic lithosphere columns indicate the thickness relationships of crust and mantle lithosphere in different parts of the diagram. Light shaded part of the columns is the crust, dark shaded part is the mantle part of the lithosphere. The point $f_c = f_l = 1$ is the reference lithosphere. **b** Some important lines in the f_c-f_l-plane. Note that f_l describes the thickening strain of the entire lithosphere (that is: the mantle part of the lithosphere *plus* the crust). *i* is the line of constant thickness of the mantle lithosphere. *ii* is the line of homogeneous thickening (thinning) of the entire lithosphere. *iii* is the line of variable crustal thickness at constant lithospheric thickness. It implies that the mantle part of the lithosphere must thin along this line as the crust thickens and vice versa. *iv* illustrates thickening of the mantle part of the lithosphere without crustal thickening. *v* limits the allowed space in the f_c-f_l-plane. Along this line, the crustal thickness and the lithospheric thickness are the same. That is, there is no mantle part of the lithosphere. Clearly, the lithosphere can not be thinner than the crust (not allowed dark shaded region). All lines are dashed in the region of thinning and continuous in the region of thickening relative to the undeformed reference lithosphere

4.1 Vertical Motion in the Crust

In the past decade, much progress in our understanding of mountain building processes has been made with studies in the interdisciplinary field between geomorphology and tectonics. In such studies, it is crucial to keep track of two different vertical velocities (Fig. 4.1). 1. The evolution of surface elevation and landforms at the earth's surface. 2. The evolution of depth and distance of rocks in the crust. Thus, the distance relative to two different reference levels must be observed simultaneously:

- The distance to the surface of an undeformed reference lithosphere. This is the reference level labeled with H on Fig. 4.1.
- The distance to the surface of the lithosphere under consideration. This is the reference level we have used for most considerations in chap. 3.

The latter tells us about depth and pressure of rocks in the crust, the former about the geomorphic evolution of the surface. In the following section we discuss vertical motions of rocks relative to both of these reference levels. In this context, it is pointed out that the vertical axis z is sometimes measured positively upwards and sometimes positively downwards, depending on the question that is being asked (s. sect. 3.7.3). As it is very easy to get confused with these different reference frames, we begin with a careful definition of our terminology.

4.1.1 Definition of Uplift and Exhumation

Motion of rocks relative to the surface of the lithosphere under consideration is called *exhumation* or *burial*, depending on the motion being *towards* the surface or away from it (Tab. 4.1). In contrast to the geomorphological use of the word "exhumation" (where it is only used to describe the surfacing of rocks that were previously at the surface, for example exhumation of a fossil or the exhumation of a river delta), tectonicists use the word "exhumation" also when describing upwards motion that has not brought rocks all the way to the surface (e. g. partial exhumation), or upwards motion of rocks that have never been on the surface previously (e. g. exhumation of a core complex).

Motion of rocks relative to the surface of an undeformed reference lithosphere is called *uplift* or *subsidence* depending on the motion being upwards or downwards. In general, the words uplift and subsidence are only used to describe the vertical motion of the surface itself. Thus, when using these terms for another level in the crust, for example the uplift of a rock relative to another rock e. g. on the other side of a fault, then the term "uplift" should only be used together with a specification of the reference level – in this case: "the other side of a fault" (England and Molnar 1990). The terminology of uplift and exhumation is not very consistently used in the literature. England and Molnar (1990) have defined these terms precisely. In the following

Table 4.1. Definition and methods of interpretation of uplift

uplift	
definition	vertical motion of the earth's surface relative to a reference level
direction of motion upwards:	called "uplift"
direction of motion downwards:	called "subsidence"
may be directly interpreted from	palaeobotany, palaeoclimatology,
may be indirectly interpreted from	sediments in the surrounding basins

Table 4.2. Definition and methods of interpretation of exhumation

exhumation	
definition	vertical motion of rocks, relative to the surface
direction of motion towards the surface:	called "exhumation"
direction of motion away from the surface:	called "burial"
may be directly interpreted from	geobarometry geothermometry (via assumption of a geotherm)
may be indirectly interpreted from	geochronology (via assumption of a geotherm)

we will follow their definition and the expansion of their logic by Stüwe and Barr (1998). Table 4.1 and 4.2 summarize these definitions and their use.

Uplift and exhumation are both measured in units of distance. Uplift and exhumation *rates* are measured in velocities: in $m\,s^{-1}$. For the following discussion we call the uplift rate v_{up} and define it positively upwards. The exhumation rate is abbreviated with v_{ex}. The rate of *uplift of rocks* v_{ro} (relative to a fixed reference frame) is given by the sum of the uplift and the exhumation rate:

$$v_{ro} = v_{ex} + v_{up} \quad . \tag{4.4}$$

The variable v_{ro} describes the vertical motion of rocks relative to a fixed reference level and is one of the most used and misused variables describing vertical motions. Ironically, v_{ro} is the only of the three vertical motions of eq. 4.4 which can not be determined directly from field or laboratory observations. It can *only* be determined from eq. 4.4. The importance of this equation can not be overemphasized. It is also emphasized that geobarometric or thermochronological data may *only* be used to infer exhumation and can not be used to interpret uplift. If there is no exhumation, then $v_{ro} = v_{up}$. In the next section we illustrate with some examples, how important the discrimination of different reference levels can be for the interpretation of tectonic features.

Vertical Motion in Convergent Tectonics. Crustal shortening will lead to downwards motion of rocks. Fig. 4.6b illustrated that – in isostatic equilibrium – only rocks above depth A will be uplifted relative to an externally fixed reference frame, while most rocks in the crustal column will move downwards. However, *all* rocks will be buried, i.e. increase their distance relative to the surface. The upwards motion of rocks can only be caused by exhumation processes as shown in Fig. 4.6c. In the literature upwards and downwards motions are often confused, for example by drawing the vertical arrows like those on Fig. 4.6c on the same diagram as the shortening arrows like those on Fig. 4.6a. While it is true that in shortening and exhumation processes occur simultaneously in many mountain belts, it is dangerous to mix vertical and horizontal motions as it obscured the responsible geodynamic processes.

Well documented examples for the simultaneous occurrence of vertical motions are found in just about all convergent orogens, for example by the fact that the highest grade metamorphic rocks often crop out near the axis of highest surface elevation. In the eastern European Alps, P-T-t-paths have been documented that show that exhumation of rocks occurred synchronously with shortening deformation phases (Cliff et al. 1985). Equivalent observations have even been made in the granulite facies roots of ancient mountain belts (Carson et al. 1997).

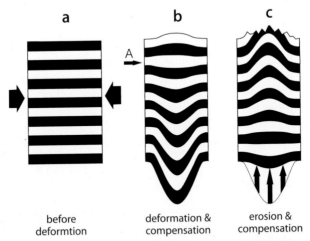

Figure 4.6. Schematic sketch of the vertical kinematics in isostatically compensated mountain belts. **a** shows a crustal column before deformation. **b** shows the crustal column after shortening. Note that all rocks below depth A are displaced downwards relative to **a**. **c** shows the crustal column after exhumation by erosion. Note that the dome strucutre that is now observed on the surface was formed by the exhumation processes only and *not* by the shortening

Figure 4.7. Schematic illustration of the vertical motion of the hanging wall (A) and the foot wall (B) of a ramp anticline, relative to the surface. Note that the thrusting process itself does *not* cause any exhumation and that it is only the subsequent erosion that brings the rocks nearer the surface

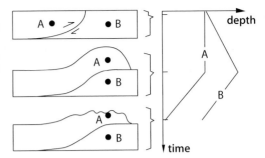

Vertical Motion on Ramp Anticlines. "Thrusting to the surface" is a commonly used term when talking of exhumation of rocks in association with a thrust. However, this use of the term is misleading. Fig. 4.7 shows that exhumation is *not* caused by the thrusting, but because the thrusting process places rocks in a position where they can easily be removed. If there is no erosion or other process removing material from the surface, then thrusting can not bring rocks closer to the surface! During thrusting, rocks in the upper plate will remain at constant depth below the surface, even though they are uplifted with respect to a reference frame fixed to an undeformed reference lithosphere (Fig. 4.7). Rocks in the foot wall on the other hand can only be buried.

Vertical Motion in Extensional Tectonics. During continental extension both *pure* and *simple* shear help to exhume rocks. However, while pure shear can exhume rocks close to the surface, ideal pure shear can never lead to complete exhumation as the complete thinned column of rock will always remain preserved (Fig. 4.8b). During extension by simple shear processes it is easy to exhume rocks as part of the column is physically removed from the top (Fig. 4.8c).

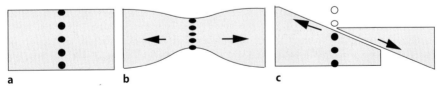

a   **b**   **c**

Figure 4.8. Exhumation during pure and simple shear of the lithosphere. **a** Starting situation with five marked depths. **b** In pure shear thinning rocks get partially exhumed, but can never exhume completely to the surface, even at very large extensional strains. The entire column above the rocks remains preserved in thinned form. **c** In simple shear it is easy to exhume rocks to the surface

Vertical Motion in Mountain Belts in General. In many mountain belts the highest grade metamorphic rocks crop out near the region of the highest topography. This suggests that the regions of the largest exhumation are also the region of largest uplift. This may not appear unusual, but there are many examples where uplift and exhumation have an opposite relationship. For example, there are small sedimentary basins on the Tibetan plateau which imply that the rocks at the base of these basins were buried at a time of rapid surface uplift. Correspondingly, there are many sedimentary basins in which the surface subsides (negative uplift) during burial of rocks. Because many of these observations are so familiar to us, *surface uplift* and *exhumation* have often not been discriminated in the past. However, the implications of these observations may often only be interpreted if the different vertical motions are discerned.

4.1.2 Kinematic Description of Vertical Motions

In the last section we have shown that *shortening* deformation can *only* lead to *burial* of rocks, while *exhumation* must (except some very special mechanisms s. p. 308) be caused by removing material from the surface. In most convergent orogens these two processes occur simultaneously. Thus, it is often not clear at what stages of the evolution rocks should exhume or get buried and if the surface uplifts or subsides at the time. In the following, we want to use eq. 4.4 to illustrate the relationships between these different vertical motions in convergent orogens subjected to erosion at the surface.

We follow the model of Stüwe and Barr (1998) and define z as the vertical distance of a rock from the surface of an undeformed reference lithosphere and z' as the vertical distance of the same rock from the surface above it, i.e. the depth of burial. For simplicity, we only consider the model one-dimensionally and assume that the convergent deformation leads to homogeneous thickening (i.e. that we can use eq. 1.5 and treat vertical strain rate as a constant). With these assumptions it is possible to describe the vertical motions of rocks with:

$$v_z = v_{\mathrm{ro}} - \dot{\epsilon}(z + H) \quad . \tag{4.5}$$

There, z is the depth of rocks relative to the undeformed reference lithosphere, v_z is the *rate* with which rocks move relative to this reference level (measured positive if they move upwards) and H is the surface elevation of the lithosphere under consideration, again measured relative to the surface of an undeformed reference lithosphere. Eq. 4.5 is fundamental for the understanding of the following considerations. If $v_{\mathrm{ro}} = 0$ and $z = -H$ (this means: rocks at the surface), then: $v_z = 0$. Relative to the reference frame of the deforming lithosphere under consideration we can write:

$$v_{z'} = v_{\mathrm{ex}} = v_{\mathrm{er}} - \dot{\epsilon}z' \quad . \tag{4.6}$$

There, v_{er} is the rate with which material is removed from the surface, e.g. the erosion rate. Note that v_{er} should not be confused with v_{ex}. The rate of exhumation depends on the difference between v_{er} (exhuming the rocks) and the deforming strain rate $\dot{\epsilon}$ (burying the rocks at a rate that is proportional to

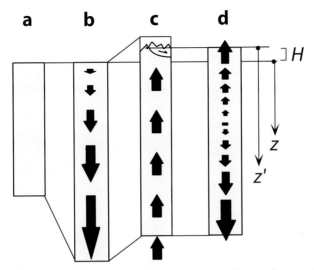

Figure 4.9. Schematic illustration of the vertical motions of rocks in the crust relative to the surface of the undeformed reference lithosphere. *a* reference lithosphere. *b* vertical motion as the consequence of thickening. *c* vertical motion as the consequence of erosion at the surface and isostatic compensation. *d* vertical motion during simultaneous thickening and erosion (after Zhou and Stüwe 1994; Stüwe and Barr 1998)

their depth z'). However, it should be clear that $v_{z'} = v_{ex}$. The relationship between the different reference levels may also be written as:

$$z' = z + H \quad . \tag{4.7}$$

The first term on the right hand side of eq. 4.5 describes the vertical motion of the lithosphere as a consequence of removal of material at the surface *and* isostatic compensation thereof. It is positive, because v_z is defined positively upwards. The second term of the equation describes the vertical thickening of the lithosphere during thickening. It is negative because thickening leads to burial of rocks. The sum of both motions is v_z or $v_{z'}$, depending on the chosen reference frame (eq. 4.5). From eq. 4.5 we can read the following: At high crustal levels, where z is small, the second term of the equation is also small. The contribution of the first term is relatively large, so that v_z is likely to be positive. Rocks move upwards in the crust during simultaneous thickening and erosion at the surface (Fig. 4.9). At deep crustal levels the second term is larger than the first term and v_z is likely to be negative. Rocks move downwards. It may be concluded that - even during homogeneous thickening - vertical motions in the crust may be very heterogeneous.

Steady State and Incremental Evolution. If we want to analyze these heterogeneous vertical motions in a bit more detail, we must pre empty some information from sect. 4.2, because we need eq. 4.29. This equation may be used to calculate the surface elevation of an isostatically compensated mountain range as a function of the thickness of crust and mantle lithosphere (expressed by $f_c \times z_c$ and $f_l \times z_l$) using some very simple assumptions on the density structure (expressed by the density terms δ and ξ). If we can find the time derivative of this equation we have a description for the *change* in elevation as a function of time: the uplift rate. The time derivative of eq. 4.29 is:

$$\frac{\mathrm{d}H}{\mathrm{d}t} = \delta z_c \left(\frac{\mathrm{d}f_c}{\mathrm{d}t} \right) - \xi z_l \left(\frac{\mathrm{d}f_l}{\mathrm{d}t} \right) \quad . \tag{4.8}$$

We can find the temporal change of f_c and f_l if we assume that the crustal and lithospheric thickening rates (the change of f_c and f_l with time) are described by the difference between *thickening* due to deformation and *thinning* due to erosion at the surface:

$$\frac{\mathrm{d}f_c}{\mathrm{d}t} = f_c \dot{\epsilon} - \frac{v_{er}}{z_c} \qquad \text{and} : \qquad \frac{\mathrm{d}f_l}{\mathrm{d}t} = f_l \dot{\epsilon} - \frac{v_{er}}{z_l} \quad .$$

Inserting these into eq. 4.8 we get the surface elevation as a function of time (the uplift rate). After some rearrangement of this equation we find this to be:

$$\frac{\mathrm{d}H}{\mathrm{d}t} = v_{up} = v_{er}b - \dot{\epsilon}(H + a) \quad . \tag{4.9}$$

In this equation, a and b summarize the constants from eq. 4.29 that are explained in eq. 4.26. They are: $a = (\delta z_c - \xi z_l)$ and: $b = (\delta - \xi)$. Eq. 4.9 describes the evolution of surface elevation of a mountain range in which there is simultaneous vertical thickening (for example due to lateral shortening) and material removal from the surface (for example due to erosion).

Eq. 4.9 may be solved, if the erosion rate v_{er} is known. So let us assume a simple erosion model for a mountain range in which the erosion rate is proportional to the elevation of the range:

$$v_{er} = \frac{H}{t_E} \quad . \tag{4.10}$$

In sect. 4.3.1 we will discuss this equation in a bit more detail and show that this model may be quite a realistic description for many mountain belts. In eq. 4.10 t_E is an erosional time constant that states in what time scale a mountain of the elevation H would be removed by erosion. According to eq. 4.10 erosion is more rapid, if t_E is small.

If we insert eq. 4.10 into eq. 4.9 it is possible to calculate the incremental uplift rates of a mountain range subject to the simple model boundary conditions assumed here. Let us do this, starting with the steady state case, where the surface elevation of the range is constant, as any uplift that may occur due to thickening is exactly balanced by subsidence due to erosion (geomorphic equilibrium). In this steady state, v_{ro} and v_{ex} at the surface have

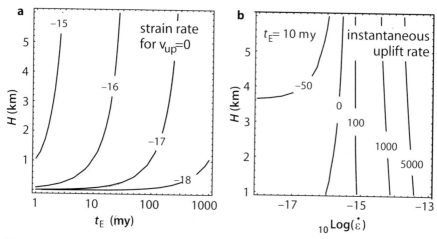

Figure 4.10. a Strain rates (contoured in $_{10}$Log($\dot{\epsilon}$)) necessary to maintain a geomorphic steady state (where the uplift rate is zero: $v_{up} = 0$) as a function of surface elevation H and erosion rate (characterized by the erosion parameter t_E). The diagram was calculated with eq. 4.11. **b** The instantaneous uplift rate (contoured in meters per million years) of an isostatically compensated mountain range as a function of surface elevation and thickening strain rate. Calculated with eq. 4.9 using eq. 4.10 to describe the erosion rate. For uplift the values are positive, for subsidence negative. Assumptions for the physical parameters are: $\rho_m = 3\,200$ kg m^{-3}, $\rho_c = 2\,700$ kg m^{-3}, $z_c = 35$ km, $z_l = 100$ km, $T_l = 1\,280\,°C$, $\alpha = 3 \cdot 10^{-5}$ K^{-1}. With these parameters the constants in the equations are: $a \approx 3545$ and $b \approx 0.14$ (s. eq. 4.26)

the same absolute value and it is true that: $dH/dt = 0$. From eq. 4.9 we get (s. Zhou and Stüwe 1994):

$$\dot{\epsilon} = \frac{bH}{t_E(H+a)} \quad \text{or} \quad H = \frac{t_E \dot{\epsilon} a}{b - \dot{\epsilon} t_E} \quad . \tag{4.11}$$

Eq. 4.11 is illustrated in Fig. 4.10a. It may be seen that - for the surface to remain constant - the erosion rate must be larger (i.e. t_E must be smaller) the *higher* the mountain range and the *faster* the thickening rate. This result is actually quite intuitive.

Let us now consider the case of a mountain range *not* in the steady state and explore the uplift or subsidence rates it will undergo. However, we need to keep in mind that we can only explore the *instantaneous* uplift rates as they will immediately change as a different elevation is reached. This may be recognized as both (dH/dt) and H occur in eq. 4.9 and we have not expressed H as a function of time. Fig. 4.10b shows the instantaneous uplift rates as a function of thickening rate and surface elevation as calculated with eq. 4.9 and eq. 4.10. The figure still shows some interesting results. For example, it shows that for thickening strain rates below about $\dot{\epsilon} = 10^{-15}$) the uplift rate is negative (subsidence occurs) and the subsidence rate increases with elevation, while this relationship is reversed for other thickening rates. Above

Figure 4.11. The evolution of surface elevation and depth of rocks in the crust in a mountain belt during simultaneous thickening and
erosion. **a** Surface elevation for four different
erosion parameters t_E (eq. 4.12). **b** Rock trajectories for the assumption of: $t_E = 0.5b/\dot\epsilon$ in **a**,
as calculated with eq. 4.13. Note that all rocks
with an initial depth of less than 30 km will
be exhumed under these assumptions. All other
assumptions are the same as in Fig. 4.10

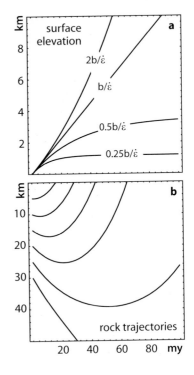

around $\dot\epsilon = 10^{-15}$ there is uplift with the uplift rate increasing for higher
mountain belts. Let us remember that such counter intuitive results are useful
to teach us about the processes controlling the evolution of surface elevation
in mountain belts, but that the model presented here is far too simplistic for
any direct application.

Evolution of Surface Elevation. In order to go beyond incremental uplift
rates and illustrate the *evolution* of surface elevation with time more directly
it is necessary to integrate eq. 4.9. Using the erosion model of eq. 4.10 integration of eq. 4.9 (using the method of Appendix A.5.1) gives:

$$H(t) = \frac{\dot\epsilon a}{\dot\epsilon - b/t_E}\left(e^{t(\dot\epsilon - b/t_E)} - 1\right) \quad . \tag{4.12}$$

Fig. 4.11a shows four examples for the evolution of surface elevation with
time as described by eq. 4.12. It may be seen that surface elevation converges
only to a steady state if $t_E < b/\dot\epsilon$ (s. Problem 4.5).

So far we have only considered the vertical motion of the surface. However, we can use our model also to describe the vertical motions of rocks
in the crust. From Fig. 4.9 we know that rock trajectories in the crust may
be divided into two groups: those paths that track upwards and those that
track downwards in the crust. The two groups are separated by the point
where $v_z = 0$ or $v_{ro} = 0$. This point is of great geological importance, as

rocks can only get exhumed by material removal from the surface if their depth z is $z < z_{v_z=0}$. We can actually calculate this point and all rock trajectories in the crust within our simple model. We can do that because we know that the vertical motion of rocks relative to an externally fixed reference frame is described by $v_{ro} = (dz/dt)$. Rock trajectories showing the evolution of depth through time may therefore be calculated by integrating eq. 4.4. After inserting eq. 4.9 and eq. 4.7 in eq. 4.4 and integrating (which we do not derive in detail here) we get:

$$z(t) = z_i e^{\dot{\epsilon}t} + \frac{a}{b}\left(1 - e^{\dot{\epsilon}t}\right) + H\left(\frac{1-b}{b}\right) \quad . \tag{4.13}$$

There, z_i is the initial depth of rocks in the crust. Fig. 4.11b shows rock trajectories calculated with eq. 4.13. Within our simple model, the depth $z_{v_z=0}$ or $z_{v_{ro}=0}$ may now be calculated reasonable easy. It is given by inserting eq. 4.9 and eq. 4.7 in eq. 4.4 assuming $v_{ro} = 0$. Solving for z gives:

$$z_{(v_z=0)} = a + \frac{H}{\dot{\epsilon}t_E}(1-b) \quad . \tag{4.14}$$

Inserting a range of realistic numerical values for the physical parameters into eq. 4.14 shows that this transition point lies at roughly 30–40 km depth. This means, that rocks can only be exhumed in orogens subject to simultaneous thickening and erosion at the surface if they lie at depths shallower than 30–40 km. This is in good correspondence with our observation that greenschist and amphibolite facies rocks with metamorphic pressures up to roughly 10 kbar (see eq. 7.1) are common in convergent orogens, while eclogites and other high pressure rocks are rare and usually confined to structures of the orogen that can only be describe with two-dimensional models (e.g. subduction zones, lithosphere scale thrusts etc.). Our model also shows that the exhumation of high pressure rocks requires consideration of other exhumation mechanisms (s. Platt 1993b).

4.2 Isostasy

Isostasy is the concept which relates the vertical distribution of mass with elevation in a state of equilibrium in which the lithosphere is considered to be floating on the underlying relatively weak asthenosphere. Isostasy does a good job of explaining the first-order variation of elevation over most of the earth's surface. Isostasy is described by a stress balance. In general, isostasy is concerned with the comparison of the surface elevation in two different places. For example, we might want to interpret the elevation difference between a mountain range and its foreland (assuming *isostatic* equilibrium) in terms of its implications for their different thicknesses. When we consider isostatic equilibrium it is useful to discriminate between:

– hydrostatic isostasy and
– flexural isostasy.

Hydrostatic isostasy is a stress balance in the vertical direction only (s. sect. 5.1.1). Thus, hydrostatic isostasy is a model that should really only be applied to regions that are large compared to the elastic thickness of the lithosphere. In other words, to geological features that are of at least several hundreds of kilometers in extent, i.e. areas like the Tibetan Plateau or the Canadian Shield. *Flexural isostasy* describes a stress balance in two or even three dimensions (s. Fig. 4.13). As a consequence, flexural isostatic considerations can be used to interpret the shape of much smaller scale features, for example foreland basins or subduction zones. While isostasy describes an equilibrium state and is therefore independent of time, many authors misinterpret the temporal evolution of isostatic rebound (for example following glaciations) as a feature inherent to isostasy. Let us begin this chapter therefore with a brief discussion of the interpretation of rebound rates before we begin with isostasy proper in sect. 4.2.1.

Isostatic Equilibration Rates. In the field we observe that isostatic equilibrium appears to be reached over finite lengths of time. For example, the uplift of Scandinavia in response to its deglaciation, occurs on a time scale of 10^4 years (e. g. Sabodini et al. 1991). Such isostatic compensation rates can be measured, for example by dating raised beaches (Fig. 4.12). This observation does *not* tell us that isostasy itself is time-dependent. Isostasy is a stress balance and as such *independent* of time. However, if a plate tries to rise or sink to reach its isotatic equilibrium state in response to a changed load, it has to displace the underlying asthenosphere. Thus, the rate of isostatic compensation can be used to estimate the viscosity of the asthenosphere (e. g. Lambeck 1993).

4.2.1 Hydrostatic Isostasy

The hydrostatic isostatic model is based on the assumption that all vertical profiles through the lithosphere may be considered independently of each other. That is, shear stresses on vertical planes are neglected (Fig. 4.13a). Then, there will be a depth at which the vertical stresses of all vertical profiles are equal. This depth is called the *isostatic compensation depth*. At this depth, the weight of all columns are equal. If we consider two profiles A and B, the *isostasy condition* may be formulated in terms of an equation (s. Fig. 4.14):

$$\sigma_{zz}^{A}|_{z=z_{\mathrm{K}}} = \sigma_{zz}^{B}|_{z=z_{\mathrm{K}}} \quad . \tag{4.15}$$

In this equation σ_{zz}^{A} and σ_{zz}^{B} are the vertical stresses of the two columns A and B and the depth z_{K} is the isostatic compensation depth. The vertical dash stands for "at the location". For most geological purposes we want to compare the elevation of two neighboring lithospheric columns in isostatic equilibrium. For this, it is useful to assume as isostatic compensation depth

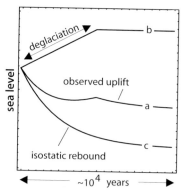

Figure 4.12. Observed and interpreted sea level changes. *a* Typical evolution of surface uplift relative to sea level in regions of recent deglaciation, for example Scandinavia (e. g. Lambeck 1991). Such curves typically contain two distinct parts and may be interpreted as the sum of sea level *rise* due to deglaciation (because of increased water mass in the oceans) and sea level *drop* because of isostatic rebound. A simple example for the former is shown in curve *b* consisting of a linear sea level rise until deglaciation ceases. Isostatic rebound (curve *c*) decreases exponentially as isostatic equilibrium is approached and therefore outlasts the deglaciation. From such curves, mantle viscosities of the order of 10^{20} Poise have been calculated

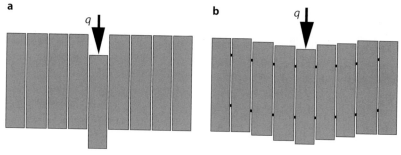

Figure 4.13. Illustration of the difference between **a** hydrostatic isostasy and **b** flexural isostasy. In **a** all vertical columns are considered independently of each other. In **b** the shear stresses between vertical columns are considered. *q* is the load

the shallowest possible depth below which there is no density differences between two neighboring columns. For most examples this can be assumed to be the base of the lithosphere of the column which reaches deepest into the asthenosphere. However, we begin by estimating the elevation of a single lithospheric column above the asthenosphere, so that $z_K = z_l$ (Fig. 4.14; s. also sect. 5.1.1).

The downward *force* that is exerted by one cubic meter of rock is given by the product of *density × gravitational acceleration* (in $N\,m^{-3}$). The downward force that is exerted by an entire vertical column per square meter (the

Figure 4.14. Illustration of isostatic equilibrium. Note that the z-axis is defined positively downwards and has its origin at the surface of the light shaded block (e.g. an iceberg or the lithosphere) that is assumed to float in a dark shaded region of higher density (e.g. water or the asthenosphere)

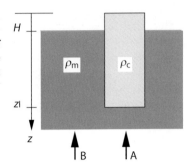

vertical normal stress) is this product integrated over the thickness of the column:

$$\sigma_{zz} = \int_0^{z_1} \rho g \mathrm{d}z \quad . \tag{4.16}$$

Inserting eq. 4.16 into eq. 4.15 gives:

$$\int_0^{z_1} \rho_\mathrm{A}(z) g \mathrm{d}z = \int_0^{z_1} \rho_\mathrm{B}(z) g \mathrm{d}z \quad , \tag{4.17}$$

where $\rho_\mathrm{A}(z)$ and $\rho_\mathrm{B}(z)$ are the densities of the two columns that are to be compared, both as a function of depth, z. The lower limit of integration 0 corresponds to the upper surface of the higher of two columns that are to be compared (this is in a coordinate system with the origin at the surface and the axis going positively downwards). The upper limit of integration is the depth z_1. At larger depths there is no density variations between the two columns (at least if we assume the geometry of Fig. 4.14). g is the gravitational acceleration.

Now we solve eq. 4.17, for the elevation of the shaded block in Fig. 4.14, relative to its surroundings. The block has a constant density ρ_c and floats in a denser medium of the constant density ρ_m. We call its elevation above the surface of the denser medium H_mat, although it is just labeled as H in Fig. 4.14. Here, we use the subscript $_\mathrm{mat}$ to emphasize that – for now – we consider only the *material* contribution to density differences between the profiles A and B. The densities and the acceleration are independent of z. Thus, they can be drawn out of the integrals on both sides of eq. 4.17 and integration is easy. By integrating the left half of the equation and splitting up the right half of eq. 4.17 we get according to Fig. 4.14:

$$\rho_\mathrm{c} g z \Big|_0^{z_1} = g \int_0^{H_\mathrm{mat}} \rho_\mathrm{air} \mathrm{d}z + g \int_{H_\mathrm{mat}}^{z_1} \rho_\mathrm{m} \mathrm{d}z \quad . \tag{4.18}$$

The density of air is negligible in comparison with ρ_m or ρ_c if these are the densities of the asthenosphere and the crust respectively (it is also negligible if we assume these two densities are that for water and ice). Thus, the first

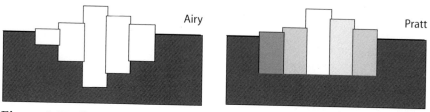

Figure 4.15. Comparison of the interpretations of the isostatic model according to Airy and Pratt. The shading indicates density. Darker shading means higher density

integral on the right hand side of eq. 4.18 is also negligible. After finishing the integration, canceling out g and inserting the integration limits we get:

$$\rho_c z_l = \rho_m z_l - \rho_m H_{mat} \quad . \tag{4.19}$$

A comparison of the two columns gives us therefore:

$$H = H_{mat} = z_l \left(\frac{\rho_m - \rho_c}{\rho_m} \right) \quad . \tag{4.20}$$

This relationship describes the hydrostatically balanced elevation of the surface of a floating body above the medium it floats in. Remember that $H = H_{mat}$ emphasizes the fact that this elevation difference is only based on the material difference between the block and the liquid. We can control this equation for some end member scenarios: If ρ_c is zero, then this equation states that $H = z_l$: the entire column floats on top of the liquid. This is the scenario given by a kids balloon floating on a lake. Alternatively if the two densities approach each other ($\rho_m = \rho_c$), then the entire body is submerged ($H = 0$). This is the scenario we observe with water soaked logs that float almost completely submerged in water. We can conclude that our observations confirm the simple model.

Isostasy According to Airy and Pratt. Two centuries ago, different models were developed to explain elevation differences observed in the mountain belts of the world in terms of the isostasy model. The two most notable models are those of *Airy* and *Pratt* (Fig. 4.15). Both earth scientists recognized that mountain belts are likely to rest in isostatic equilibrium and that their elevation is proportional to the density contrast between crust and mantle, as expressed by eq. 4.20. Pratt observed that many low lying Proterozoic shields are made up of high grade metamorphic rocks of high density, while mountain belts are often made up of hydrated, low grade metasediments. He concluded that most continental crusts extend to roughly similar depths and that the observed differences in surface elevation are the consequence of horizontal density variations in the crust.

In contrast, Airy estimated that the density of the crust is largely the same in all continental regions and therefore concluded that topographically higher regions, must be compensated by crustal roots at depth. The models of Airy and Pratt still bear their names. Seismic studies in many mountain belts show that most regions of high surface elevation are indeed compensated by significant roots at depth. On the other hand, it is true that there is a relationship between surface elevation and density of rocks. In short, the truth lies between the models of Airy and Pratt, although much success has been made by following Airy's model.

The Elevation of Mountain Belts. Gravimetric data tell us that many active orogens are *not* in isostatic equilibrium, but that their topography is dynamically supported. In other words, the surface elevation is actively held up or pushed down and is *out* of isostatic equilibrium. Dynamically supported topography may generally be found on length scales that are comparable to the elastic thickness of the lithosphere and will be discussed there (e. g. Forsyth 1985; Lyon-Caen and Molnar 1983; Molnar and Lyon-Caen 1989) (sect. 4.2.2). However, we note here that the model of hydrostatic isostasy should only be used for topographic features that are at least some hundreds of kilometers in lateral extent. For example, the European Alps are barely 200 kilometers across and are only partly compensated isostatically. This limitation of the hydrostatic model should be kept in mind when we develop it in the next sections.

When considering the hydrostatically supported elevation of a mountain belt, it is useful to divide the density variations in the lithosphere into two parts:

- density variations that are due to material differences and
- density variations that are caused by thermal expansion.

The fact that both types of density variations may form significant contributions to the density structure of the lithosphere is familiar to us from Fig. 2.15. Let us now consider the elevation of a lithosphere with the thickness z_l and a crustal thickness of z_c above its surroundings considering both the influence of the different materials and the influence of thermal expansion.

- *Isostasy due to thermal expansion.* In order to calculate the contribution of thermal expansion to surface elevation we need to introduce α: the *coefficient of thermal expansion*. α has the units of strain per temperature increment, which is K^{-1} (s. sect. 5.1). For most rocks the coefficient of thermal expansion is of the order $\alpha = 3 \cdot 10^{-5}\,K^{-1}$. Using α and the density of the mantle ρ_m, the density of colder rocks of the same material as a function of temperature may be calculated with:

$$\rho(T) = \rho_m(1 + \alpha(T_l - T)) \ . \tag{4.21}$$

There, T_l is the temperature at the base of the lithosphere at $z = z_l$. According to eq. 4.21: $\rho = \rho_m$, where $T = T_l$. At lower temperatures, the density increases linearly. At the surface, where we can assume that the temperature is $T_s = 0\,°C$, eq. 4.21 becomes:

$$\rho_{(T=T_s)} = \rho_0 = \rho_m (1 + \alpha T_l) \quad . \tag{4.22}$$

If the density of the mantle is about $\rho_m=3\,200$ kg m^{-3} at T_l, then the density at the surface is: $\rho_0=3\,300$ kg m^{-3}. Assuming a linear geotherm in the lithosphere, we can describe the mean density of the lithosphere with:

$$\overline{\rho} = \rho_m \left(1 + \alpha \frac{T_l + T_s}{2}\right) \quad . \tag{4.23}$$

In order to estimate which proportion of the elevation of a mountain belt is due to thermal expansion (H_{therm}), we insert eq. 4.23 into the right hand side of eq. 4.17. The left side remains as in eq. 4.18 and eq. 4.19. After integration according to the same principles as we did to get H_{mat} we get here:

$$H_{therm} = -z_l \alpha (T_l + T_s)/2 \quad . \tag{4.24}$$

The negative sign arises because $\overline{\rho}$ is larger than ρ_m.

• *Isostatically supported elevation of mountain belts.* The higher density of the cold lithosphere provides a *negative* contribution to the overall buoyancy (eq. 4.24). The *material* contribution of the crust to the elevation, on the other hand, is positive and was derived in eq. 4.20. Density variations within the mantle part of the lithopshere are neglected here (s. however p. 35). Then, the isostatically supported surface elevation relative to the surroundings is given be the sum of the thermal and the material contributions:

$$H = H_{mat} + H_{therm} = z_c \left(\frac{\rho_m - \rho_c}{\rho_m}\right) - z_l \alpha (T_l + T_s)/2 \quad . \tag{4.25}$$

If we summarize all the material parameters into the constants:

$$\delta = (\rho_m - \rho_c)/\rho_m \qquad \text{and}: \qquad \xi = \alpha (T_l + T_s)/2 \quad , \tag{4.26}$$

then this eq. 4.25 simplifies to:

$$H = \delta z_c - \xi z_l \quad . \tag{4.27}$$

If we insert meaningful numbers into eq. 4.25 (e. g. $\rho_m = 3\,200$ kg m^{-3}, $\rho_c = 2\,700$ kg m^{-3}), we get:

$$\delta \approx 0.15 \qquad \text{and}: \qquad \xi \approx 0.018 \quad . \tag{4.28}$$

This implies that the influence of material difference between crust and mantle, per meter of lithospheric column, is about ten times more important to the isostatically supported surface elevation than the influence of the thermal expansion. However, because the crust constitutes only about one third of the whole lithosphere, the crustal material contribution to the elevation is in total only about 3 times larger than the contribution of thermal contraction, which applies to the whole lithosphere. In total, H is about $3\,600$ m.

This is the elevation of the upper surface (of a lithosphere with z_c and z_l as above) above the hypothetical surface of a liquid mantle, as we illustrated in Fig. 4.14. Mid-oceanic ridges are the only place on the globe where we can measure the depth of this reference level. It turns out that mid-oceanic ridges lie indeed about 3 600 m below the average elevation of the continents and lie at a very constant depth below sea level (Turcotte et al. 1977; Cochran 1982).

In most geological problems it is much more interesting to know the elevation of a mountain belt above its surroundings, rather than above the mid-oceanic ridges. For this purpose, it is useful to reformulate eq. 4.25, so that the elevation is given as the elevation difference between a thickened (or thinned) lithosphere and an undeformed reference lithosphere:

$$H = (\delta f_c z_c - \xi f_l z_l) - (\delta z_c - \xi z_l) = \delta z_c (f_c - 1) - \xi z_l (f_l - 1) \ . \qquad (4.29)$$

The parameters f_c and f_l describe the thickening strains of the crust and the mantle lithosphere and were discussed in detail in sect. 3.4.1 and 4.0.2 (also: sect. 6.1.4, eq. 6.10). The elevation of isostatically supported mountain belts above the undeformed reference lithosphere is shown in Fig. 4.16 (for the concept of an undeformed reference lithosphere see: Le Pichon et al. 1982). More detailed assumptions about the thermal expansion have no influence on the surface elevation (e. g. Zhou and Sandiford 1992). Fig. 4.16 shows clearly that homogeneous thickening of the entire lithosphere (a diagonal line from bottom left to top right in this diagram) causes relatively small changes of the surface elevation, because the two contributions in eq. 4.27 and eq. 4.29 have opposite signs. Accordingly, the negative buoyancy caused by the thickening of the mantle part of the lithosphere is largely compensated by the positive buoyancy of the thickened crust. It may also be read from this figures, that doubling of the crust, without thickening of the lithosphere would imply an

Figure 4.16. Isostatically supported surface elevation of mountain belts in the f_c-f_l-plane (after Sandiford and Powell 1990). Contours were calculated with eq. 4.29 and are labeled in km. Following assumptions were used: $\rho_m = 3\,200$, $\rho_c = 2\,750$, $\alpha = 3 \cdot 10^{-5}$, $z_c = 35$ km, $z_l = 100$ km. Using these values, the two constants are: $\delta \approx 0.14$ and $\xi = 0.018$. The thick line in the lower right hand corner of the diagram delineates the not allowed region of this parameter space (s. sect. 4.0.2)

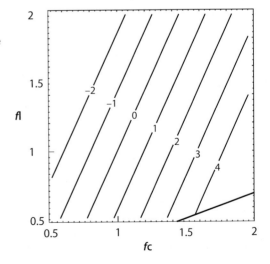

isostatic uplift of about 3–4 km (compare the paths in Fig. 4.5b, s. sect. 2.4.1).

The Depth of the Oceans. The water depth of the oceans is a direct function of the distance to the mid-oceanic ridges (Fig. 2.1). The functional relationship between water depth and distance from the mid-oceanic ridge was described with a fantastically simple model by Parsons and Sclater (1977). Their model is one of *the* largest successes of the theory of heat conduction (sect. 2.1, 3.5.1). It can be derived using the principles of hydrostatic isostasy.

Oceanic lithosphere consists (except for a thin 7 km thick crust) largely of asthenosphere material that has cooled to form lithospheric mantle. Because of the small and constant thickness of the crust, material contributions to density variations may be neglected and thermal expansion (contraction) is the governing factor for variations in the density structure of the oceanic lithosphere. In order to use this density variation to estimate the isostatically supported elevation of the ocean floor, we use the model sketched in Fig. 4.17. According to eq. 4.17 the vertical normal stresses of the columns A and B must be the same in the compensation depth $z = z_1$. For column A the vertical normal stress at depth $z = z_1$ is given by:

$$\sigma_{zz}^{A}|_{z=z_1} = \rho_w g w + \int_0^{z_1} \rho_{(z)} g \mathrm{d}z \quad . \tag{4.30}$$

There, w is the water depth in column A, ρ_w is the water density, g is the gravitational acceleration and $\rho_{(z)}$ is the density of the lithosphere as a function of depth which we shall discuss on the next page. For column B we can formulate:

$$\sigma_{zz}^{B}|_{z=z_1} = \rho_m g w + \rho_m g z_1 \quad . \tag{4.31}$$

It should be possible to follow eqs. 4.30 and 4.31 by considering Fig. 4.17. After inserting eqs. 4.30 and 4.31 into eq. 4.17, the isostasy condition of gets the following form:

$$\rho_m z_1 + w(\rho_m - \rho_w) = \int_0^{z_1} \rho_{(z)} \mathrm{d}z \quad . \tag{4.32}$$

With foresight to the following steps in our calculation we bring the first term of this equation to the right hand side, finds its derivative with respect to z and write it therefore into the integral. Eq. 4.32 gets the form:

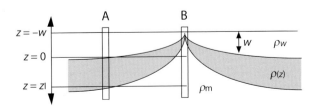

Figure 4.17. Schematic profile through a mid-oceanic ridge

$$w(\rho_m - \rho_w) = \int_0^{z_1} (\rho_{(z)} - \rho_m)dz \quad . \tag{4.33}$$

This equations states that the water depth is dependent on the density structure as a function of depth $\rho_{(z)}$. In oceanic lithosphere this density function is a direct function of the temperature profile (s. sect. 3.5). Thus, if we know the temperature as a function of depth, then $\rho_{(z)}$ in eq. 4.33 is known, because we know already the relationship between density and temperature from eq. 4.21. Thus we can begin by inserting eq. 4.21 into eq. 4.33:

$$w(\rho_m - \rho_w) = \int_0^{z_1} \rho_m \alpha(T_1 - T(z))dz \quad . \tag{4.34}$$

The variable $T(z)$ is the only unknown in this equation, but we determined it in sect. 3.5.1. It is well-described by the half-space cooling model. Thus, the temperature profile of eq. 3.81 may be directly inserted into eq. 4.34:

$$w(\rho_m - \rho_w) = \int_0^{z_1} \rho_m \alpha \left(T_1 - T_s - (T_1 - T_s)\mathrm{erf}\left(\frac{z}{\sqrt{4\kappa t}}\right)\right) dz \quad . \tag{4.35}$$

After simplification using eq. 3.17 we get:

$$w(\rho_m - \rho_w) = \int_0^{z_1} \rho_m \alpha(T_1 - T_s)\mathrm{erfc}\left(\frac{z}{\sqrt{4\kappa t}}\right) dz \tag{4.36}$$

or, after taking the constants out of the integral:

$$w = \frac{\rho_m \alpha(T_1 - T_s)}{(\rho_m - \rho_w)} \int_0^{z_1} \mathrm{erfc}\left(\frac{z}{\sqrt{4\kappa t}}\right) dz \quad . \tag{4.37}$$

If we introduce the variable $n = z/\sqrt{4\kappa t}$, we can take all the constants out of the integral (s. Appendix B) and get:

$$w = \sqrt{4\kappa t}\frac{\rho_m \alpha(T_1 - T_s)}{(\rho_w - \rho_m)} \int_0^{z_1} \mathrm{erfc}\,(n)\,dn \quad . \tag{4.38}$$

The definite integral of the error function is not know for integration limits of 0 and z_1, but it *is* known for integration with limits at infinity. It is:

$$\int_0^\infty \mathrm{erfc}(n)dn = \frac{1}{\sqrt{\pi}} \quad .$$

This is a close enough approximation, in particular since $\rho \to \rho_m$ at the base of the lithosphere. Thus, the integral of eq. 4.38 may be substituted by the integral from above. The water depth as a function of distance from the mid-oceanic ridge may thus be described with this model by:

$$w = \frac{2\rho_m \alpha(T_1 - T_s)}{(\rho_m - \rho_w)}\sqrt{\frac{\kappa t}{\pi}} \quad . \tag{4.39}$$

If we insert standard values for all the constants in this equation we get:

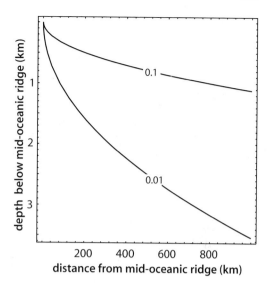

Figure 4.18. Profiles of water depth as a function of distance from the mid-oceanic ridge as calculated with eq. 4.39. The curves are shown for different rifting rates in $m\,y^{-1}$. Following constants were used: $\rho_m = 3\,200\ kg\,m^{-3}$, $\rho_w = 1\,000\ kg\,m^{-3}$, $\alpha = 3 \cdot 10^{-5}\ K^{-1}$, $T_l = 1\,280\,°C$, $T_s = 0\,°C$

$$w \approx 5.91 \cdot 10^{-5}\sqrt{t}\ . \tag{4.40}$$

In words, the depth of the water is proportional to the square root of age of the oceanic lithosphere. Note that this water dept is only the *additional* water depth on top of the water depth *at* the mid-oceanic ridge (Fig. 4.17). We can convert this into water depth as a function of distance from the mid-oceanic ridge if we substitute age by the ratio of distance to rifting rate: x/u; which is also age. Fig. 4.18 shows some water depth profiles calculated with this equation. The fantastic coincidence of these curves with bathymetric measurements in the oceans of the world confirm the model (see also its correspondence with heat flow data in sect. 3.5.1).

4.2.2 Flexural Isostasy

We can measure that most topographic features of our planet that are less than many hundreds of kilometers across are *not* completely in hydrostatic isostatic equilibrium. This includes whole mountain ranges like the European Alps (Karner and Watts 1983; Lyon-Caen and Molnar 1989). This can be measured gravimetrically: Gravimetry measures mass and in isostatic disequilibrium the total mass above the isostatic compensation depth is not everywhere the same. Thus there are gravity anomalies that may be interpreted in terms of isostatic disequilibrium. Isostatic disequilibria may form in response to a large range of processes. For example, a continental plate may be *actively* pushed downwards by the load of another plate, or it may be *actively* held up by mantle convection exerting an upwards force to the bottom of a plate. Topographic features that are created by non-isostatic processes are called *dynamically supported* topography.

Flexural isostasy is a stress balance that also considers *horizontal* elastic stresses (Fig. 4.13b). Flexural isostasy is therefore at least a *two-dimensional* stress balance. It may be used to interpret surface topography in terms of both, hydrostatic isostasy and elastic flexure. In flexural isostasy, lithospheric plates are viewed as elastic plates that are bent by vertical loads. Interestingly, this model describes a large number of observations extremely well, although it is not at all trivial that the lithosphere should behave elastically at all. For example, we will show in sect. 5.2, that deformation of the lithosphere on geological time scales may be best described by brittle and ductile deformation mechanisms (sect. 5.1.2). Nevertheless, we observe a number of large scale features that are well-described by elastic lithospheric models. In the following we summarize a few of these observations.

Examples of Elastic Deformation. *Oceanic* lithosphere is characterized by little internal deformation. This is because it is rheologically much stronger than continental lithosphere. It has a uniform thickness (on the scale of 10^2 km), plane surface. As a consequence, elastic features that develop in response to vertical loads may spectacularly be seen without much disturbance by features created by other deformation mechanisms. The best known example for elastic deformation of the oceanic lithosphere are the valleys around *sea mounts*, for example around the Hawaii-Emperor chain. Sea mounts are volcanoes that have formed far from mid-oceanic ridges. They were created by *hot spots* that have their origin deep inside the mantle (Fig. 4.19). Thus, hot spot volcanoes that have formed on the surface of the oceanic lithosphere have no compensating root at the base of the plate. The volcano may be considered as an external load to a plate of more or less constant thickness that bends it downwards.

Another example of elastic deformation of oceanic lithosphere is the bending of the plates at subduction zones. The shape of trenches and the fore

Figure 4.19. Flexure of oceanic lithosphere due to the loading of a sea mount

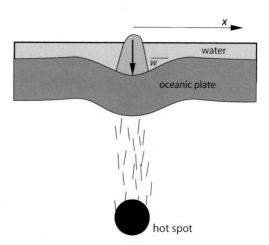

bulge on the seaward side of the trench are also the consequence of elastic bending of the plate.

The elastic bending of *continental* plates may be observed in the foreland of many collisional orogens, where molasse basins form as the consequence of the elastic deflection of the plate in response to the load of the mountain belt. One of the best know examples is the northern molasse of the European Alps. There, the European plate is bent downwards under the load of the alpine mountain chain. The deepest point of the deflection is the valley of the river Donau. However, in collisional orogens the *external* load applied by the weight of the mountain belt is partly compensated by an *internal* loads: the root of the mountain belt. Thus, the distribution of loads is not as clear or as easily interpreted as the examples of oceanic plates.

Passive continental margins also show often evidence for elastic bending of continental lithosphere (Fig. 4.23). The best known examples for this are the great escarpments along the coasts of southern Africa and Australia (Tucker and Slingerland 1994; Kooi and Beaumont 1994). There, the unloading of the plate that is caused by the asymmetric erosion of the continental margin is compensated by elastic updoming of the coastal foreland. The Australian Great Barrier Reef, for example, may be interpreted as an elastic fore bulge similar to those observed in the vicinity of subduction zones (Stüwe 1991, Fig. 6.10).

Interestingly, many geological structures that may be explained with the theory of elasticity – as for example the shape of the Australian escarpment – are preserved for many tens of millions of years. In sect. 5.3 we will show that on such a large time scale most stresses should be compensated by viscous deformation.

The Flexure Equation. In order to describe flexural isostasy quantitatively, we need to preempt some information on elastic deformation from sect. 5.1.2. Elastic deformation describes an empirically derived *constitutive relationship* in which stress and strain are proportional to each other (eq. 5.20). The proportionality constant between stress and strain is called the modulus of elasticity or *Young's modulus* E. How much a plate bends under an applied stress depends on E and its compressibility, which is described by the *Poisson ratio* ν.

Let us now consider the bending of a simple, ideal elastic plate like the one sketched in Fig. 4.20. We also neglect buoyancy forces for now. When integrating the horizontal normal stresses σ_{xx}, over the thickness of the elastic plate h, then it may be shown (or even intuitively seen) that the bending moment M is proportional to the curvature of the plate (s. Fig. 4.20):

$$M = -D\frac{\mathrm{d}^2w}{\mathrm{d}x^2} \quad . \tag{4.41}$$

In this equation, w is the vertical deflection of the plate and the constant of proportionality D is called the *flexural rigidity* of the plate. The bending moment M is the integrated torques on both sides of the load. The derivation

Figure 4.20. Bending of an ideal elastic plate in a simplified model view which is useful for the description of bending lithospheric plates

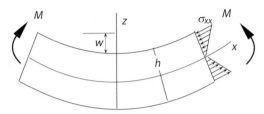

of eq. 4.41 and the following steps are explained in some detail by Turcotte (1979) and Turcotte and Schubert (1982) and will not be repeated here.

Eq. 4.41 is in itself not yet very useful. However, it may be coupled with a force balance equation that places bending moments, the vertical load q, any applied horizontal forces F and the shear forces (s. Fig. 4.20) into a relationship to each other (s. Turcotte and Schubert 1982; Ranalli 1987). When coupled with eq. 4.41, one arrives at the one-dimensional flexure equation:

$$D\frac{\mathrm{d}^4 w}{\mathrm{d}x^4} = q_x - F\frac{\mathrm{d}^2 w}{\mathrm{d}x^2} \quad . \tag{4.42}$$

There, q_x is the vertical load as a function of horizontal distance x and has the units of force per area: stress. Thus, if the distribution of loads is known, this equation may be solved for either the deflection of the plate w or for its flexural rigidity D (in N × m). Usually, the deflection is well known from bathymetric or topographic observation and eq. 4.42 is used to derive the rigidity or "stiffness" of the plate. This flexural rigidity is a direct function of the elastic material properties of an ideal elastic plate of thickness h and is related to these by:

$$D = \frac{Eh^3}{12(1 - \nu^2)} \quad . \tag{4.43}$$

Thus, if the material constants are known and the flexural rigidity of a plate was derived from its shape using eq. 4.42, then this may be converted directly into an elastic thickness of the lithosphere using eq. 4.43. All descriptions of the bending of elastic plates are based on the integration of eq. 4.42, or its two-dimensional equivalent.

Application to the Lithosphere. Eq. 4.42 may be directly applied to describe flexural isostatic equilibrium, i.e. the elastic bending of lithospheric plates under external and internal loads. When we do this, we need to consider at least three important points:

1. We have to keep in mind that the flexure equation is based on completely different deformation mechanisms from those that we will discuss in sect. 5.2, where we consider the rheology of the lithosphere. In other words: eq. 4.42 is only a model that describes some field observations very well, but may be quite useless for the description of many other observations.

2. The flexural rigidity D must be interpreted correctly. Field observations tell us that the rigidity of lithospheric plates is of the order of $D \approx 10^{23}$ Nm

($\pm$ about one order of magnitude) and laboratory experiments show that the material constants are about $E \approx 10^{11}$ Pa and $\nu \approx 0.25$. According to eq. 4.43 these parameters imply that the elastic thickness of the lithosphere h is only some tens of kilometers. Thus, the elastic thickness of the lithosphere is much thinner than the lithosphere according to thermal or mechanical definitions. The elastic thickness must be considered as the theoretical thickness of a plate with homogeneous elastic properties. Considering that the brittle strength of the upper crust as well as the ductile strength of the lower most lithosphere are likely to be very small, it is only the central part of the lithosphere that is dominated by elastic behavior on time scales that are short compared to the viscous response. However, it is important to note that the concept of an elastic thickness remains a theoretical one. Ranalli (1994) showed that the elastic thickness of the lithosphere is largely dependent on the depth of the 900 °C-isotherm (s. sect. 5.2.1, Fig. 5.16). As the rigidity is proportional to h^3 (eq. 4.43), it might therefore be expected that D is indirectly proportional to the cube-root of the geothermal gradient (Molnar and Lyon-Caen 1989). This, however, is not the case. For example, the rigidity of the Indian and Adriatic plates varies by about *three* orders of magnitude but the difference in geothermal gradients is less than *one* order of magnitude. Thus, the elastic thickness is likely to be not only dependent on the 900 °C-isotherm.

3. The distribution of loads on the plate must be clear. The load as a function of distance q_x as used in eq. 4.42 is the sum of a series of *internal* and *external* loads that act upwards and downwards onto a plate. In order to clarify *which* different forces act on the plate, it is useful to divide the plate under consideration according to the scheme illustrated in the right hand

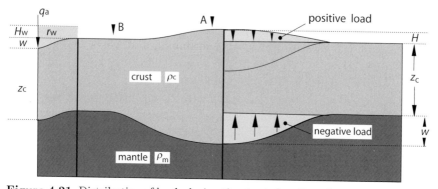

Figure 4.21. Distribution of loads during the elastic bending of lithospheric plates. The shown example illustrates the displacement of the mantle by crustal material. At the left margin of the diagram the case of water covered oceanic lithosphere is illustrated (Problem 4.12). The central part of the figure illustrates the case of a continental plate. The right hand part of the figure shows how this plate can be split up into parts in order to consider the different forces acting on the plate. Points A and B are marked for Problem 4.12

part of Fig. 4.21. There, it may be seen that the downward force exerted by the mountain range on the plate is given by the vertical normal stress $q_{ext} = \rho_c g H$. This is the *external* or the *positive* load. This load is opposed by a buoyancy force in the region of the displaced mantle. This is the *internal* or *negative* load shown on Fig. 4.21 with the upwards arrows. This internal load has the magnitude $q_{int} = (\rho_m - \rho_c) g w$. The net load that is applied to the plate is therefore:

$$q(x) = q_{ext} - q_{int} = \rho_c g H(x) - (\rho_m - \rho_c) g w \quad . \tag{4.44}$$

When eq. 4.44 is inserted into eq. 4.42, this may be solved for w numerically or – for some simple boundary conditions - also analytically. When considering multi-layered plates, the same principle may be followed in deriving the net load on the plate (s. Problem 4.12).

Application Examples. For many plate tectonic questions there is no analytical solution of eq. 4.42, and numerical methods must be used to solve it (s. Problem 4.11). However, there are also a range of important geological applications for which there are closed solutions of eq. 4.42. Some of these are discussed below.

• *Examples in oceanic lithosphere.* A series of elastic bending problem in the oceanic lithosphere may be well-described with eq. 4.42 using two simplifying assumptions:

– 1. We assume that there are no horizontal forces applied to the plate. Then, the entire last term of eq. 4.42 is zero.
– 2. We assume that the vertical load is only applied at a single location at the end of the plate; i.e. there is no dependence of the load on x.

Based on the second assumption, and assuming that the downwards deflected region is filled with water, eq. 4.44 simplifies to:

$$q = q_a - (\rho_m - \rho_w) g w \quad . \tag{4.45}$$

as illustrated on the very left hand edge of Fig. 4.21 (ρ_w is the water density). Both the above assumptions are a fair approximation to problems like the deflection of plates under long, narrow island chains on large oceanic plates, or to the load of a continent onto the edge of a subducting oceanic plate. Eq. 4.42 simplifies to:

$$D \frac{d^4 w}{dx^4} = -(\rho_m - \rho_w) g w \quad . \tag{4.46}$$

Eq. 4.46 describes a range of geological features surprisingly well and has the great advantage that it may be integrated analytically for a range of geologically relevant boundary conditions. After integration, the constants D, g, ρ_m and ρ_c often occur in the following relationship:

$$\alpha = \left(\frac{4D}{g(\rho_{\mathrm{m}} - \rho_{\mathrm{w}})}\right)^{1/4} \quad . \tag{4.47}$$

α is called the *flexure parameter* of the lithosphere (s. Problem 4.13). The first example we want to discuss is that of a line-shaped load of islands on a continuous plate of constant thickness. For appropriately formulated boundary and initial conditions (e.g. the load applies only at $x = 0$, symmetry of the deflection so that $dw/dx = 0$ at $x = 0$ and others) a solution of eq. 4.46 is:

$$w = w_0 e^{-x/\alpha} \left(\cos(x/\alpha) + \sin(x/\alpha)\right) \quad . \tag{4.48}$$

There, w_0 is the maximum deflection of the plate directly underneath the load and w is normalized to this value (we can see from eq. 4.48 that $w \to w_0$ for $x \to 0$). Eq. 4.48 is a good approximation for the description of the water depth around the Hawaii and Emperor island chains (s. Fig. 4.22a). The equation is also historically important, as is was one of the first models used to estimate the elastic thickness of the lithosphere using the bathymetric surveys around Hawaii.

The second example that may be described with the approximation of eq. 4.46 is the shape of oceanic lithosphere near trenches. There, the loading of the subducting oceanic plate may be viewed as a line-loading by the margin of the upper plate. For this case, boundary conditions must be assumed that

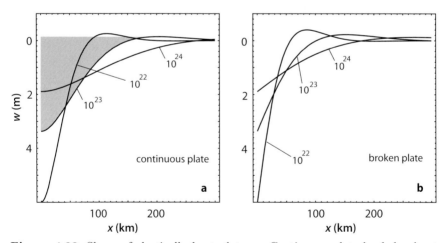

Figure 4.22. Shape of elastically bent plates. **a** Continuous plate loaded only at $x = 0$: the left margin of the diagram (eq. 4.48). Only half of the plate is shown. **b** Broken plate, also loaded only at $x = 0$ (eq. 4.49). The curves are labeled with the flexural rigidity of the plates in Nm. For one example In **a**, the volume of the formed basin is shown by the shaded region. The basin volume is for all rigidities the same! Note the elastic bulges that occur outwards of the basins before the plates return to their normal position in the far field. These bulges occur because – according to eq. 4.42 – the mean curvature of the plate must be minimal

describe a broken half plate which is subjected to a load at its end. For appropriately formulated boundary conditions a solution of eq. 4.46 is:

$$w = w_0 e^{-x/\alpha} \left(\cos(x/\alpha) \right) \quad . \tag{4.49}$$

The shape of plates as described by eq. 4.49 is illustrated in Fig. 4.22b. A comparison of these curves with bathymetric measurements shows that most subduction zones are steeper near the trench than what is described by the curves at the left margin of Fig. 4.22b. It is interpreted that this indicates that subducted plates are not only loaded by the upper plate but that convection in the mantle wedge and other forces exert a additional torques on subducting plates.

• *Examples in continental lithosphere.* Continental lithosphere deforms internally much easier than oceanic lithosphere by pervasive ductile mechanisms. Thus, elastic features are often not so clearly exposed and loads of mountain ranges and the like are distributed over large parts of the plates. As a consequence, continental lithosphere does not lend itself so easily to description with analytical solutions of eq. 4.42. Nevertheless, it should be said that the load of long mountain chains on homogeneous continental plates is analogous to the problem of long island chains on oceanic lithosphere. Thus, eq. 4.48 can – in principle – also be used to describe foreland basins, but care must be taken by accounting for sedimentary fill of foreland basins, compensating crustal roots etc (s. Turcotte and Schubert 1982). For example, ρ_w must be replaced by ρ_c in the formulation of the flexural parameter (s. eq. 4.44). However, much more progress has recently been made in foreland basin modeling by using numerical approximations, where more realistic scenarios may be described (Garcia-Castellanos et al. 1997).

The topography of *passive continental margins* is probably the example in the continental lithosphere that is most obviously described by elastic flexure (Fig. 4.23). While the geometry of passive margins may be characterized with a very simple geometry, there is no analytical solution of the flexure equation to describe it and models in the literature rely on numerical approximations. Passive margins feature a range of interesting morphological relationships between the position of the drainage divide, the shape and direction of the drainage networks, the position of a characteristic great escarpment and the slope of the range (s. p. 171). Many of these relationships indicate that the relief of the plate margin is characterized by elastic bulges in the foreland of the escarpment that are interpreted to have formed in response to the unloading of the plate margin by erosional retreat of the escarpment. The Great Barrier Reef in Australia has been explained as such a forebulge (e. g. Stüwe 1991; Gilchrist et al. 1994; Tucker and Slingerland 1994). Other curious features of passive continental margins, for example the inland drainage of rivers, are consistent with such interpretations.

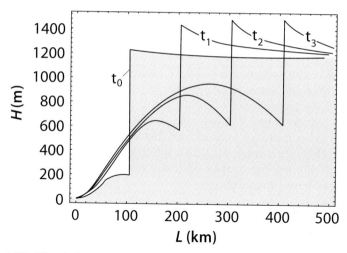

Figure 4.23. Elastic flexure at passive continental margins. The figure is a cross section through an idealized passive margin showing surface elevation H as a function of distance from the continental shelf L at four different time steps t_0 to t_3 during the successive erosional retreat of a $1\,000$ m high escarpment. The land region of the cross section is shaded for time step t_0. Topographic profiles similar those at t_1 to t_3 can be found in southern Africa and eastern Australia (after Stüwe 1991)

4.3 Geomorphology

The interpretation of geomorphological features in terms of an integrated geodynamic understanding of plate tectonic processes has become an important topic in the earth sciences. This new direction has become known by the name "tectonic geomorphology" and the appearance of a range of recent excellent textbooks testifies of the importance of this new field (e. g. Burbank and Anderson 2001, Summerfield 1991, Keller and Pinter 1996, Schumm et al. 2000, Kirkby 1994, Julien 2000).

For the description of geomorphic processes it is often useful to discriminate between:

– geomorphic shaping by tectonic ("endogenic") processes, and
– geomorphic shaping by erosion and sedimentation ("exogenic") processes.

Most of both process groups may be described using some very basic principles of which some are introduced on the following pages. From the viewpoint of a Eulerian observer, both process groups may be described as a material transport into or out of the system. *Tectonic processes* ("endogenic" processes) move material by faulting, by uplift or by subsidence. *Erosion* and *sedimentation* ("exogenic") processes move material by removing or depositing material. In many cases the material transport may be described with the same principles as the transport of heat, that is, by:

– diffusion,
– advection or
– production

of material in the system. Accordingly, many geomorphic processes may be
described with the same principles discussed in sect. 3.1, 3.2 and 3.3, by
considering the transport of *mass* instead of *energy*. However, there are also
processes that are unique to landscape formation, for example, the hydrolog-
ical processes in drainage networks or the threshold mechanisms governing
landslides. Before we discuss the description of individual processes in some
detail, the following paragraphs illustrate the difference between *endogenic*
and *exogenic* landscape formation.

• *Geomorphic shaping by tectonic processes.* Most tectonic processes are spatially discontinuous: faulting moves coherent packages of rock, while others
remain fixed in space. The geometry and kinematics of individual landscape
forming tectonic processes depends on the mechanical anisotropy of the rock
and on the deformation mechanism. Thus, relief development by tectonic
("endogenic") processes is generally described with mechanical and kinematic
models and – while undoubtedly an integral process of landscape formation –
are usually not the subject of the modeling of geomorphic processes as such.
Rather, tectonic processes form *boundary conditions* to geomorphic modeling, which generally concentrates on the description of the exogenic processes.
Fortunately, many tectonic processes have one of the following two characteristics that help to integrate them in geomorphological models:

– 1. Some tectonic processes important in landscape evolution are very rapid
     compared to the subsequent erosion/sedimentation processes, for example
     faulting by earthquakes. Thus, in model descriptions, such processes can
     often be used as the starting conditions of a subsequent erosional process
     that is modeled explicitly.
– 2. Others are slow, but spatially very extensive, for example the uplift of
     a mountain range. Such processes can be described in geomorphological
     models as boundary conditions to the model.

• *Geomorphic shaping by erosion and sedimentation.* Erosion and sedimentation processes are governed by a large range of different physical processes
including soil creep, solution, rain splash, chemical- and aeolian weathering,
down- and sideways cutting of drainages, debris flows, as well as discontinuous
processes like landslides and many others (e.g. Carson and Kirkby 1972). In
order to describe these processes with simple models it is useful to summarize
them into three groups:

– short range continuous transport,
– long range continuous transport,
– discontinuous processes.

Short range transport describes the local redistribution of mass on a hill slope scale and is discussed starting on p. 171. *Long range transport* describes the erosion and sedimentation processes in rivers (p. 182) and *discontinuous processes* are processes that are episodic in time, like landslides (p. 187). Fortunately, geological time scales are long enough so that many local *discontinuous* processes may be described by *continuous* models. However, as always, the chosen model description depends critically on the question being asked. For example, when modeling the first order morphological features of orogens as a whole, the subdivision discussed above is unnecessary. We begin therefore with a summary of geomorphic modeling on the largest scale.

4.3.1 Erosion Models on Orogenic Scale

Erosion models are not only a tool of geomorphologists. Many questions typically thought about by tectonicists and metamorphic petrologists also require the consideration of erosion in some simplified way. To illustrate this need, consider the following example: Metamorphic rocks may exhume from deep in the crust by either *erosion* or *extension*. Which of these two processes plays a more important role is a much debated question in the literature (Ring et al. 1999). The studies of England (1981), Summerfield and Hutton (1994) or Harrison (1994) show that at least in some orogens, erosion is the principle exhumation mechanism. Thus, it is crucial to consider erosion processes as an integral part of most geodynamic models (e.g. sect. 6.2.1). In most geodynamic models one of the following four erosion models is used (Fig. 4.24):

a Erosion rate is constant through time and space,
b erosion rate is a function of elevation,
c erosion rate is a function of slope,
d erosion rate is a function of surface curvature.

Which of these four models should be used, depends on the nature of the problem. In models that consider only a single point of the surface (for example one-dimensional vertical sections through the crust, e.g. sect. 3.4.1 or 4.1) are confined to one of the first two model descriptions of erosion.

Constant Erosion Rate. Assuming that the erosion rate during a given orogenic process is constant through space and time is the most dramatic thinkable simplification of real erosion processes. However, let us recall that a good model must find the right balance between accurate description and simplicity (p. 5). For example, erosion models assuming constant erosion rate have been very successful to explain the causes of the clockwise shape of metamorphic PT paths during regional metamorphism (sect. 6.2.1). Thus, for many purposes and problems of metamorphic petrologists, this model is the best. The model can be formulated as:

$$v_{\mathrm{er}} = -\frac{\mathrm{d}H}{\mathrm{d}t} = \text{constant} \ . \tag{4.50}$$

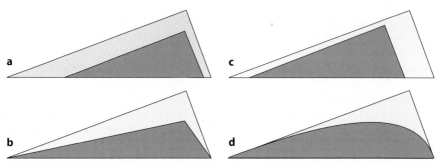

Figure 4.24. The influence of four different erosion models on the shaping of an asymmetric mountain belt. The light shaded area shows the mountain before the onset of erosion. The dark shaded region shows the shape of the mountain after some time. **a** constant erosion rate, **b** erosion rate proportional to elevation, **c** erosion rate proportional to slope, **d** erosion rate proportional to surface curvature. In all four erosion models, the erosion rate is the rate of removal of vertical section of the topography, measured in $\mathrm{m\,s}^{-1}$. Note that only in models **a** and **b** the highest point of the topography remains laterally fixed during erosion

where v_{er} is the erosion rate, H is elevation and t is time. Note that the sign must be negative for erosion to *decrease* surface elevation (this is different from other vertical reference frames we have used elsewhere in this book). Remember that the erosion rate can only be set equal to the rate with which the surface goes downwards if the uplift rate of rocks is $v_{\mathrm{ro}} = 0$ (s. eq. 4.4).

Erosion Rate Proportional to Elevation. Assuming that the erosion rate of a mountain belt is proportional to its elevation is the next closer approximation of nature. In fact, this model describes many aspects of real erosion processes very well and we have used it extensively in the section around page 146. For example, we observe that high mountain ranges like the European Alps erode much quicker than low lying hills in the German or Italian foreland. However, there are also examples where the opposite is true. For example, the Tibetan Plateau is 5 000 m high, but erosion rate is practically zero. Nevertheless, this erosion model has found much application in the literature. In the most simple case, the proportionality between elevation and erosion rate is linear and may be described by the relationship:

$$v_{\mathrm{er}} = -\frac{\mathrm{d}H}{\mathrm{d}t} = -\frac{H}{t_{\mathrm{E}}} \quad . \tag{4.51}$$

Note again that the sign convention used here is consistent with the convention used in the last paragraph, but opposite to that used in eq. 4.10, where we used a different reference frame. The erosion parameter t_{E} (in units of time) describes how long it takes to erode a mountain of the elevation H. Note, however, that the model implies that the erosion rate drops immediately as the first increment of erosion has decreased the elevation of a mountain. If eq. 4.51 is integrated (e.g. using the principle explained on p. 372), the model describes an exponential decrease of the elevation through time:

$$H = H_0 \times e^{\left(\frac{-t}{\tau_E}\right)} \quad , \tag{4.52}$$

where H_0 is the original elevation at the onset of erosion.

Erosion Rate Proportional to Slope. A proportionality between erosion rate and slope may be described by:

$$v_{er} = -\frac{dH}{dt} = -u\frac{dH}{dx} \quad . \tag{4.53}$$

There, x is a horizontal spatial coordinate and dH/dx is the topographic gradient: the slope. The proportionality constant u is the horizontal rate of displacement of the slope. As with eq. 4.50, it is important to note that erosion rate only corresponds to the rate of elevation change, if all other uplift or subsidence processes are zero. Equation 4.53 should remind us of eq. 3.43. Both are one-dimensional transport equations, which may be solved with the methods discussed in sect. 3.3 (s. Fig. 3.12 or Fig. A.8). The model is a good description for the evolution of many landforms, for example the motion of sand dunes. The model has also been applied to describe the geomorphic evolution of passive continental margins.

- *Advection at passive margins.* Markedly asymmetric mountain ranges are developed at several locations around the globe along the passive margins of continents. The best developed examples are the *Great Dividing Range* along the east coast of Australia and the coastal ranges along much of the southern African continent (Fig. 1.1). In both these examples, the inland side of the mountain range has a very small slope angle, while the coastal side is characterized by much steeper slope, often referred to as *Great Escarpment* (Ollier 1985). Because of this marked asymmetry, King (1953) concluded that erosion occurs largely along the escarpment which causes that the shape of the range remains largely self similar through time and that the great escarpment retreats inland (s. Fig. 4.23). In southern Africa this process is aided by the fact that the escarpment is often made up by the very resistant Karoo-basalts. This erosion model and the implied morphological evolution of passive continental margins has found much interest in the past decade. The simple model of King (1953) is now largely superceded (e. g. Stüwe 1991; Kooi and Beaumont 1994, Tucker and Slingerland 1994), but it still must be acknowledged that it describes a number of features quite well.

4.3.2 Short Range Transport

On a hill slope scale local mass transfer is well-described by the process of diffusion which we have already met and discussed in much detail on p. 50. We will recall from there, that diffusion describes a proportionality between *temporal* change and *spatial* curvature (eqs. 3.1, 3.4). That this model can also be applied to describe the geomorphic shaping of landforms was established by Culling (1960), Ahnert (1970) or Andrews and Bucknam (1987). Many observations in nature lend themselves to description with this proportionality: We can observe that ragged, pointy mountains erode much quicker than flat plateaus (even if they lie very high), that sharp escarpments erode quicker than smoothly curved hills and many more. Among diffusion models we discern:

– linear diffusion,
– non-linear diffusion.

Before we describe and apply diffusion models on the next pages we want to recall that the diffusion model is only a simple description summarizing a range of physical processes (including even discontinuous processes like landslides if they may be temporally averaged).

• *Linear diffusion.* Linear diffusion of mass is completely analogous to the logic discussed on p. 48. It is based on the assumption that the rate of down slope transport of mass (described by mass flux q) is proportional to the hill slope:

$$q = -D\frac{\mathrm{d}H}{\mathrm{d}x} \ .$$
(4.54)

where $\mathrm{d}H/\mathrm{d}x$ is the topographic gradient, (i. e. the slope) and the erosional diffusivity D corresponds to κ in the theory of heat conduction and has the units of $\mathrm{m^2\,s^{-1}}$. This equation is directly analogous to Fourier's first law (eq. 3.1). Note that the mass flux q has the units of $\mathrm{m^2\,s^{-1}}$ which may be interpreted as the volumetric flow normalized to the width of a profile. In other parts of this chapter we will encounter fluxes that are normalized to the channel cross section and have therefore the units of $\mathrm{m^2\,s^{-1}}$ (e.g. p. 185). The diffusivity D may be interpreted as the product of horizontal rate of mass transport v and thickness of an erodable near surface layer h_s (Beaumont et al. 1992; Carson and Kirkby 1972):

$$D = vh_\mathrm{s} \ .$$
(4.55)

From this equation, we can recognize the origin of the variability of D: h_s is dependent on rock type, but v is not. As in the theory of heat transfer, the flux equation eq. 4.54 may be combined with a one-dimensional mass balance of the form:

$$\frac{\partial H}{\partial t} = -\frac{\partial q}{\partial x} \ .$$
(4.56)

This equation is analogous to eq. 3.2 and is not derived in detail here. By inserting eq. 4.54 in eq. 4.56 we obtain the mass diffusion equation in one dimension:

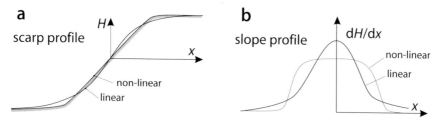

Figure 4.25. Comparison between linear and non-linear diffusion in the evolution of a degrading scarp. Note that in the non-linear case slope curvatures are more localized than for the linear case

$$\frac{\partial H}{\partial t} = D\frac{\partial^2 H}{\partial x^2} \quad . \tag{4.57}$$

Eq. 4.57 corresponds to eq. 3.6. If we insert concentration instead of elevation into eq. 4.57 then this equation may also be used to model the distribution of elements in minerals (s. eq. 7.2). In mineralogy and geochemistry, the equations governing diffusion are known as *Fick's laws*.

• *General conservation of mass.* The diffusion equation describes conservation of mass on a local scale. However, just as we have combined conduction, advection and production of heat in a complete thermal energy balance in eq. 3.53, we can write a more general form of a mass conservation equation in two dimensions as:

$$\frac{\partial H}{\partial t} = D\left(\frac{\partial^2 H}{\partial x^2} + \frac{\partial^2 H}{\partial y^2}\right) + u_x\frac{\partial H}{\partial x} + u_y\frac{\partial H}{\partial y} + v \quad . \tag{4.58}$$

This equation describes the change in elevation at a given point of the topography as the sum of diffusion (first term on right hand side), sidewards advection of a topographic profile (e.g. dune motion or retreat of an erosional escarpment) at the horizontal rates u_x and u_y (second and third term) and material "production", i.e. upwards advection at the rate v. This "upwards advection rate" v may be seen as the sum of uplift and sediment deposition and is particularly important for descriptions in Lagrangian reference frames.

Note that this equation describes a two-dimensional model, although all three spatial dimensions x, y and H occur in it and although the model can be represented as a three-dimensionally (e.g. as a plaster model). However, H is evaluated only on the basis of two model variables and landscape models are therefore in general two-dimensional models (s. p. 7).

• *Non-linear diffusion.* In linear diffusion it is assumed that the mass transport is *directly* proportional to slope. However, this need not be so. In fact, it may be intuitively followed that the rate of material transport increases more rapidly than linear with slope angle, for example because the material on the slope is loose, in contrast to outcropping material near the top or consolidated debris at the base. A more general formulation of diffusion may be made that looks like:

$$\frac{\partial H}{\partial t} = D \frac{\partial}{\partial x^2} \left(\frac{\partial H}{\partial x^2} \right)^n \quad .$$
(4.59)

If $n = 3$ this equation describes what is called *cubic diffusion*. (For an explanation of the definition of a non-linear differential equation see p. 352). If the rate of down slope transport is related to some power of slope, then this will have the consequence that more even slope profiles develop with sharper edges as illustrated in Fig. 4.25. Non-linear diffusion was discussed by Newman (1983) and established for the modeling of scarps by Andrews and Bucknam (1987) and most recently discussed by Roering et al. (2001). Hanks and Andrews (1989) suggested a "linear plus cubic" diffusion model as a good fit to some field data and use a diffusion equation of the form:

$$\frac{\partial H}{\partial t} = D \left(1 + 3c \left(\frac{\partial H}{\partial x} \right)^2 \right) \left(\frac{\partial^2 H}{\partial x^2} \right) \quad ,$$
(4.60)

where c is a constant. Modern discussion of non-linear versus linear diffusion in landform processes are found by Avouac (1993) or Avouac and Peltzer (1993). Non-linear diffusion describes in effect a down-slope change of the erosional diffusivity D (Pierce and Coleman 1986). However, we will not discuss non-linear diffusion further in the context of this chapter.

Scarp Degradation. As a first example for the application of the short range diffusion model, let us discuss the morphological dating of fault scarps. As scarps are often small compared to the extend of the slope they occur on, the diffusive decay of scarp-like landforms may be described between boundary conditions at infinity and the problem is therefore analogous to the thermal diffusion problem discussed on p. 102 (Fig. 3.27, eq. 3.85). Using H for elevation, x for a spatial coordinate system normal to the fault scarp and with its origin in the center of the scarp (as shown in Fig. 4.25), the boundary conditions may be formulated as $H = a$ at $x = \infty$ for $t \geq 0$ and $H = -a$ at $x = -\infty$ for $t \geq 0$. These boundary conditions describe a scarp displacing a flat surface by the height $2a$. For these conditions, eq. 4.57 may be integrated to give:

$$H = a \times \text{erf} \left(\frac{x}{\sqrt{4Dt}} \right) \quad ,$$
(4.61)

which is completely analogous to eq. 3.85, except that we shifted the co-ordinate system to the middle of the scarp. If the vertical scarp occurs on a slope with the slope b, ($b = \tan(\beta)$, where β is the slope angle), then the factor bx can simply be added to the right hand side of eq. 4.61. However, field observations show that many fault scarps are not vertical, in part because they did not form on vertical faults and in part because they sag gravitationally after a faulting event and assume a stable angle of repose (angle α on Fig. 4.26c) only some time thereafter. Fortunately, this "gravitational phase" is usually short compared to the subsequent degradation and the decay of the landform may still be described with the diffusion equation. However, an

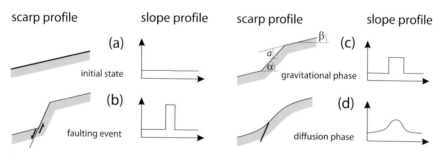

Figure 4.26. Stages in the evolution of a fault scarp (after Avouac 1993). Note the similarity of the slope profile during the diffusion process from **c** to **d** to the cooling of intrusions shown in Fig. 3.29. Parameters labeled in **c** are those used in eq. 4.62

analytical solution of eq. 4.57 for the initial geometry shown in Fig. 4.26 is unlike more complicated than eq. 4.61. It is:

$$H = \left(\frac{c}{a}\sqrt{\frac{Dt}{\pi}}\right) \times \left(\exp\left(-\frac{(x+c)^2}{4Dt}\right) - \exp\left(-\frac{(x-c)^2}{4Dt}\right)\right)$$

$$+ \left(\frac{c}{2a}\right) \times \left((x+c)\mathrm{erf}\left(\frac{x+c}{\sqrt{4Dt}}\right) - (x-c)\mathrm{erf}\left(\frac{x-c}{\sqrt{4Dt}}\right)\right) + bx \quad , \qquad (4.62)$$

where $c = a/(\alpha - b)$, the half height of the scarp is a and $b = \tan(\beta)$, as shown in Fig. 4.26c.

While it is simple to use this equation for the morphological dating of fault scarps (e.g. by plotting it up using MATHEMATICA), it is often possible date the decay of scarps by using a single value: the maximum scarp slope $\tan(\theta)$. That may be found by taking the derivative of eq. 4.62 with respect to x and evaluating this at $x = 0$. This gives much more simple expressions which have been used to date scarps by a variety of authors (e.g. Avouac 1993; Avouac and Peltzer 1993). The slope distribution shown in Fig. 4.25 and 4.26 shows that diffusive decay of scarps is characterized by a Gaussian distribution of slope. Diffusion of landforms may therefore also be described by Gaussian smoothing (Avouac 1993).

Mass Diffusion With Fixed Boundary Conditions. While the degradation of scarps may be described with boundary conditions fixed at infinity, most geomorphological diffusion problems are characterized by spatially fixed boundary conditions. For example, the rounding of hill slopes is usually spatially confined to a hill between two drainages from where material is efficiently transported out of the system. Similarly, the weathering and erosion of granitic boulders occurs between joints from where material is removed out of the system. There are many related problems that may be described with analytical solutions of eq. 4.57 (or its two dimensional equivalent) us-

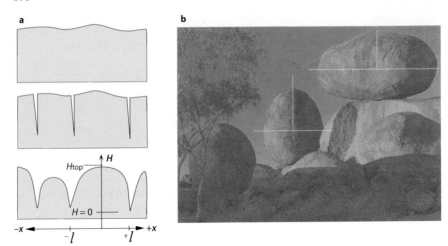

Figure 4.27. Illustration of spheroidal weathering. **a** Schematic illustration. The top diagram shows the starting geometry, the middle diagram shows how jointing of the rock surface occurs rapidly due to climatic influence and occurs rapidly in comparison with the subsequent erosion. The bottom diagram shows the typically rounded shapes that form during the subsequent weathering and the coordinate system is shown that is used for the formulation of eq. 4.63. **b** shows an example from the Devils Marbles, Central Australia. The coordinate systems on the photograph are drawn for Problem 4.16

ing spatially fixed boundary conditions, but we focus below on two specific examples.

• *Spheroidal weathering.* Mechanically isotropic rocks (for example granite) often weather in typical rounded shapes (Fig. 4.27). This *spheroidal weathering* occurs because individual blocks are separated relatively quickly by jointing, but round off by erosion on a much longer time scale. The weathering process is most effective on surfaces that have a high spatial curvature: it is a diffusion process and may be described with eq. 4.57. In order to formulate appropriate boundary conditions for the integration of eq. 4.57 it is useful to choose a coordinate system with an origin at the center of the block (Fig. 4.27). If the diameter of the block is $2l$, then the joints on either sides of the block lie at $x = l$ and $x = -l$. Initial and boundary conditions may be formulated as:

– Initial condition: $H = H_{top}$ in the region $-l < x < l$ at time $t = 0$.
– Boundary conditions: $H = 0$ at $x = l$ and $x = -l$ at $t > 0$.

H_{top} is the height of the block as shown in Fig. 4.27a. A solution of eq. 4.57 subject to these boundary conditions requires the use of Fourier series (for reasons explained on p. 368 and p. 111) and therefore contains infinite summations and trigonometric functions. It is:

$$H = \frac{4H_{\text{top}}}{\pi} \sum_{n=0}^{\infty} \frac{-1^n}{2n+1} \exp\left(-D(2n+1)^2\pi^2 t/4l^2\right)$$

$$\times \cos\left(\frac{(2n+1)\pi x}{2l}\right) . \tag{4.63}$$

The shape of weathering profiles as a function of time calculated with this equation are shown in Fig. 4.28. The mismatch between the curves shown in this figure and the photograph in Fig. 4.27b is predominantly because eq. 4.63 is one-dimensionally, while the boulders in the photograph also round off in the plane normal to the photograph, plus from below. Fortunately, a two-dimensional equivalent to eq. 4.63 is easily formulated as the product of two solutions in orthogonal directions, as we discussed on p. 110.

Figure 4.28. The shape of an initially rectangular ridge that erodes by mass diffusion processes after 1 000, 10 000 and 30 000 years. Streams transporting eroded material out of the system must be imagined in the bottom left and right corners to run normal to this page. Calculated with eq. 4.63 using $l=1$ km, $H_{\text{top}}=1$ km and $D=10^{-6}$ m^2 s^{-1}

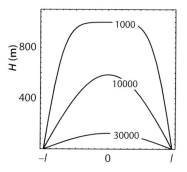

• *Hill slope profiles.* Eq. 4.63 and its graph on Fig. 4.28 may not only be used to date spheroidally weathered granite boulders. It can also be used to model the shape of ridge profiles between parallel drainages (Fig. 4.29). However, when modeling hill slopes between drainages we need to be careful, as the boundary conditions we have formulated in the last paragraph describe stationary (non-incising) river beds. Thus, the model applies only to *graded rivers* where the material transport in the stream equals the amount of material fed into them by the ridge. If the streams incise into the landscape simultaneously with the diffusion on the hill slope, we must formulate the problem with variable boundary conditions.

Mass Diffusion With Variable Boundary Conditions. If the incision rate of drainages is comparable to the rate of diffusion, then the incision itself is part of the hill slope shaping process and the boundary conditions of the previous sections may not be used. The description of simultaneous diffusion and incision at the model boundaries depends on the reference frame used. Two possibilities offer themselves.

• *Eulerian description.* For an observer fixed to an external reference frame, the incision of drainages may be formulated as follows:

Figure 4.29. Example of hill slope formation due to mass diffusion with incising boundaries. Western Mac Donnell Range, Central Australia. The length of the visible part of the principle ridge is about 500 m. Different stages of hill slope development may be seen. Near the principle drainage divide, (where head waters in the gullies have incised last), profiles are similar to the youngest profile of Fig. 4.30 (profile *a*), while hill slopes have adapted a steady state by *b* which is maintained in *c* before the slope disintegrated into smaller landforms

– Initial condition: $H = 0$ in the region $-l \geq x \leq l$ at time $t = 0$.
– Boundary conditions: $H = v_\text{fl} t$ for $x = l$ and $x = -l$ at time $t > 0$.

In this formulation we used $H = 0$ at the surface before the onset of erosion as the origin of a vertical axis going positively downwards and a horizontal coordinate system with the origin half way between two parallel streams that are a distance $2l$ apart. v_fl is the vertical incision rate of the rivers. The elevation H increases linearly at the model boundaries with time. Note that we changed the direction of the vertical axis from the last section, for clearer illustration. Solving eq. 4.57 subject to these temporally varying boundary conditions is difficult and will not discuss here. However, an example of a result from a corresponding solution is shown in Fig. 4.30a.

• *Lagrangian description.* From the observation point in the river bed it appears as if the landscape uplifts between the river beds (Fig. 4.30b). We can describe this using the uplift rate v_ro for which it is true that: $v_\text{ro} = -v_\text{fl}$. The uplift applies to the entire model, not only the boundaries. Thus we can not use eq. 4.57 for a description in a Lagrangian reference frame. The equation we need to solve includes a term describing the rock uplift. It is:

$$\frac{\partial H}{\partial t} = D \frac{\partial^2 H}{\partial x^2} + v_\text{ro} \; . \tag{4.64}$$

which is equivalent to the heat conduction equation with a heat production term (see eq. 3.24, also see eq. 4.58). The initial and boundary conditions that apply for the solution of this equation are:

– Initial condition: $H = 0$ in the region $-l \geq x \leq l$ at the time $t = 0$.
– Boundary condition: $H = 0$ at $x = l$ and $x = -l$ at all times $t > 0$.

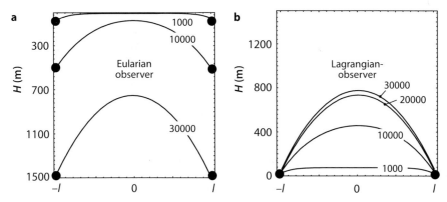

Figure 4.30. Landscape evolution of a ridge subject to mass diffusion processes between vertically incising drainages. **a** Within in a Eulerian reference frame (spatially fixed) and **b** in a Lagrangian reference frame (fixed to the incising drainages). The shape of the ridge is shown after 1 000, 10 000 and 30 000 years and was calculated using eq. 4.65, $l = 1$ km, $v_{ro} = 0.05$ m per year and $D = 10^{-6}$ m^2 s^{-1}

For these boundary conditions a solution of eq. 4.64 is given by Crank (1975):

$$H = v_{ro} \left(\frac{l^2 - x^2}{2D} \right) - \frac{16 l^2 v_{ro}}{D \pi^3}$$

$$\times \sum_{n=0}^{\infty} \frac{-1^n}{(2n+1)^3} \exp\left(-D(2n+1)^2 \pi^2 t / 4l^2 \right) \cos\left(\frac{(2n+1)\pi x}{2l} \right) \quad .(4.65)$$

A model example of a hill slope profile calculated with this solution is shown in Fig. 4.30b. Note that the landscape profiles in Fig. 4.30a and b have the same shape at the same times. They differ only in the reference frame and thus the profiles occur at different absolute elevations.

On Fig. 4.30b we can see that the landscape evolution described here approaches a steady state shape after about 20 000 years. This is what is called an *erosional steady state* (Ahnert 1984). In the steady state, the shape of the hill slope profile does not change anymore. This steady state is reached when the curvature of the landscape profile is exactly large enough, so that the diffusive mass transport in vertical direction is exactly as much as the mass production described by the rock uplift v_{ro} (s. p. 61 to estimate when such steady states are reached). The attainment and preservation of an erosional steady state can be observed on Fig. 4.29. We have discussed an equivalent steady state in the theory of heat conduction when discussing the steady state temperatures in subduction zones (p. 99). A steady state solution of eq. 4.64 is discussed in Problem 4.17.

Diffusion Age, Time Constants and Diffusivities. For the construction of the curves in Fig. 4.28 or 4.30 we have assumed that $D = 10^{-6}$ m^2 s^{-1}. We have assumed this value in order to retain some analogy to chapter 3 where we have discussed that the thermal diffusivity is $\kappa \approx 10^{-6}$ m^2 s^{-1} and rarely varies by more than a factor of two from this value. However, the diffusivity of mass in erosion processes is much more variable. It depends on climate, material and many other parameters. Mass diffusivities have been reported to range between $D = 1 \times 10^{-10}$m^{2}s^{-1} or $D = 1.7 \times 10^{-10}$m^{2}s^{-1} for largely unconsolidated materials in different regions in China and the Tien Shan (Tapponier et al. 1990; Avouac and Peltzer 1993), to $D = 5.3 \times 10^{-12}$m^{2}s^{-1} and $D = 1.8 \times 10^{-11}$m^{2}s^{-1} as estimates for the in-strike and cross-strike diffusivities for vertically bedded sandstone at Ayers Rock, Australia (Stüwe 1994). Mass diffusivities may vary by many orders of magnitude and unless we have measured D in our region of interest, we can usually not know if a given landform formed over a long time with a small diffusivity or a short time with a high diffusivity.

However, we can see that in eqs. 4.63 and 4.65, *time* and *diffusivity* occur always as a linear product and profiles are therefore identical if the product of time and diffusivity is a constant. Thus, it is often useful to define a "degradation coefficient", t_a, as the product of time and diffusivity:

$$t_a = Dt \quad . \tag{4.66}$$

Degradation coefficients are also known by the name "diffusion age", although we can see from eq. 4.66 that it has the units of m^2. Note however, that degradation coefficients are only defined by the simply linear product of eq. 4.66 if the diffusivity is constant in time. If the diffusivity changes with time (as often is the case if climate or the state of consolidation change during erosion), then the degradation coefficient is defined by a more complicated function discussed and explained in detail by Avouac (1993).

• *Diffusive time versus length scale.* The degradation coefficient corresponds to the product $\tau \times \kappa$ in eq. 3.18 and can therefore be used to convert between time and length scale in diffusion processes (s. also eq. 7.3). The diffusive time scale argument used on p. 57 can directly be applied to estimate the rough time scale of erosion processes and is therefore a useful tool for the field geologist. As the analogy between energy and mass diffusion is straight forward, the discussion from p. 57 is not repeated here.

4.3.3 The Shape of Volcanoes

Many strato-volcanoes have an intriguingly similar shape and surface elevation suggesting that they formed by similar processes (Turcotte 1997). For example, Etna in Sicily, Mt. Fuji in Japan as well as many volcanoes in Indonesia and in Alaska are all about 3 500 m high and have conical profiles with concave flanks. Two very simple models but different may be used to explain this shape. One relies on the principals of mass production and diffusion as we have discussed in several parts of this book; the other is a hydrostatic model.

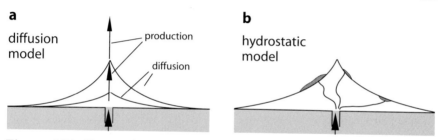

a diffusion model

b hydrostatic model

production

diffusion

Figure 4.31. Two models for the formation of strato volcanoes. **a** This model assumes that volcanoes get their shape from the interaction of mass *production* at a single point at the center of the volcano and mass *diffusion* distributing the material (s. also Fig. 3.8). **b** This model assumes that volcanoes describe surfaces of constant hydrostatic pressure. Magma does not necessarily extrudes at the tip of the volcano, but follows the way of least resistance giving the mountain its shape without erosive influence. While both models lead to similar shapes, detailed comparison between measured shapes of volcanoes with the two models may help to provide constraints on future eruption sites

- *Diffusion model.* The diffusion model is based on the assumption that magma erupts always from the same point on the surface and that the erupted material is distributed from there by mass diffusion (Fig. 4.31a). This model may be described with the equations we have used in sections 3.1 and 3.2 for the simultaneous production and diffusion of heat. If we assume that the erupted magma is distributed concentrically around the eruption point, then we can describe this problem in cylindrical coordinates using the eruption point as the origin and using an equivalent of eq. 3.24 in cylindrical coordinates (s. p. 54). Initial and boundary conditions for the problem may be formulated as:

   – Initial condition: $H = 0$ for all r at $t = 0$ as well as v_{ro} = magma production rate at $r = 0$ and $v_{ro} = 0$ at $r > 0$.
   – Boundary condition: $dH/dr = 0$ at $r = 0$ and $H = 0$ at $r \to \infty$ for $t > 0$.

   v_{ro} is the rate of magma production at the coordinate origin and replaces the heat production rate in eq. 3.24. These conditions are equivalent to those we have used on p. 67 for the description of the influence of frictional heating around shear zones. The only difference lies in that we have used Cartesian coordinates there (as shear zones are generally planar) and will require cylindrical coordinates for the description of volcanoes. However, the curves in Fig. 3.8 give a qualitative indication of the shape obtained with this model.

- *Hydrostatic head model.* The second model that has been used to describe the shape of volcanoes relies on the assumption that the surface of a volcano corresponds to a surface of constant hydrostatic head over a point source where the magma erupts (Fig. 4.31b). Thus, it is assumed that magma will always erupt to the surface where it finds the least resistance, even if this is not the crater vertically above the point source. This second model is discussed in detail by Turcotte and Schubert (1982).

Both models discussed above lead to cone shapes with concave surfaces, just like we observe volcanoes to look like. However, there is sufficient difference in the details of the two models, so that it may be possible to use the observed shape of a volcano in comparison with both model to predict what processes govern the magma distribution in the chosen example. It may be possible to use the models described here to make predictions about the likely points of future eruptions.

4.3.4 Long Range Transport: Drainages

Fluvial erosion is one of the most important landscaping processes and the transport of material by rivers is an efficient mechanism for redistribution of mass on a large scale. Fluvial erosion is therefore often called *long-range* transport. In the field, this is documented by the enormous incision rates of some rivers and their extensive sedimentary fans elsewhere; for example the Indus or the Tsangpo in the Himalayas. In fluvial erosion processes it is often useful to discern between:

− supply limited erosion,
− transport limited erosion.

In supply-limited processes the transport of material out of the system is much more efficient than the sediment production rate, while transport limited processes are limited by the fluvial transport of rapidly supplied sediment (e.g. Tucker and Slingerland 1994). In the following we discuss some methods of characterizing fluvial erosion.

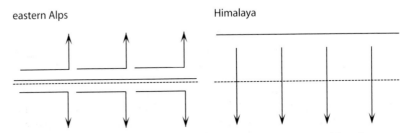

Figure 4.32. Schematic sketch of two different spatial relationships between drainage divide (continuous line), axis of the highest topography (dashed line) and direction of drainages (arrows). In the eastern European Alps the principle drainage divide corresponds to the region of the highest topography and the principle drainages are parallel to this axis. In the Himalayas, the principle drainage divide is some hundreds of kilometers north of the line of the highest topography and the drainages are perpendicular to this axis

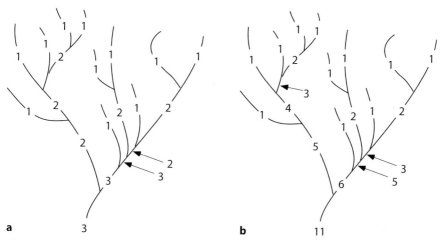

Figure 4.33. Stream order according to two different schemes: **a** according to Horton (1945) and Strahler (1964); **b** according to Shreve (1967)

Networks. The geometry of drainages – both with respect to a longitudinal profile along the drainage, as well as in plan view – usually have spatial patterns that are characteristic of tectonic and erosion processes (s. Summerfield 1991, Burbank and Anderson 2001), (Fig. 4.32). It is therefore useful to discuss some model tools that can be used to describe the spatial pattern of drainages. In *plan view*, the spatial characteristics of drainage networks may be described by their:

– topological properties,
– geometrical properties.

Both properties are often largely independent of scale. They are *self-similar* and lend themselves to a description as fractal shapes (s. sect. 4.3.6).

• *Network rules.* The topological properties of drainage networks may be characterized by allocating each stream section an order. There are different rules, how this may be done. According to Horton (1945), Strahler (1964) and Schumm (1956) the first stream after the spring has the order 1. When two streams of different order merge, the subsequent stream has the order of the higher order stream at the confluence. If both streams have the same order, then the order of the subsequent stream is larger by one (Fig. 4.33a). According to the scheme of Shreve (1967) the order of a stream is the number of contributing springs (Fig. 4.33b). If one uses the scheme of Horton (1945), then the topology of most natural networks appear to follow simple exponential laws. For example, the law of stream number states that the number of streams of order i may be described by:

$$N_i = a_1 e^{b_1 i} \quad .$$

$$(4.67)$$

There, N_i is the number of streams of the order i and a_1 and b_1 are constants. Similar laws apply to the length of streams of different order l_i and the size of the catchment of each stream A_i:

$$l_i = a_2 e^{b_2 i} \qquad \text{and} \qquad A_i = a_3 e^{b_3 i} \; . \tag{4.68}$$

There, a_2, a_3, b_3 and b_3 are constants. These network laws can be used to characterize the topology of an entire network with only a small portion of the network. However, Kirchner (1993) showed that practically all networks follow these laws and that it is therefore difficult to discern between artificial, random and natural networks using these rules (s. also Tarboton 1996).

• *Bifurcation and length-order ratios.* The bifurcation ratio R_b and the length-order ratio R_r of a drainage network are network parameters that are useful for the characterization of networks using fractal logic (p. 188). The bifurcation ratio is the ratio of the number of stream of a given order to the number of streams of the next higher order. The length-order ratio may be formulated correspondingly:

$$R_b = \frac{N_i}{N_{i+1}} \qquad \text{and :} \qquad R_r = \frac{r_{i+1}}{r_i} \; . \tag{4.69}$$

For many natural drainage networks, the bifurcation ratio is between three and five and it has been suggested that there is a climatic dependence (Turcotte 1997).

• *Grade river beds.* If we plot the surface elevation of a river bed against distance from the spring (a cross section along the river) many drainages have a shape that reminds of an exponential function: The drainage bed is steep at first and then gets shallower with increasing distance from the spring. This natural development often leads to the development of *graded river beds* (Mackin 1948). In a graded river, the slope of the drainage and the flow of water are in equilibrium so that neither erosion nor sedimentation takes place. The flow rate is exactly large enough to transport the sediment from the drainage basin above. Because of this, a graded river can maintain its shape and is in geomorphic equilibrium.

Models Describing Fluvial Erosion. In order to describe landscaping by rivers, a series of elegant models have been designed in the last few years (Ahnert 1976; Kooi and Beaumont 1994; Beaumont et al. 1992; Willgoose et al. 1991; Chase 1992; Tucker and Slingerland 1994, 1996). In many of these models two different types of fluvial channels are distinguished which are thought to form due to different physical processes:

– bedrock channels,
– alluvial channels

In *bedrock* channels, the rate of incision of a channel due to bedrock erosion may simply be assumed to be proportional to discharge and channel gradient

$$\frac{\partial H}{\partial t} \propto q_r^\alpha \left(\frac{\partial H}{\partial l} \right)^\beta . \tag{4.70}$$

where t is time, H is the elevation of the channel bed, l is the horizontal distance along the channel and the discharge q_r has the units of discharge volume per drainage width and time ($\mathrm{m\ s^{-1}}$) and is usually in some way proportional to the size of the drainage basin. α and β are exponents defining the non-linearity.

In *alluvial* channels it is often useful to consider the erosion/sedimentation processes in terms of the sediment carrying capacity of the stream

$$q_f^{eq} = -K_f q_r^m \left(\frac{\partial H}{\partial l} \right)^n . \tag{4.71}$$

where K_f is the constant of proportionality (Beaumont et al. 1992; Begin et al. 1981; Willgoose et al. 1991). There is considerable uncertainty about the exponents defining the non-linearities in eq. 4.70 and eq. 4.71 and most modern models assume therefore $\alpha = \beta = m = n = 1$ (see appendix of Kooi and Beaumont 1994).

Beaumont et al. (1992) and Kooi and Beaumont (1994) have presented an elegant model in which both, *bedrock* channels and *alluvial* channels may be modeled by comparing the sediment carried by a drainage q_f with the *equilibrium carrying capacity* of the river q_f^{eq} given by eq. 4.71. Whether the river erodes- or sediments onto its bed, depends on whether the sediment flow q_f is larger or smaller than the sediment equilibrium carrying capacity. q_f is given by the following two assumptions:

- 1. The rate of change of the sediment content of a river is proportional to the magnitude of the disequilibrium $dq_f/dt \propto (q_f^{eq} - q_f)$. If $q_f < q_f^{eq}$, then the amount of sediment in the river increases and the river will erode its bed. If $q_f > q_f^{eq}$, then there is more sediment in the stream than it can hold and there is sedimentation in the river bed.
- 2. The change of the sediment content dq_f/dt is inversely proportional to a *length scale of reaction*, l_f, which may be interpreted as the distance that water needs to flow along a river bed in order to do work on the river bed. If the reaction length scale is large, then the change of sediment freight of the river occurs slowly and over larger length scales and vice versa.

Using these two assumptions the rate of vertical lowering (by erosion) or rising (by sedimentation) of the river bed may be described by:

$$\frac{\partial H}{\partial t} = -\frac{dq_f}{dl} = -\frac{1}{l_f} (q_f^{eq} - q_f) . \tag{4.72}$$

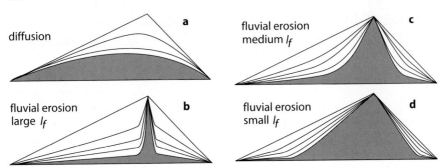

Figure 4.34. Erosion of a model landscape. **a** Relief formation by diffusion. If the starting profile is asymmetric, then diffusion will lead to a lateral shift of the drainage divide. **b**, **c** and **d** Relief formation by fluvial erosion (eq. 4.72). The model rives flow on both sides of the drainage divide along the surface of the gray shaded region outwards. the water shed (drainage divide) itself is not affected by this. The three examples differ in the magnitude of the reaction length scale l_f. In **b**, l_f is much smaller than the length of the river, in **c** it is comparable and in **d** l_f is much longer than the length of the river

In order top use eq. 4.72, q_f must be evaluated by numerical integration along the river bed using the two assumptions above. Figs. 4.34b, c and d show schematic illustrations of the temporal evolution of drainages as calculated with eq. 4.72. All drainages may be divided into two sections: a lower part where the slope of the river *decreases* and an upper part where the slope of the river *increases*. In the lower section of the rivers a *graded river bed* develops. The section of the river bed which is *graded* grows with the evolution of the drainage upwards. The length of the section of the grade river bed and the rate with which it grows drainage upwards, depends on the reaction length scale.

• *Integrated landscape models.* Eq. 4.72 describes many field observations made about the geometry of river profiles very well and has therefore been used by a range of authors to describe the temporal evolution of drainages and drainage networks. However, it fails to describe processes at the drainage divide for which different models must be invoked. Within the model of eq. 4.72 the parameters q_r and therefore q_f^{eq} and q_f are zero at the watershed and drainage divides will therefore remain preserved as steep ridges. The steepness of the ridge is proportional to the reaction length scale. However, the decay of the drainage divide itself may be well-described by diffusive mass transport discussed in the previous sections. Beaumont et al. (1992), Tucker and Slingerland (1994) and others have integrated combinations of diffusive short range transport and long range transport to describe two-dimensional landscape evolutions. This model was transferred onto irregular grids by Braun and Sambridge (1997) which is today probably one of the most elegant model for the integrated description of landscape evolutions.

4.3.5 Discontinuous Landscape Formation

Not all geomorphic processes may be described using continuous models like those discussed in the previous sections. For example, the occurrence of landslides is a classic discontinuous process that requires its own class of model description. How *continuous* geological processes may cause *discontinuous* processes will be discussed in some detail in sections 6.3.6 and 6.3.6. We will show there that discontinuous processes may be triggered by *threshold mechanisms*, or be the direct cause of *non-linear feedback* between different processes. The particular example of landslide occurrence is a beautiful example illustrating the meaning of *self-organized criticality*.

Consider an incising valley. As a river incises into the landscape, the slopes will steepen until they reach a critical angle where a steady state geomorphic profile is reached (s. p. 179, Fig. 4.30b). However, in many real landforms, this steady state will not be maintained by continuous incremental transport of material into the drainage, but by discontinuous landslides that cause a temporal fluctuation of slope around the steady state angle. The size-frequency distribution of such land slides is fractal and the state of the slope is said to be in a state of self-organized criticality (Turcotte 1997).

4.3.6 Fractals

Many morphological forms on the earth's surface have a fractal shape, for example coast lines (Mandelbrot 1975) or the shape of the earth's surface itself (Chase 1992). Fractals are usually brought in connection with non-linear feed back and chaotic processes because many geometric representations of non-linear phenomena do indeed render fractal shapes (e. g. Turcotte 1997; s. sect. 6.3.6). However, fractals themselves have nothing to do with chaos or feedback. Rather, they are simply defined as a geometric object that has the following properties (Fig. 4.35):

- The shape of the object must be able to be characterized without having to give a scale. This property is called *self similarity* or *scale independence*.
- It must be possible to characterize the object with a *fractal dimension*.

- *Fractal dimension.* The fractal dimension of an object is defined as follows:

$$D = \frac{\log(m)}{\log(n)} \quad . \tag{4.73}$$

There, m is the number of objects and n is a characteristic linear dimension. This may be illustrated with the fractal dimension of a simple geometric object (called a *geometrical fractal*). Consider a square that is subdivided into four sub-squares. If the edge of the square is subdivided into n pieces, then the total number of small squares is $m = n^2$. If one magnifies the length of the edge of each sub-square by the factor n we return to the original square. Using eq. 4.73 we arrive at $D = \log(m)/\log(n) = 2$. For

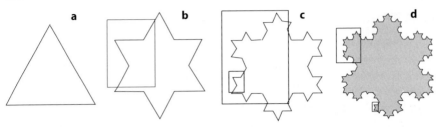

Figure 4.35. The first four stages of Koch's snow flake, a typical example for the development of a fractal geometry. The final form at **d** has a fractal geometry as it is self similar. The sections inside the box in **b** and **c** or in **c** and **d** are identical

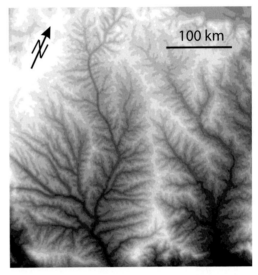

Figure 4.36. Fractal drainage patterns in the Chinese Loess Plateau. Similar fractal patterns can be found in the shape of the Grand Canon and countless other examples around the world. However, the Chinese loess plateau is probably the largest and most spectacular example in the world (because of the isotropy and high erodability of the material). Because of its spectacularity it was also chosen by Turcotte (1997) as a title image for his textbook on fractals

a square (having obviously two dimensions) this is quite trivial. However, for the snow flake of Koch things are not so clear (Fig. 4.35). Each edge of a triangle with length a corresponds to four edge sections each of which has a length of $a/3$ in each subsequent figure in Fig. 4.35. We can write: $D = \log(4)/\log(3) = \log(16)/\log(9) \approx 1.262$.

In natural landscapes there are two important fractal dimensions that may be defined and used to characterize the landscape:

− the fractal dimension of the drainage network,
− the fractal dimension of the topography itself.

Both can be obtained *statisitically* from natural landscapes and they are therefore called "statistical fractal" (in contrast to the geometrical fractal dimension of Koch's snow flake). The fractal dimension of *drainage networks* is defined as the ratio of the logarithms of bifurcation ratio and length-order ratio (s. eq. 4.69):

$$D = \frac{\log(R_b)}{\log(R_r)} \quad . \tag{4.74}$$

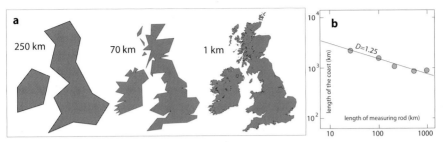

Figure 4.37. a The coast of Britain in three different resolutions (labeled in km). The figure was constructed using the generic mapping tool GMT. The length of the coast line segments are determined by the resolution of the digital elevation models used and are therefore not exactly of the same length. Nevertheless, the figure illustrates the relationship between segment length and coast line length used to determine the fractal dimension in **b** (after Mandelbrot 1967)

Typical drainage networks (like that shown in Fig. 4.36) have fractal dimensions that are somewhat less than space filling with $D \approx 1.8$. The fractal dimension of *coast lines* is similar to that of drainage networks inasmuch as it is smaller than 2. The fractal dimension of coast lines may be measured by taking m as the length of a measured coast line and n as the length of the measuring rod (Figs. 4.37). For the classic example of the west coast of Britain it is $D = 1.25$ (Mandelbrot 1967).

Corresponding to the measurement of the fractal dimension of coast lines, the fractal dimension of entire landscapes may be defined as the ratio of the logarithm of the relief and the logarithm of length scale. Natural landscapes have a fractal dimension around $D \approx 2.1$–2.7 (Mandelbrot 1982).

Using Fractal Dimension. Consider a landscape that was created by tectonic processes, for example the idealized horst and graben structure shown for time step t_0 in Fig. 4.38a. If the tectonic processes created a more or less random topography, then such a landscape may be characterized by a single fractal dimension as shown by the straight line for time t_0 in Fig. 4.38b. Subsequent erosion and sedimentation processes will destroy landforms at a rate that is proportional to their length scale. Small landforms will be quickly destroyed by mass diffusion processes, while intermediate landforms are only rounded and the largest landforms will preserve their original shape the longest. The length scale of the landform that separates *smaller* landforms (that are now characterized by "erosive shapes") from *larger* landforms (that still retain the shape created by the initial tectonic event) may be used to derive a characteristic diffusive length scale of the landscape. At time t_1 this is roughly the length scale of the landform e for the example in Fig. 4.38a. At time t_2 erosion has proceeded further and only the valley located to the right of landform f is large enough so that its original shape is still dominant. The landscape becomes multifractal as indicated by the curves for t_1 and t_2 in Fig. 4.38b. As we have discussed on p. 180, the diffusive length scale may now be converted into a time scale and therefore can be used to date the time

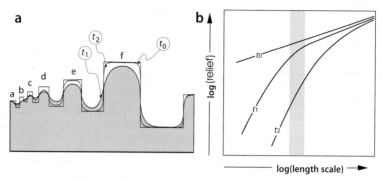

Figure 4.38. a Schematic cartoon of an idealized landscape made up of horsts and grabens which were subsequently eroded and sedimented upon by mass diffusion processes. Curve t_0 describes the landscape at the end of the tectonically induced horst and graben formation. The other two curves show the landscape at two subsequent times t_1 and t_2 **b** shows the fractal dimension of the landscape at times t_0, t_1 and t_2 in a plot of relief against length scale l. There, relief is the elevation difference between two points separated by a distance l (after Braun and Sambridge 1997)

of the last landscape forming tectonic event. Such qualitative considerations can be considerably improved by using statistic evaluation of length scales over an entire landscape (Chase 1992). Braun and Sambridge (1997) used this approach to date the age of the eastern Australian highlands.

4.4 Problems

Problem 4.1. *Vertical reference frames (p. 135):*
The summit of Chimborazzo in Equador is 6 180 m above sea level. The summit of Mt Everest is 8 848 m above sea level. Use eq. 4.2 to calculate the elevation difference between the two summits measured from the center of earth rather than from sea level. Chimborazzo lies practically at the equator, Mt Everest at roughly $\lambda = 28°N$.

Problem 4.2. *Understanding the f_c-f_l plane (p. 139):*
Draw an f_c-f_{ml}-diagram, where f_{ml} is the thickening strain of the *mantle part* of the lithosphere only. Draw into this diagram lines that describe the following deformation paths of the lithosphere: a) homogeneous thickening of the whole lithosphere, b) thickening of the mantle lithosphere without thickening of the crust and c) thickening of the crust at constant thickness of the whole lithosphere (i.e. with thinning of the mantle lithosphere corresponding to thickening of the crust). Compare your diagram with Fig. 4.5, where f_l describes the thickening strain of the *entire* lithosphere.

Problem 4.3. *Uplift versus exhumation (p. 141)*:
a) How much was a rock uplifted relative to a fixed external reference level if it was exhumed from 10 km depth to 5 km depth by erosion and the eroding mountain range decreases in this time from 8 000 m to 3 000 m elevation. b) What is the amount of exhumation a rock experienced if the surface uplift was 5 km and the rock was uplifted 4 km relative to a fixed reference level.

Problem 4.4. *Vertical kinematics (p. 148)*:
Use eq. 4.9, eq. 4.11 and eq. 4.12 to calculate the uplift and the exhumation history of a simple model mountain range that may be described with these equations. Assume that the range is in geomorphic steady state, so that its surface elevation remains constant because $v_{er} = v_{ro}$. a) What is the magnitude of the erosion parameter t_E of this range? Use the following constants: $a = 2\,500$ m and $b = 0.13$ and $\dot{\epsilon} = 10^{-15}$ s^{-1}. b) What was the depth of rocks that reach the surface after 40 my at the start of their evolution? Use eq. 4.13. c) How long is the hiatus between the onset of surface uplift and the onset of exhumation of these rocks? Use eq. 4.13.

Problem 4.5. *The principle of isostasy (p. 152)*:
We are familiar with the fact that 90% of icebergs are under water and 10% above the surface. Use Fig. 4.14 and eq. 4.17 to determine the density of ice from this observation. The density of water is $\rho_w = 1\,000$ kg m^{-3}.

Problem 4.6. *Relationship between relief and uplift (p. 152)*:
Parallel rivers incise an isostatically compensated plateau with a vertical incision rate of: $v = 5$ mm y^{-1} (Fig. 4.39). The slopes on both sides of the rivers are 45° steep. All rivers have a distance of $l = 10$ km from each other. Draw the surface elevation of the ridges and the river beds as a function of time assuming the plateau remains isostatically compensated during the incision process. Assume the following numerical values for the necessary physical parameters: $\rho_m = 3\,200$ kg m^{-3} and $\rho_c = 2\,700$ kg m^{-3}. (An interesting discussion of this problem may be found by Montgomery 1994.)

Figure 4.39. Illustration of Problem 4.6

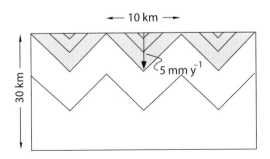

Problem 4.7. *Controls on surface elevation (p. 152):*
Calculate the surface elevation changes of an isostatically compensated mountain belt that occur in response to the following processes: a) Continental crust of thickness $z_c = 30$ km ($\rho_c = 2\,700$ kg m^{-3}) is being underplated by a 10 km thick basaltic layer of density $\rho_u = 2\,900$ kg m^{-3}. The density of the underlying mantle is $\rho_m = 3\,200$ kg m^{-3}. b) Continental crust is being thinned to half its thickness ($f_c = 0.5$) and underplated by a 5 km thick layer of basaltic underplate of density $\rho_u = 2\,950$ kg m^{-3}. c) What is the difference in surface elevation between a) and b)?

Problem 4.8. *Controls on surface elevation (p. 155):*
Consider an isostatically compensated mountain range. What are the relative contributions of *thermal contraction* and *material* to the surface elevation of the lithosphere above the asthenosphere? Do these relative contributions change for mountain belts of different elevation? Use eq. 4.25 and the following numerical values for the physical parameters for your considerations: $\rho_m = 3\,200$ kg m^{-3}; $\rho_c = 2\,700$ kg m^{-3}; $\alpha = 3 \cdot 10^{-5}$ K^{-1} and $T_l = 1\,200\,^\circ$C.

Problem 4.9. *Age depth relationship of oceanic lithosphere (p. 157):*
a) Is the rifting rate at the mid Atlantic ridge 1.5 cm y^{-1}, 2.5 cm y^{-1} or 3.5 cm y^{-1}? Calculate the answer assuming that the bathymetry of the Atlantic may be described with the model of eq. 4.39 (Fig. 4.18), and use the following bathymetric data from the north Atlantic ocean. The data are given in pairs for water depth/distance from the mid Atlantic ridge: 300 m/50 km; 500 m/175 km; 800 m/250 km; 1300 m/500 km; 1500 m/700 km; 1800 m/825 km; 2300 m/1300 km; 2600 m/1575 km ; 2800 m/1775 km; 2900 m/1950 km; 3200 m/2500 km; 3200 m/3125 km; 3300 m/3375 km; 3200m/3625 km. $\rho_m = 3\,200$ kg m^{-3}; $\rho_w = 1\,000$ kg m^{-3}; $\alpha = 3 \cdot 10^{-5}\,^\circ$C^{-1}; $T_l = 1\,200\,^\circ$C; $\kappa = 10^{-6}$ m^2 s^{-1}. b) From a certain distance from the ridge (age) onwards, these data do not correspond very well with the model, even for the correct answer of a). What is this distance (age) and why does the model not work for older oceanic lithosphere?

Problem 4.10. *Understanding the flexure equation (p. 162):*
Perform a dimensional analysis of eq. 4.43 and eq. 4.42. a) What are the units of D, F and q? b) Typical continental lithosphere may be described with a Poisson's ratio around $\nu = 0.25$ and a Young's modulus of $E = 10^{11}$ Pa. What is the flexural rigidity of a continent that may be described with an elastic thickness of $h = 10$km to $h = 70$ km?

Problem 4.11. *Understanding the flexure equation (p. 162):*
(a) Integrate the flexure equation (eq. 4.42) to describe the shape of a fishing rod of length L held horizontally at its end and bent downwards under its own weight. Use the geometry shown in Fig. 4.40. (Equivalent curves are described by a piece of paper hanging off the edge of a table, or a bridge under construction extending half over a valley).

Some hints: 1.) Since there is no horizontal force applied to the rod, the last term of eq. 4.42 is zero. 2.) Since the load per unit length is everywhere the same, the load is independent of x. Eq. 4.42 simplifies to $d^4w/dx^4=q/D$. 3.) Since there is no torque *or* change in torque applied to the free-hanging end of the rod, it is true that $d^2w/dx^2 = 0$ at $x = L$ as well as that: $d^3w/dx^3 = 0$ at $x = L$. 4.) Since the rod is held out horizontally, it is also true that $w = 0$ at $x = 0$ and that: $dw/dx = 0$ at $x=0$. (b) Use your result to estimate the flexural rigidity of a piece of paper hanging $L = 10$ cm off the edge of a table with the lowest point of the paper tip hanging $w_{x=L} = 3$ cm down. The weight of the paper corresponds to a load of $q = 2$ N m^{-2}.

Figure 4.40. Illustration of Problem 4.11

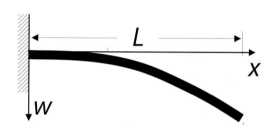

Problem 4.12. *Internal and external loads (p. 163):*
a) Repeat the derivation of eq. 4.44 using the principles of eq. 4.15. In other words, derive eq. 4.44 by comparing the weights of the vertical columns in Fig. 4.21 at the points A and B. b) Derive an equation for the load on oceanic plates that corresponds to eq. 4.44. This case is illustrated in Fig. 4.21 on the very left. The variable q_a is a line load. That is, it has to be imagined to extend infinitely in the direction normal to this page.

Problem 4.13. *Understanding the flexural parameter α (p. 165):*
What are the units of the flexure parameter α in eq. 4.47? How large is α roughly for the continental lithosphere? Use $D \approx 10^{23}$ Nm; $\rho_m = 3\,200$ kg m^{-3} and $\rho_w = 1\,000$ kg m^{-3}.

Problem 4.14. *Plate flexure near Hawaii (p. 165):*
The water depth of the Pacific in the vicinity of the Hawaii- and Emperor island chains is somewhat deeper than in the remainder of the abyssal planes of the Pacific. Also, there is a high point in the ocean floor, about 250 km off the coast of Hawaii. (Fig. 4.19). This water depth profile corresponds well with the shape of an elastically flexed plate loaded by the weight of the island chain with the high point in the ocean floor being the flexural bulge. What is the thickness of the elastic lithosphere in the Pacific? Use eq. 4.48, 4.47, 4.43 and the following values for the physical parameters: $E = 70$ GPa, $\rho_m = 3\,200$ kg m^{-3}; $\rho_w = 1\,000$ kg m^{-3}; $g = 10$ m s^{-1} and $\nu = 0.25$. For the

most elegant solution of this problem we need not even know the details of the shape of the flexed plate or the maximum water depth at the point of maximum flexure.

Problem 4.15. *Fault scarp degradation (p. 174):*
Avouac and Peltzer (1993) report a fault scarp from China with the following profile: (27,1; 51,1.1; 68,1.2; 77,1.8; 90,1.8; 110,1.9; 111,2.0; 120,2.2; 122,2.45; 125,2.6; 128,3.1; 130,4; 133,4.5; 135,5; 140,6.6; 143,7.5; 147,8.8; 150,10; 155,11; 160,12; 170,12.8; 180,13.5; 190,13.5; 220,13.7). This list shows the scarp co-ordinates $(x_1,H_1; x_2,H_2 ...)$, where x horizontal distance from a fix point in the far field and H is the surface elevation above the base, both in meters. Assuming that the surface is flat in the far field and that the scarp was ini-tially vertical, use eq. 4.61 to date the earth quake that formed it. Assume that the mass diffusivity is 5.5 m^2 per 1000 years.

Problem 4.16. *Spheroidal weathering (p. 176):*
Measure the topographic profiles off the two boulders with the marked co-ordinate systems on Fig. 4.27 and use eq. 4.63 to estimate whether the two boulders commenced weathering at the same time, or if the smaller one has been subjected to weathering longer.

Problem 4.17. *Steady state hill slopes (p. 178):*
Calculate the steady state $(dH/dt=0)$ hill slope profile of a ridge develop-ing between two parallel incising drainages by integrating eq. 4.64. Use the same boundary conditions used to derive eq. 4.65. Compare your result with Fig. 4.30b to judge if the curve for 30 000 years is already near the steady state profile or not.

Figure 4.41. Illustration for Problem 4.18

Problem 4.18. *Fractals and fractal dimension (p. 187):*
What is the fractal dimension of the triangle in Fig. 4.41? Use eq. 4.73

5. Force and Rheology

In this chapter we discuss the basics of dynamic descriptions of geodynamic processes. Dynamic descriptions, for example force balance estimates for orogens, provide a useful independent constraint on models based on field observations. Let us consider an example: A field geologist finds in a Precambrian shield folds and thrusts that he interprets to have formed as the consequence of crustal shortening. A detailed strain analysis shows that 80% shortening occurred and the geometry of shortening indicates that this resulted in fourfold thickening of the crust. He therefore further infers (using the principle of isostasy) that - at the time, a mountain range of some 15 km elevation existed above the metamorphic terrain. While this interpretation may be consistent with the field observations in the terrain, it has no independent test. In this example we could argue that we have no knowledge of any present day mountain range on this planet that is this high and that, therefore, this interpretation is unlikely. However, in many less obvious examples there are no direct analogies and the resulting models - albeit perfectly imaginable and fully consistent with field observations - are wrong. One way to provide an independent test of such models is to make a rough estimate of the involved force balance. In the next chapters we want to perform such estimates. In order to do so, it is necessary to commence with a brief repetition of the basics of stress and strain. However, this repetition remains brief and the interested reader is referred to a range of excellent text books on the subject. For example:

- Jaeger and Cook (1979) Fundamentals of Rock Mechanics.
- Twiss and Moores (1992) Structural geology.
- Pluijm and Marshack (1997) Earth Structure: An Introduction to Structural Geology and Tectonics.
- Ramsay and Huber (1983) Modern Structural Geology. Volume 1: Strain Analysis
- Ramsay and Huber (1987) Modern Structural Geology. Volume 2: Folds and Fractures
- Ramsay and Lisle (2000) Modern Structural Geology. Volume 3: Applications of Continuum Mechanics in Structural Geology
- Weijermars (1997) Principles of Rock Mechanics.

5.1 Stress and Strain

The state of stress of a rock is described by a symmetrical tensor with nine components. However, in deformed rocks we can usually measure (using geobarometric methods) only a single scalar quantity: pressure (for details on the scalar properties of pressure s. p. 365). Only with in situ stress measurements and some palaeopiezometric methods is it possible to resolve the stress state in more detail than this single value. In connection with eq. 5.7 we will show that pressure may be viewed as the mean principal stress. For now we only want to note that our geodynamic interpretations of ancient metamorphic terrains often rely on the interpretation of *pressure* even though this is only a mean quantity of the principal stresses that define the complete state of stress. Because many field geologists remain uncertain about the relationship of the pressure they measure to the complete stress tensor, we begin in sect. 5.1.1 with a very brief repetition of this tensor.

Strain (often imprecisely called deformation) is *also* described by a tensor. However, we will not discuss the strain tensor in this book in much detail, as just about all problems we treat here are kept very simple so that only *one* component of the strain tensor is non-zero. The whole strain tensor can therefore be usually described by a single scalar quantity (s. sect. 1.2.1, A.3). Nevertheless we need a few different terms describing features of strain. We define the following common terminology: The *stretch* of a rock s is the ratio of its length after deformation l to that before deformation l_0. Its *elongation* is the ratio of the change in length and the original length. We call this e.

We can write the relationship between stretch (uniaxial strain), elongation and length in short:

$$s = \frac{l}{l_0} = 1 + e = 1 + \left(\frac{l - l_0}{l_0}\right) \ . \tag{5.1}$$

Both s and e have been referred to as "strain" by different workers in the past. In this book we refer to s as strain.

5.1.1 The Stress Tensor

There are many excellent descriptions of stress in an abundance of good text books (e. g. Jaeger and Cook 1979; Means 1976; Suppe 1985; Twiss and Moores 1992; Engelder 1993). However, confusion of terminology is still common in the literature (Engelder 1994). Thus, we begin by defining a few terms related to stress. For any more detail, the reader is referred to the literature, in particular to the excellent summaries by Engelder (1994) and Twiss and Moores (1992).

Term definitions. *Force* is a vector and - like all vectors - is described by a *magnitude* and a *direction*. It has the units of mass × acceleration: 1 N = 1 kg m s^{-1}. A related vector quantity is *traction*. Traction is a force (with magnitude and direction) *per area*, where the orientation of this area is not defined. Tractions may be subdivided into normal and a parallel components called *normal traction* and *shear traction*. It is important to note that tractions are vectors, although they have the same units as stress. In contrast, *stress* is a tensorial quantity described by all the tractions acting on a unit cube (s. sect. A.3). We will now discuss the stress tensor in a bit more detail.

- *The stress tensor.* In three-dimensional space, the state of stress of a single point inside a rock (i.e. a unit cube) is given by nine numbers, all of which have the units of force per area. These nine numbers define the tensor as follows:

$$\sigma = \begin{pmatrix} \sigma_{11} & \sigma_{12} & \sigma_{13} \\ \sigma_{21} & \sigma_{22} & \sigma_{23} \\ \sigma_{31} & \sigma_{32} & \sigma_{33} \end{pmatrix} = \begin{pmatrix} \sigma_{xx} & \sigma_{xy} & \sigma_{xz} \\ \sigma_{yx} & \sigma_{yy} & \sigma_{yz} \\ \sigma_{zx} & \sigma_{zy} & \sigma_{zz} \end{pmatrix} \qquad (5.2)$$

(s. sect. A.3). The two different notations of subscript used in eq. 5.2 are both common in the literature. The first of the two specifying spatial indices x, y and z or 1, 2 and 3 indicates the normal to the plane on which this stress component acts. The second index indicates the direction in which the stress component acts. We can see that the three tensor components in the diagonal of this matrix have two identical indices. They are called *normal stresses* because the surface onto which the stresses act are *normal* to the direction in which they act, (i.e. the indices for "direction in which it acts" and "plane onto which it acts" are the same). In the following we abbreviate normal stresses with σ_n. The remaining six components of stress in eq. 5.2 are *shear stresses*. In these, the stress components they describe act *parallel* to the plane onto which they are exerted. The stress components in a given row of eq. 5.2 act on the same plane, but in different directions. The vertical columns contain stress components oriented in the same direction, but acting on different planes.

In the literature, shear stresses are often abbreviated with τ and normal stresses with σ. Thus, eq. 5.2 is often written as:

$$\sigma_{ij} = \begin{pmatrix} \sigma_{xx} & \tau_{xy} & \tau_{xz} \\ \tau_{yx} & \sigma_{yy} & \tau_{yz} \\ \tau_{zx} & \tau_{zy} & \sigma_{zz} \end{pmatrix} . \qquad (5.3)$$

However, the notation of eq. 5.2 is mathematically more correct, as its nine components are all components of the same tensor with the same units, so they should be abbreviated with the same symbol. Therefore, we retain the description used in eq. 5.2, in particular using x and y rather than 1 and 2 as subscripts and describe shear stresses with $\sigma_{i \neq j}$ or σ_s and normal stresses with $\sigma_{i=j}$ or σ_n. In the following sections we will also introduce τ to describe deviatoric stresses, but this should not be confused with shear stress as used in eq. 5.3 or as done in many places in the literature.

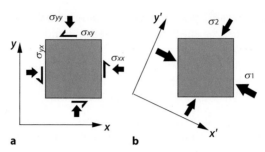

Figure 5.1. a The state of stress of a unit square within the two dimensions of this book page. In two dimensions, the stress tensor has only four independent components illustrated here by the four labeled arrows. In **b** the coordinate system x', y', z' was chosen in such an orientation so that the shear stresses become zero. The normal stresses become therefore *principal stresses*. The state of stress of the square in **a** and **b** is identical

The stress tensor is symmetrical, that is, each component above the diagonal has an equivalent component of equal magnitude below it. It is true that: $\sigma_{yx} = \sigma_{xy}$, $\sigma_{zx} = \sigma_{xz}$, $\sigma_{yz} = \sigma_{zy}$. If this were not the case, the body described by this state of stress would rotate. Thus, the stress tensor consists of only six independent numbers: three normal stresses (written in the diagonal) and three shear stresses (the off diagonal terms).

The state of stress described by eq. 5.2 can be expressed a bit more simply in a differently oriented coordinate system. It is always possible to assume a coordinate system with the coordinates x', y' and z', in which all shear stresses (all off diagonal terms in eq. 5.2) become zero. The diagonal components in this new coordinate system are called *principal stresses*. Principal stresses are denoted with a single subscript as σ_1, σ_2 and σ_3. In the earth sciences it is common to use the subscript "1" for the largest principal stress and "3" for the smallest. Thus, the state of stress at a point may always be characterized by only three principal stresses (Fig. 5.1):

$$\sigma'_{ij} = \begin{pmatrix} \sigma'_{xx} & 0 & 0 \\ 0 & \sigma'_{yy} & 0 \\ 0 & 0 & \sigma'_{zz} \end{pmatrix} = \begin{pmatrix} \sigma_1 & 0 & 0 \\ 0 & \sigma_2 & 0 \\ 0 & 0 & \sigma_3 \end{pmatrix} . \tag{5.4}$$

The order in which σ_1, σ_2 and σ_3 appear in eq. 5.4 implies that the new coordinate system was chosen here so that the x'-axis is parallel to the largest of the three principal stresses. Of course this need not be the case. Also note that the numbers denoting the three principal stresses have nothing to do with the spatial subscripts we briefly used in eq. 5.2.

In the stress diagrams of this book the coordinates are generally drawn parallel to the principal stress directions so that we can write: $\sigma_{ij} = \sigma'_{ij}$. In other words, the principal stresses are parallel to the coordinate axes x', y' and z', but their magnitude is independent of the orientation. Fortunately, in the earth's crust the principal stresses are often oriented roughly parallel to the vertical and horizontal, because there is no shear stresses on the earth's surface and shear stresses at the base of the lithosphere are negligible.

The sum of the three principal stresses is called the *first invariant of the stress tensor*. It is called *invariant* because its value is independent of the choice of the coordinate system:

$$I_1 = \sigma_1 + \sigma_2 + \sigma_3 \ . \tag{5.5}$$

Note that pressure is directly related to the first invariant (see eq. 5.7). Two more invariants of the stress tensor are defined as:

$$I_2 = -\left(\sigma_2\sigma_3 + \sigma_3\sigma_1 + \sigma_1\sigma_2\right) \ ,$$

and:

$$I_3 = \sigma_1\sigma_2\sigma_3 \ . \tag{5.6}$$

The 2nd and 3rd invariants are relevant for the understanding of some fundamental geodynamic processes (e. g. sect. 6.2.2).

Derived Quantities of the Stress Tensor.

• *Mean stress.* The *mean stress* σ_m is given by the mean of the three principal stresses. It is therefore independent of the coordinate system:

$$\sigma_m = P = \frac{\sigma'_{xx} + \sigma'_{yy} + \sigma'_{zz}}{3} = \frac{\sigma_1 + \sigma_2 + \sigma_3}{3} \ . \tag{5.7}$$

The mean stress is also called pressure P. Strictly speaking, the mean stress is the mechanical definition of pressure, while a chemist or thermodynamicist would say that work is the product of pressure and volume change and that, therefore, pressure has the units of energy per volume (1 Pa = 1 J m^{-3}). The most common place where geologists encounter these non-intuitive units for pressure is when looking up the molar volumes of mineral phases. These are generally quoted in the units Joule per bar. As the volume change may be highly anisotropic in an anisotropic stress state, chemically defined pressure may be determined by integrating the volume change over the surface of a unit volume. *Chemical* and *mechanical* pressure only correspond in an isotropic state of stress. Earth scientists measure pressure using geobarometers (sect. 7.2.1). Many geobarometers rely on the pressure sensitivity of chemical equilibria. It is therefore not clear if we measure mechanical or chemical pressure with them. This is some of the reason why the relationship between geobarometrically measured pressure and depth is often discussed in the literature (e. g. Harker 1939; Wintsch and Andrews 1988). However, in the following we assume that the differences between chemical and mechanical pressure are so small (if any) that we need not be concerned with them.

• *Differential stress.* Differential stress is a scalar value defined as the difference between the largest and the smallest principal stress:

$$\sigma_d = \sigma_1 - \sigma_3 \ . \tag{5.8}$$

It is a measure of how far the stress state is from isotropic. During viscous (ductile) deformation, the application of any differential stress will cause permanent deformation. However, in cold rocks, the application of a differential stress will cause minimal elastic deformation unless it is large enough so that the rock breaks. Thus, there may be differential stresses that are supported by the internal strength of rocks and that remain without evidence in the field. Geologists often refer to differential stress only when they see rocks deform. We want to keep in mind that *any* differential stress will cause deformation in the elastic and ductile regime. However, in the brittle regime, deforming rocks will only record the differential stress that causes failure (i.e. if it is large enough to touch the failure envelope on the Mohr circle).

• *Deviatoric stress.* Unlike mean stress, pressure or differential stress, *deviatoric stress* is not a single number, but a tensor, denoted commonly with τ. This tensor is defined by the deviation of the stress tensor in a general coordinate system (i.e. eq. 5.3) from pressure, or from mean stress:

$$\tau = \begin{pmatrix} \tau_{xx} & \tau_{xy} & \tau_{xz} \\ \tau_{yx} & \tau_{yy} & \tau_{yz} \\ \tau_{zx} & \tau_{zy} & \tau_{zz} \end{pmatrix} = \begin{pmatrix} \sigma_{xx} - P & \sigma_{xy} & \sigma_{xz} \\ \sigma_{yx} & \sigma_{yy} - P & \sigma_{yz} \\ \sigma_{zx} & \sigma_{zy} & \sigma_{zz} - P \end{pmatrix} \ . \tag{5.9}$$

It can be seen that the total stress tensor is the sum of the isotropic stress tensor plus the deviatoric stress tensor:

$$\begin{pmatrix} \sigma_{xx} & \sigma_{xy} & \sigma_{xz} \\ \sigma_{yx} & \sigma_{yy} & \sigma_{yz} \\ \sigma_{zx} & \sigma_{zy} & \sigma_{zz} \end{pmatrix} = \begin{pmatrix} P & 0 & 0 \\ 0 & P & 0 \\ 0 & 0 & P \end{pmatrix}$$

$$+ \begin{pmatrix} \sigma_{xx} - P & \sigma_{xy} & \sigma_{xz} \\ \sigma_{yx} & \sigma_{yy} - P & \sigma_{yz} \\ \sigma_{zx} & \sigma_{zy} & \sigma_{zz} - P \end{pmatrix} \ . \tag{5.10}$$

The total stress tensor (left hand side of eq. 5.9) is important for calculating viscous deformation, whereas the deviatoric stress tensor (last term in eq. 5.9) is important for calculating viscous deformation (Fig. 5.2).

For a coordinate system parallel to the principal stress directions the deviatoric stress tensor may simply be written as (in analogy to eq. 5.4):

$$\tau' = \begin{pmatrix} \tau'_{xx} & 0 & 0 \\ 0 & \tau'_{yy} & 0 \\ 0 & 0 & \tau'_{zz} \end{pmatrix} = \begin{pmatrix} \sigma_1 - \sigma_m & 0 & 0 \\ 0 & \sigma_2 - \sigma_m & 0 \\ 0 & 0 & \sigma_3 - \sigma_m \end{pmatrix} \ . \tag{5.11}$$

In this book, just about all discussed examples are for orientations of the stress field where $\tau = \tau'$. Also, we usually use both σ and τ as if it were a

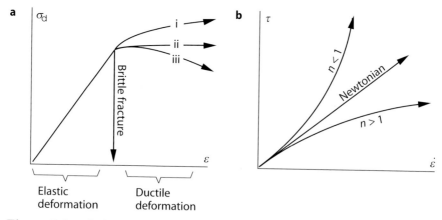

Figure 5.2. a Relationship between stress σ and strain ϵ for elastic, plastic and brittle deformation. Curve *ii* is for the ideal case of plastic (ductile) deformation; *i* is with strain hardening; *iii* with strain softening. **b** Relationship between stress and strain rate $\dot{\epsilon}$ for three different viscous materials. For a Newtonian fluid this relationship is linear. The number n is the power law exponent which we will discuss in some detail in sect. 5.1.2

scalar, i. e. we use them in equations that involve no tensor calculus. This is because most problems discussed in this book are reduced – for simplicity – to mere one-dimensional scenarios where there is only a single direction in which stresses are non-zero. We can therefore say usually that: $\sigma = \sigma_d = \sigma_1$.

The deviatoric stress tensor is important as its components cause viscous deformation. The absolute magnitude of the deviatoric stress tensor components indicates how the rock will strain in ductile deformation. For example, a rock will *extend* in the direction in which the deviatoric stress components are negative (negative is tensional in the earth science convention), even if all the principal stresses indicate compression (s. Fig. 5.3). Thus, when making cartoons of a field terrain it is always most instructive to sketch arrows for the principal components of the deviatoric stress tensor onto them, as their magnitude and direction corresponds to what is observed kinematically in the field (s. Fig. 6.25). For example, two rocks from different crustal levels may suffer the same deviatoric stresses and therefore deform similarly, but they may be in completely different states of total stresses, for example if they experience deformation at different crustal levels. In short, it is useful to calculate the deviatoric stresses when trying to compare seemingly different states of stress (Problem 5.4).

Let us consider a simple two dimensional example (there are only two principle stresses σ_1 and σ_2) where both σ_1 and σ_2 are positive (the rock is under compression) and the mean stress is $\sigma_m = (\sigma_1 + \sigma_2)/2$. There, the principle components of deviatoric stress are: $\tau_1 = \sigma_1 - \sigma_m = (\sigma_1 - \sigma_2)/2$ and $\tau_2 = \sigma_2 - \sigma_m = -(\sigma_1 - \sigma_2)/2$. We can see that the largest possible principal component of the deviatoric stress tensor is $\tau_1 = \sigma_d/2$ and the

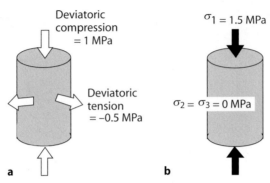

Figure 5.3. Cartoon illustrating a typical uniaxial deformation experiment. The state of deviatoric stress of the cylinder in **a** is identical to that of the cylinder in **b**, although the total state of stress is undefined in **a**. However **a** is consistent with the pressure inside both cylinders being 0.5 MPa, although both σ_d and σ_1 are 1.5 MPa. In **a** the state of stress is illustrated in terms of the components of the deviatoric stress tensor. In **b** in terms of the uniaxially applied stress. Because the experiment is uniaxial, all illustrated stress components are also principal stress components both for the deviatoric stresses in **a** and the applied stress in **b** (after Engelder 1994)

smallest is $\tau_2 = -\sigma_d/2$. We can summarize this information with a more applied example of a continent that is under horizontal compression where the principle components of stress σ_{xx} and σ_{zz} are parallel to the horizontal (x) and vertical (z) directions. There, the principle horizontal deviatoric stress is:

$$\tau_{xx} = \sigma_{xx} - \sigma_m = \frac{\sigma_{xx} - \sigma_{zz}}{2} = \frac{\sigma_d}{2} \quad . \tag{5.12}$$

Strength. *Strength, failure strength* or *shear strength* are terms used to describe the critical value which the differential stress must reach to cause permanent deformation. Elastic stresses lead *not* to permanent deformation, but the strain that is caused by elastic processes is so small that elastic processes are often ignored in geodynamics (however it *is* important in seismology, as well as in some important geomorphological questions, s. sect. 4.2.2). For most problems of the field geologist however, we are interested in the stresses that lead to the deformation that we can observe in the field. In the *brittle* regime, this value depends directly on the magnitude of the principal stresses and is given by the stress where the curve on Fig. 5.2a deviates from its linear course. In the *viscous* regime, all differential stresses will lead to permanent deformation (Fig. 5.2). In this book we use the terms "strength" and "differential stress" synonymously as we assume that viscous deformation dominates the deformation mechanisms in the lithosphere. We therefore use the terms in the following way:

$$\text{strength} = \sigma_d \quad . \tag{5.13}$$

Figure 5.4. Different surface forces acting on a unity cube in the z direction. If z is the vertical, then there is also a body force due to gravity of the magnitude ρg. At rest, each of the three labeled forces is compensated by a force of equal magnitude but opposite direction. Other forces in the y- and x-directions are not labeled for clarity

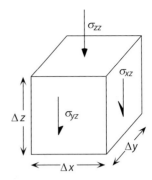

In the literature both terms are often confused with deviatoric stress (s. Engelder 1994). In order to avoid this confusion and the misuse of the term *deviatoric stress* it is useful to study the example in Fig. 5.3.

Stress Balance. The equations describing the balance of stresses are the basics for all mechanical descriptions of deformation. A stress balance is a generalized form of Newton's second law:

$$\text{force} = \text{mass} \times \text{acceleration} \; . \tag{5.14}$$

Eq. 5.14 has its only complications in that it is a vector equation, because force is a vector. That is, it consists of three equations each of which describe a force balance in one of the three spatial directions. Also, within each of these equations, several *surface* and *body forces* must be summed up and set equal to the product of mass times acceleration on the right hand side of the equations. Also note that the equations of force balance are generally considered per unity volume so that eq. 5.14 is usually written in terms of force/volume = density × acceleration. The different forces acting in the z direction can be summed up from Fig. 5.4. The sum is:

$$\left(\sigma_{zz} + \frac{\partial \sigma_{zz}}{\partial z}\Delta z\right)\Delta x \Delta y - \sigma_{zz}\Delta x \Delta y + \left(\sigma_{xz} + \frac{\partial \sigma_{xz}}{\partial x}\Delta x\right)\Delta y \Delta z$$

$$-\sigma_{xz}\Delta y \Delta z + \left(\sigma_{yz} + \frac{\partial \sigma_{yz}}{\partial y}\Delta y\right)\Delta x \Delta z - \sigma_{yz}\Delta x \Delta z - \rho g \Delta x \Delta y \Delta z$$

$$= \rho \frac{\partial u_z}{\partial t}\Delta x \Delta y \Delta z \; , \tag{5.15}$$

where x, y and z are the three spatial directions, ρ is density, g is acceleration and u is velocity. Even if this equation appears enormously complicated, it should be easy to follow it using Fig. 5.4. We can see that the equation has six similar looking terms on the left hand side. Every group of two terms describes a difference between the force on one side of the cube (e.g. the 2nd term in eq. 5.15: $\sigma_{zz}\Delta x \Delta y$) and the force on the opposite side of the cube (e.g. the 1st term in eq. 5.15: $(\sigma_{zz} + (\partial \sigma_{zz}/\partial z)\Delta z)\Delta x \Delta y$). If these are equally large, then the body does not accelerate. If this difference is finite, then the body accelerates with the rate written on the right hand side of the equation: $\rho \frac{\partial u_z}{\partial t} \Delta x \Delta y \Delta z$.

In most geodynamic problems, acceleration is assumed to be negligible or equal to zero as any change in velocity occurs over a very long time scale. In the horizontal cases then the term $\rho \frac{\partial u_x}{\partial t} = \rho \frac{\partial u_y}{\partial t} \to 0$ and in the vertical case $\rho \frac{\partial u_z}{\partial t} \to \rho g$, as gravitational force is still felt as a body force. Thus, eq. 5.15 simplifies to the following:

$$\frac{\partial \sigma_{zz}}{\partial z} + \frac{\partial \sigma_{xz}}{\partial x} + \frac{\partial \sigma_{yz}}{\partial y} - \rho g = 0 \quad . \tag{5.16}$$

Eq. 5.16 describes the equilibrium of stresses in the vertical direction and is generally applicable in the earth sciences. The first three terms of this equation are the sum of the *surface forces* acting in the z-direction, the fourth term is the *volume* or *body force* downwards. In analogous equations for the x- and y-direction this fourth term does not appear. The relationships for the x- and y-directions are:

$$\frac{\partial \sigma_{xx}}{\partial x} + \frac{\partial \sigma_{yx}}{\partial y} + \frac{\partial \sigma_{zx}}{\partial z} = 0 \tag{5.17}$$

and:

$$\frac{\partial \sigma_{yy}}{\partial y} + \frac{\partial \sigma_{xy}}{\partial x} + \frac{\partial \sigma_{zy}}{\partial z} = 0 \quad . \tag{5.18}$$

Eqs. 5.16, 5.17 and 5.18 are the basics of all mechanical equilibria discussed in this book (e. g. sect. 4.2.1; 6.2.2, s. Problem 5.5).

The Difference Between Lithostatic and Non-Lithostatic. The pressure measured by petrologists with geobarometers in metamorphic rocks is generally interpreted as the "burial pressure", that is, the pressure is directly correlated with the depth of the rocks at the time of metamorphism according to eq. 7.1 (s. sect. 7.1.1). This interpretation is based on the assumption that rocks have negligible strength, i.e. they cannot support any differential stress. This state of stress is called *lithostatic*. In this state, the *lithostatic pressure* is of the same magnitude as each of the principal stresses (eq. 5.7). The state of stress is isotropic. An every day example is the state of stress inside fluids which have negligible strength, for example a glass of beer. The force exerted by the beer onto the outside glass is exactly as large as the weight of the vertical section of beer lying above this point (s. Problem 5.6). However, if we consider a more general state of stress (i.e. a material that *can* support differential stresses), then we can see that only part of this pressure is caused by depth. How large the non depth related contribution to pressure is depends on the directions of σ_1, σ_2 and σ_3 as well as on the magnitude of the difference $\sigma_1 - \sigma_3$ and has long been a much debated topic of geodynamics. In the widest sense, this discussion can be summarized under the term "tectonic overpressure" (Rutland 1965, Ernst 1971) and will be discussed in a bit more detail in s. sect. 6.3.5.

For some special orientations of a general stress field it is possible to divide pressure into a lithostatic and a non lithostatic component. Such a division helps to illustrate the different contributions to pressure and allows us to estimate the magnitude of differential stresses under different boundary conditions. In a stress field where σ_1 and σ_3 are the maximum and minimum principle stresses we can write:

$$P = \frac{\sigma_1 + \sigma_3}{2} = \sigma_3 + \frac{\sigma_1 - \sigma_3}{2} = \sigma_{\mathrm{lith}} + \frac{\sigma_{\mathrm{d}}}{2} = \rho g z + \tau_1 \quad . \tag{5.19}$$

There, σ_{lith} is the component of pressure caused by the weight of the overlying rock column, and the non-lithostatic component is given by the largest principal component of the deviatoric stress tensor. Of course, eq. 5.19 is only valid if $\sigma_2 = (\sigma_1 + \sigma_3)/2$. For other values of σ_2, or for differently oriented stress fields, this simple subdivision in lithostatic and non-lithostatic terms of pressure is not possible and non-lithostatic components of pressure can only be calculated from the complete tensor. In sect. 6.3.5 we discuss how large such tectonic contributions to pressure might be.

5.1.2 Deformation Laws

When describing deformation of rocks mechanically we need to invoke a mathematical rule that relates stress (or force) to strain (or strain rate). Such a relationship is called a *flow law, deformation law* or *constitutive relationship*. If we know the flow law for a given rock, then we can use the relationships of mechanical equilibrium (eqs. 5.15 to 5.18) to describe the deformation of rocks in response to an applied force.

On a microscopic scale, structural geologists discriminate between a large number of deformation laws. However, on geological time scales and lithospheric length scales, deformation mechanisms may be summarized into two major groups (Table 5.1):

1. elastic deformation and
2. viscous (ductile) deformation.

These two deformation mechanisms are fundamentally different:

- In elastic deformation the *strain* of a rock is proportional to the applied *total stress.*
- In viscous deformation the strain *rate* of a rock is proportional to the applied *deviatoric stress.*

Viscous deformation is often also called "ductile" deformation. However, the term "ductile" is also often used for other non-fracturing deformation mechanisms, for example for plastic deformation. Thus, it is more precise to

Table 5.1. Deformation laws

brittle		no deformation law but a stress state; usually described with plastic law	
plastic	(ductile)	constant stress; example: sand	
viscous	(ductile)	stress and strain rate are proportional	- linear (Newtonian) - non linear (power law)
elastic		stress and strain are proportional	

use the term "viscous", although in this book we use "viscous" and "ductile" synonymously.

We all have every day encounters with these two deformation mechanisms, namely with rubber bands (elastic) and mixing cake dough (viscous). When stretching a rubber band, the amount of stretch depends on how hard we pull. The more pull, the more stretch. The applied stress and the resulting strain are proportional. On the other hand it does not matter at all whether we pull fast or slow. The stretch is always the same for the same applied force, independent on the speed (strain rate) with which we do the experiment. With mixing dough its exactly opposite. It does not matter at all whether we mix it only a bit or very thoroughly (little or much strain in the dough), the needed force is always the same. However, how much force we need depends very strongly on the mixing rate. If we mix it rapidly, we need much more force than if we mix slowly.

Aside from elastic and viscous deformation there are *brittle* deformation and *plastic* deformation that are often listed as deformation mechanisms. However, strictly speaking, brittle and plastic deformation only describe a state at which deformation occurs, but not the degree of deformation (neither strain nor strain rate). Therefore, strictly speaking, plastic and brittle deformation describe no constitutive relationship. For example, the Mohr Coulomb criterion only describes the stress state at which failure occurs, but not the amount of strain that occurs. *Plastic* deformation describes the stress state at which flow occurs but not the strain rate (s. e. g. Weijermars 1997; Twiss and Moores 1992, Jaeger and Cook 1972). In the following we present a brief repetition of the deformation laws needed for the understanding of the rheology of the lithosphere and the flexural equilibria as discussed in chapter 4.2.2.

Elastic Deformation. Elastic deformation is characterized by a proportionality between stress and strain (Fig. 5.2). Both these parameters are described by tensors which each include 6 independent values. However, if this proportionality is ideally linear, and one-dimensional, (i.e. during uni-

Figure 5.5. Stretch of a cube as the consequence of compression in the vertical direction.
The Poisson constant is defined as $\nu = -e_3/e_1$

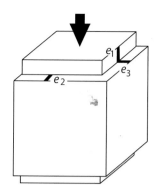

axial loading) then this relationship is called *Hook's law* and may be simply
written as:

$$\sigma = E\epsilon \ . \tag{5.20}$$

When the deformation is uni-axial, then ϵ is the (dimensionless) longitudinal normal strain and is defined as the change in length during deformation
relative to the original length (s in eq. 5.1). The proportionality constant E
is called the *Young's modulus* and has the units of stress ($\mathrm{N\,m^{-2}}$). The Young
modulus corresponds to the slope of the elastic section of the curve in Fig. 5.2.
For rocks, the Young's modulus is of the order of 10^{10} to 10^{11} Pa.

If more than one of the three principal stresses is larger than zero, then it
is also important to consider that rocks are compressible. This is described
by the Poisson constant ν. For the largest principal stress we can write:

$$\sigma_1 = \epsilon_1 E + \nu\sigma_2 + \nu\sigma_3 \tag{5.21}$$

or, if strain is written as a function of stress:

$$\epsilon_1 = \frac{1}{E}\sigma_1 - \frac{\nu}{E}\sigma_2 - \frac{\nu}{E}\sigma_3 \ . \tag{5.22}$$

For the other two spatial directions equivalent equations may be formulated. The Poisson constant is given by the ratio of two stretches, namely the
infinitesimal strain normal to the applied stress and the stretch *in* direction
of the applied stress (Fig. 5.5).

During compressive deformation, a rock will shorten in the direction of the
applied stress. Thus, the incremental stretch e_1 in Fig. 5.5 is negative. If the
rock is isotropic, then this shortening is distributed evenly between expansion in the other two spatial directions. Thus, $\nu = +0.5$ for incompressible
materials. For example, rubber is almost incompressible and has a Poisson
constant of almost $\nu = +0.5$. In contrast, the Poisson constant of rocks is
of the order of 0.1–0.3. We can see that rocks are quite compressible in the
elastic regime. However, the total strains of rocks in the elastic regime are
quite small, because the Young's modulus of rocks is very large.

While the Poisson constant is directly related to how compressible a rock is, it should not be confused with the *compressibility* β. Under isotropic stress (with pressure P) a rock will compress isotropically (i.e. $\epsilon_1 = \epsilon_2 = \epsilon_3$) and

$$P = \frac{1}{\beta}(\epsilon_1 = \epsilon_2 = \epsilon_3) \quad . \tag{5.23}$$

β is related to Young's modulus and the Poisson constant by:

$$\beta = \frac{3 - 6\nu}{E} \quad . \tag{5.24}$$

We can see from this equation that, for incompressible materials where $\nu = +0.5$ the compressibility becomes $\beta = 0$. The inverse of the compressibility is called the *bulk modulus*: $K = 1/\beta$. The derivation of eq. 5.24 as well as a more rigorous treatment of the theory of elasticity will not be performed here and the reader is referred to the literature.

Brittle Fracture. When the stresses applied to rocks cannot be compensated elastically, permanent deformation will occur. This may occur by ductile or brittle processes. Among brittle processes, two different modes of brittle deformation may be discerned: rocks deform either by creating new cracks, or by friction along existing fractures. In both cases the friction along the failure planes plays a critical role. Brittle failure is commonly described with the *Mohr-Coulomb-criterion*. However, it should be said here that – strictly speaking – the Mohr-Coulomb-criterion describes only a state of stress, namely the critical state at which failure occurs. It does not place stress and strain in a relationship to each other and is therefore not a constitutive relationship or flow law.

• *Mohr-Coulomb-criterion.* Coulomb (1773) was the first to recognize that the brittle strength of materials is largely a linear function of the applied normal stress σ_n and that it depends only to the second order on a material constant called cohesion σ_0. At geological stresses cohesion is largely negligible. According to the Coulomb criterion, failure occurs when the shear stress on a given plane reaches a critical value τ_c that is a function of the normal stress acting on that plane σ_n, as:

$$\tau_c = \sigma_0 + \mu\sigma_n \quad . \tag{5.25}$$

The coefficient that relates shear stress and normal stress on a failure plane is called the *internal coefficient of friction*. This coefficient is dimensionless. According to the Coulomb criterion (eq. 5.25) brittle deformation is a nearly linear function of total stress. It is independent of temperature or strain rate $\dot{\epsilon}$ and almost independent of the material as the cohesion is almost negligible and the internal coefficients of friction are very similar for most rocks (Byerlee's law).

Mohr (1900) then discovered that the failure criterion of Coulomb may be elegantly portrayed graphically. His graphical analysis is called the Mohr diagram. In the Mohr diagram shear stresses are plotted against normal stresses and the stress state in a rock is plotted as a circle. Fig. 5.6 shows that the shear stresses in a rock σ_s are a function of the angle between the considered plane and the principal stresses. This may be formulated in terms of the equation:

$$\sin(2\theta) = \frac{2\tau_c}{(\sigma_1 - \sigma_3)} \quad, \tag{5.26}$$

where the angle θ is the angle between any considered plane in a rock and the principal stress directions. From eq. 5.26 we can see that σ_s is the largest on planes that lie at an angle of 45° to the principal stress direction:

$$\tau_{\max} = \frac{\sigma_1 - \sigma_3}{2} \quad. \tag{5.27}$$

Thus, the maximum shear stress a rock can support is half as large as the applied differential stress. However, it is important to note that the *largest* shear stress is *not* where failure occurs. From Fig. 5.6 we can see that the normal stress at $\tau_{\max}$ is so much larger than the normal stress at τ_c that the plane is inside the failure envelope.

The slope of the tangent to the Mohr circles in Fig. 5.6 is given by the internal *angle of friction*. This angle of friction ϕ and coefficient of friction μ are related by:

$$\tan\phi = \mu \quad. \tag{5.28}$$

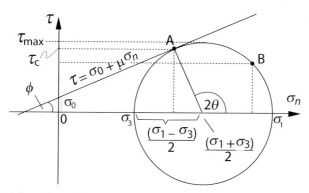

Figure 5.6. The relationship between normal stress (horizontal axis) and shear stress (vertical axis) in the Mohr circle. Note that we now use τ to denote shear stress and not for the deviatoric stress tensor as we did in eq. 5.9. The normal stresses are compressive (positive) to the right of the origin and tensional (negative) to the left. Eq. 5.25 describes the tangent to the Mohr circles drawn around centers at $(\sigma_1 + \sigma_3)/2$ with the radius $(\sigma_1 - \sigma_3)/2$. In the ductile regime shear stresses do not increase linearly with normal stresses anymore. However, for many rocks the curve is even in the brittle regime not completely linear, but slightly concave against the normal stress axis. For failure planes in a rock it is true that: $\tau_c = \sin 2\theta(\sigma_1 - \sigma_3)/2$ and $\sigma_n = (\sigma_1 + \sigma_3)/2 - \cos(2\theta)(\sigma_1 - \sigma_3)/2$

For most rocks this angle is about 30–40°, which is equivalent to an internal coefficient of friction between roughly 0.6 and 0.85. This relationship is called Byerlee's law as he was the first to measure μ on a crustal scale and derived ϕ from μ.

If fluid pressure plays a role, then the Mohr-Coulomb criterion is often formulated as:

$$\tau_c = \sigma_0 + \mu\,(\sigma_n - P_f) \quad , \tag{5.29}$$

where P_f is the pore fluid pressure. This can be approximated as $\tau_c = \sigma_0 + \mu\sigma_n(1 - \lambda)$, where $\lambda = P_f/\sigma_L$ is the ratio of pore fluid pressure to lithostatic stress, σ_L, if $\sigma_n \approx \sigma_L$. If both are of the same magnitude, then $\lambda = 1$ and the shear stress necessary for failure is only a function of the cohesion. If there is no fluid, then eq. 5.29 reduces to eq. 5.25.

• *Byerlee's and Amonton's laws.* If preexisting cracks occur in a rock then there is no cohesion. To be more precise: the remaining cohesion is negligible compared to the cohesion of an intact rock. The shear stresses needed to deform a rock only need to overcome the coefficient of friction and the normal stresses applied to the rock. Eq. 5.25 simplifies to:

$$\tau_c = \mu\sigma_n \quad . \tag{5.30}$$

This equation is usually called *Amonton's law.* Byerlee (1968, 1970) showed empirically that, at pressures below 200 MPa, (roughly less than 8 km) the crust may be characterized by an internal coefficient of friction around 0.85:

$$\tau_c = 0.85\sigma_n \quad . \tag{5.31}$$

At larger depths, but above the brittle ductile transition brittle failure in the crust appears to be best described by:

$$\tau_c = 60\,\mathrm{MPa} + 0.6\sigma_n \quad . \tag{5.32}$$

These empirical relationships are called *Byerlee's laws* (Fig. 5.7). Because of the fact that $\mu \approx 0.85$, most faults occur at 30 degrees angle to the maximum principle stress. Byerlee's laws state that rocks at 5 km depth will fail at roughly 110 MPa, in 10 km depth at roughly 230 MPa and in 15 km depth at about 300 MPa. If we want to consider fluid pressure as well we can reformulate Byerlee's laws to:

$$\tau_c = 0.85\sigma_n(1 - \lambda) \quad \text{and} \quad \tau_c = 60\,\mathrm{MPa} + 0.6\sigma_n(1 - \lambda) \quad . \tag{5.33}$$

• *Anderson's theory.* Byerlee's laws describe the relationship between shear stresses (in particular τ_c) and normal stresses (which relate to depth in the crust) in general, but they do not explain the spatial orientation of failure planes relative to the *principal stresses.*

Anderson reformulated the Mohr Coulomb law in terms of differential stress $(\sigma_1 - \sigma_3)$ and lithostatic stress $(\sigma_L = \rho g z)$ instead of shear stress

Figure 5.7. Brittle failure as a function of depth and normal stress or "lithostatic pressure", as calculated with 5.34, 5.35 and 5.36

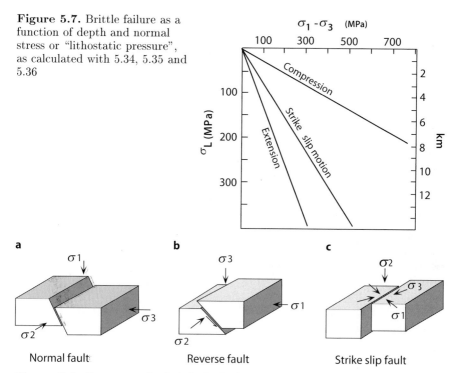

Figure 5.8. Geometry of a brittle fault during: **a** extension, **b** compression and **c** strike slip motion

and normal stress. He took three cases to represent reverse faults, normal faults and strike slip faults and assumed that $\sigma_L = \sigma_3$, $\sigma_L = \sigma_1$ and $\sigma_L = \sigma_2 = 0.5(\sigma_1 + \sigma_3)$ for each of these cases, respectively (Fig. 5.8). The relationship for each case can be directly derived from Fig. 5.6 using trigonometric relationships (e. g. Weijermars 1997). They may be written as follows:

$$\text{reverse faults}: \quad \sigma_1 - \sigma_3 = \frac{2(\sigma_0 + \mu\sigma_L(1 - \lambda))}{\sqrt{\mu^2 + 1} - \mu} \quad , \tag{5.34}$$

$$\text{normal faults}: \quad \sigma_1 - \sigma_3 = \frac{-2(\sigma_0 - \mu\sigma_L(1 - \lambda))}{\sqrt{\mu^2 + 1} + \mu} \quad , \tag{5.35}$$

$$\text{strike slip faults}: \quad \sigma_1 - \sigma_3 = \frac{2(\sigma_0 + \mu\sigma_L(1 - \lambda))}{\sqrt{\mu^2 + 1}} \quad . \tag{5.36}$$

Fig. 5.7 shows these three linear relationships for $\mu = 0.85$ and $\sigma_0 = 0$.

• *Interpretation of fault orientations.* The spatial orientation of faults in the earth's crust is one of the most important pieces of evidence for the interpretation of the magnitude and geometry of the stress field in the lithosphere

(s. sect. 6.2; Zoback 1992). If the orientation of faults can be measured directly in outcrop, then there is a series of statistical methods that may be used to derive the palaeostress field at the time of fault formation (Angelier 1984; Angelier 1994). For the interpretation of the *present day* stress field of the lithosphere the most important methods are:

- Interpretation of seismic data,
- Interpretation of bore hole break outs (e. g. Bell and Gough 1979; Mastin 1988; Wilde and Stock 1997),
- direct in-situ- measurement of stresses (Zoback and Haimson 1983),
- Interpretation of GPS data (Global Positioning System; e. g. Argus and Heflin 1995),
- interferometric methods (e. g. Molnar and Gibson 1996).

The application of interferometric methods has only been possible since they are accurate enough to measure plate motions directly. However, strictly speaking, these methods (and those using GPS data) measure motions rather than stresses.

The most reliable data source for the interpretation of the intra plate stress field remains the interpretation of seismic data, in particular because it allows to characterize processes that occur deep inside the lithosphere (Fig. 2.2). Thus, they can not only be used to determine the stress field in two dimensions, but also its depth dependence on a plate scale. Seismic data are commonly interpreted with the aid of *fault plane solutions* (s. e. g. Michael 1987; McKenzie 1969a). In the following paragraphs we explain fault plane solutions to the extent that a tectonically interested field geologist can read them.

Fault plane solutions illustrate the qualitative direction of motion (*polarity*) of seismic P-waves (primary longitudinal waves) on a Schmidt net (s. Figs. 2.10, 5.11). In order to be able to do this, a seismic event must be recorded by a large number of seismic stations around the globe. Fig. 5.9 illustrates how such data are then used to plot a fault plane solution using a two-dimensional example. The gray shaded bar and the arrows in the middle of this figure symbolize a dextral fault, and we assume that the center of the seismic event lies in the middle of the figure. Points A to G are seismic stations in the region that have recorded the event. The very first motion along the fault as measured by the seismic stations is either tensional or compressional. According to their position relative to the epicenter, the stations A, B and F will register a *tensional* first motion, while stations G and D will register a *compressional* first motion. Stations C and E will record no P-waves at all. These results are illustrated by the clove leaf around the center of the diagram. Tensional regions are colored white and characterized by a "–", compressional regions are drawn black and characterized by a "+". The distance of the clove leaf circumference from the middle indicates the inten-

Figure 5.9. Polarity and intensity of P-waves in two dimensions. The two planes along which there are no P-waves are called nodal lines or nodal planes

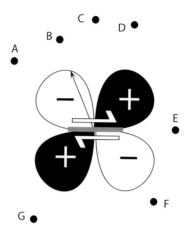

sity of the measured P-waves. The arrow from the origin towards station B is a vector indicating direction and intensity of the P-waves measured at B.

For the interpretation of a fault in three-dimensional space, we cannot restrict ourselves to the two-dimensional cartoon of Fig. 5.9 and it is necessary to use data from seismic stations around the globe. Fig. 5.10a shows a schematic cross section through earth. Now consider a seismic event that occurs at point P. From there, P-waves will begin to propagate in all directions. Because of the refraction of waves through the interior of the globe, the wave measured at point S will have departed from point P almost straight downwards. In order to know *exactly* which direction this wave has left from point P, we need to know the curvature of the wave through earth. However, this is known and may be looked up in seismic tables. It is therefore possible to plot the relative position of all seismic stations that have recorded the event into a Schmidt net. All stations in the vicinity of the epicenter (i.e. on a tangential plane to earth at the epicenter) will plot near the circumference of the net. For these stations we could use the two-dimensional illustration of Fig. 5.9. Fig 5.10b is an enlarged section from Fig. 5.10a. The polarity of the P-waves on the projection hemisphere is indicated by the light and dark shading.

Fig. 5.11a shows the fault plane solution for the seismic event in the three-dimensional cartoon in Fig. 5.10b. Note that the solution has two alternative interpretations. Firstly, Fig. 5.11a may be interpreted as a normal fault with a steep eastwards dip. Then, the boundary between the right hand black colored region and the central white region is the fault plane (plane A in Fig. 5.10b) and the boundary between the left hand black region and the white region is the normal to this plane (*auxiliary plane*; plane B in Fig. 5.10b). However, the figure may alternatively be interpreted as a normal fault with a shallow westward directed dip. All fault plane solutions have two alternative interpretations for the orientation of the fault they represent. Fig. 5.11b is the fault plane solution of the reverse fault of the same orientation as in a and

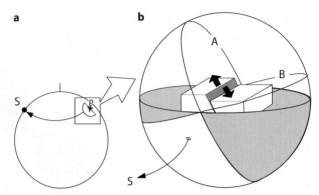

Figure 5.10. a Schematic cross section through the earth. Point P marks the occurrence of a seismic event. Point S marks a seismic station that has recorded the event. **b** Enlarged section from **a**. The polarity of all P-waves emanating downwards from point P is plotted in a Schmidt net surrounding the fault. The fault itself intersects the Schmidt net along line A. Plane B is the normal to this plane and to the movement direction. A and B are called nodal planes

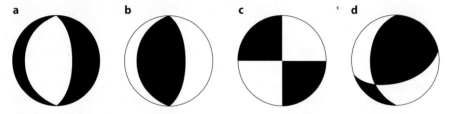

Figure 5.11. Some examples of fault plane solutions. **a** Fault plane solution for a normal fault that dips steeper than $45°$ towards the right (east) *or* shallower than $45°$ towards the left (west). **b** Fault plane solution for a reverse fault of the same orientations as in **a**. **c** Fault plane solution of a north south striking dextral- *or* an east-west striking sinistral strike slip fault. **d** Fault plane solution for a reverse fault that is inclined with roughly $45°$ towards the left (west) and has a dextral strike slip component in its motion. Alternatively, this fault plane solution could be for a reverse fault that is inclined towards the south east and contains a sinistral component. Note that the polarity of the center of the fault plane solution always indicates the overall kinematics of the structure (i.e. extensional vs. compressional)

Fig. 5.11c is the fault plane solution of a strike slip fault. Fig. 5.11d is the fault plane solution for a fault containing reverse thrust as well as strike slip components in its motion. Fault plane solutions as illustrated in Fig. 5.11 are a common means of interpretation of geodynamic processes in active orogens (Molnar and Lyon-Caen 1989) and plate boundary processes (s. e. g. Frohlich et al. 1997).

Viscous Deformation. On the scale of a thin section, rocks behave not viscously but according to a large range of deformation mechanisms. The dependence of deformation mechanism on the physical conditions may be portrayed in *deformation mechanism maps* (Frost and Ashby 1982). In such maps parameters like temperature, grain size, stress and viscosity are plotted against each other and the diagrams a divided into different fields where different mechanisms apply. However, on a larger scale, for example the scale of the crust, it is useful to average different deformation mechanisms and assume that rocks behave like viscous fluids and deform according to the laws of viscosity. Viscous deformation of ideal fluids is described by a proportionality between deviatoric stress and strain rate:

$$\tau = 2\eta\dot{\epsilon} \ . \tag{5.37}$$

The proportionality constant η is called the *dynamic viscosity*. It has the units of Pascal times second (Pa s) or $kg\,m^{-1}\,s^{-1}$. The viscosity of air is roughly 10^{-5} Pa s, the viscosity of water roughly 10^{-3} Pa s, the viscosity of ice roughly 10^{10} Pa s, of salt 10^{17} Pa s and of granite it is roughly 10^{20} Pa s. If η is constant with respect to strain rate then eq. 5.37 is linear. A fluid that behaves according to a linear relationship is called a *Newtonian fluid*. Eq. 5.37 states that the larger the deviatoric stress that is applied, the faster the rock will deform. Note that, in the orientation of the *maximum* shear strain rate, the stress in eq. 5.37 will be $\tau_{\max}$, which is equivalent to the differential stress $\sigma_d = (\sigma_1 - \sigma_3)$, (s. eq. 5.27).

• *A factor two.* The factor 2 in eq. 5.37 results from the definition of strain rate that is generally used in geodynamics. It is worth noting that in fluid dynamics a definition of strain rate $\dot{\gamma}$ is used that is twice that used here: i. e. $\dot{\gamma} = 2\dot{\epsilon}$. Using this definition would result in the equation describing viscous deformation as: $\tau = \eta\dot{\gamma}$. Both definitions may be found in the literature. A more detailed discussion of different definitions for strain rate may be found in Ranalli (1987).

• *The Arrhenius relationship.* Viscosity is extremely strongly temperature dependent. This temperature dependence is described by the *Arrhenius relationship*:

$$\eta = A_0 e^{Q/RT} \ . \tag{5.38}$$

In this relationship the constants A_0 and Q are material-specific constants called the *pre exponent constant* and the *activation energy* (in $J\,mol^{-1}$), respectively. The parameter R is the universal gas constant and T is the absolute temperature. If we try to read eq. 5.38 we can see that it states that the viscosity of any material will trend towards infinity at absolute zero and will decrease exponentially from there to asymptotically approach the value A_0 at very high temperatures. The Arrhenius relationship will be discussed in some more detail in sect. 7.2.2 in connection with eq. 7.4. In the Arrhenius relationship absolute temperature must be always used, to be compatible with the way the material constants are usually formulated. The *kinematic viscosity* is the ratio of dynamic viscosity and density and has the units of diffusivity, namely $m^2\,s^{-1}$.

Rocks rarely deform as a Newtonian fluid (i.e. there is rarely a linear relationship between the applied deviatoric stress and strain rate). In general, rocks deform roughly 8 times as rapid if the applied force is doubled. This may be written in terms of the power law:

$$\tau^n = A_{\text{eff}}\dot{\epsilon} \ . \tag{5.39}$$

There, the exponent n is called the *power law exponent* which is a material constant and is between 2 and 4 for many rocks types. The parameter A_{eff} is a material constant. It is analogues to η in eq. 5.37, but does not have the units of viscosity and we therefore use a different symbol. A_{eff} has the units $\text{Pa}^n\,\text{s}$. However, in analogy to a Newtonian fluid, it is possible to derive an *effective viscosity* from eq. 5.39 which is given by the ratio of deviatoric stress and strain rate. This is:

$$\eta_{\text{eff}} = \frac{\tau}{2\dot{\epsilon}} = \frac{1}{2}A_{\text{eff}}^{-1/n} \times \dot{\epsilon}^{((1/n)-1)} \ . \tag{5.40}$$

If we want to apply eq. 5.39 to rocks it is useful to couple it with the Arrhenius relationship (eq. 5.38). However, because of the difference between *dynamic* viscosity (eq. 5.38) and *effective* viscosity (eq. 5.40) this coupling is not trivial.

In general, the rheology of rocks that do not behave like a Newtonian fluid is described empirically. Such an empirical relationship is called *Dorn's law* and is constrained with deformation experiments. Because of the experimental set up that is generally used (where the maximum shear stress is measured; s. eqs. 5.37, 5.27), Dorn's law is often formulated in terms of differential stress rather than deviatoric stress:

$$\dot{\epsilon} = \sigma_d^n A e^{-\frac{Q}{RT}} \qquad \text{or}: \qquad \frac{\sigma_d}{2} = \frac{1}{2}\left(\frac{\dot{\epsilon}}{A}\right)^{(1/n)} e^{\left(+\frac{Q}{nRT}\right)} \ . \tag{5.41}$$

For exponents larger than 1, Dorn's law is also called simply *power law*. Note that the constant A has - in contrast to A_{eff}, the units of $\text{Pa}^{-n}\,\text{s}^{-1}$. Many rheological experiments have been calibrated to obtain material constants in terms of the three constants Q, A and n in eq. 5.41 (e.g. Gleason and Tullis 1995).

For mechanical models in which temperature is not considered explicitly, it is useful to summarize the temperature dependent terms of eq. 5.41. We will discuss this in some more detail in sect. 6.2.2. Fig. 5.12a shows the differential stress as a function of temperature and strain rate for the material constants of quartz. Fig. 5.12b shows differential stress as a function of activation energy and pre exponent constant at fixed temperatures and strain rates.

Dorn's law is an empirical deformation law and in some cases it is necessary to modify it empirically. One example where this is necessary is the deformational behavior of olivine. Fig. 5.12b shows that olivine deforms at $500\,^\circ\text{C}$ (which may be a realistic assumption for the Moho-temperature) only at unrealistically high stresses around 10^7 MPa if it were described with eq. 5.41.

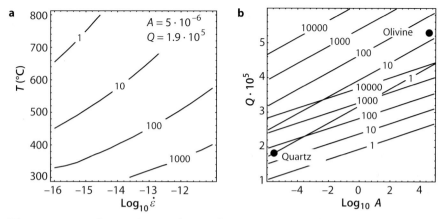

Figure 5.12. Differential stress (in MPa) during viscous deformation as a function of a range of parameters as calculated with eq. 5.41. A power law exponent of $n = 3$ was assumed. **a** Differential stress as a function of temperature and strain rate for the material constants of quartz. **b** Differential stress as a function of activation energy Q (in $J\,mol^{-1}$) and pre exponential constant A (in $MPa^{-3}\,s^{-1}$). Continuous lines are for $500\,^\circ C$, dashed lines are for $1\,000\,^\circ C$. The assumed strain rate is $\dot\epsilon = 10^{-13}\ s^{-1}$. The rheological data for quartz and olivine from sect. 5.2.1 are plotted

Table 5.2. Dependence of brittle and viscous deformation on some physical parameters

dependent on	brittle	viscous
total pressure (depth)	yes (linear)	no
material	no	yes ($\approx$ power of 3)
strain rate	no	yes
temperature	no	yes (exponential)

Thus, Goetze (1978) suggested that a better description of the behavior of olivine above 200 MPa is given by the relationship:

$$\sigma_1 - \sigma_3 = \sigma_D \left(1 - \sqrt{\frac{RT}{Q_D}\ln\left(\frac{\dot\epsilon_D}{\dot\epsilon}\right)}\right) \quad . \tag{5.42}$$

There, Q_D is again an activation energy, σ_D a critical stress that must be exceeded and $\dot\epsilon_D$ is the critical strain rate (s. Table 5.3). Comparing eq. 5.42 with eq. 5.41 shows that this law is by far not as temperature dependent as the power law. Combinations of eqs. 5.41, 5.42 and Byerlee's laws form the basics of many simple quantitative models describing the rheology of the lithosphere as a whole (Brace and Kohlstedt 1980).

Eq. 5.41 shows that the stresses during viscous deformation are strongly dependent on *temperature, strain rate* and *material constants*, but are independent of the confining pressure. Thus, ductile deformation is subject to completely different laws than brittle deformation (Table 5.2).

5.2 Rheology of the Lithosphere

Rheology is the science of the *flow characteristics of materials*. In a more general sense, *rheology* is often used as a term to describe the deformational behavior of material (in the case of geologists: rocks), independent of whether the deformation is actually by *flow* or rather by brittle fracture or other deformation mechanisms. Rheology describes the relationships between forces and motions, and between stress and strain. As such, constitutive relationships form the basics for the all rheological questions (s. p. 205).

In the previous section we showed that two of the most important deformation mechanisms in the lithosphere, namely brittle and viscous deformation, depend on very different physical parameters (Table 5.2). This is the reason for strong rheological heterogeneieties in the lithosphere and also for the different rheological behavior of continental versus oceanic lithosphere. This will be the subject of the next sections.

5.2.1 Rheology of the Continental Lithosphere

In the late seventies of last century Brace, Goetze and others summarized much of the information from the previous sections to formulate a simple rheologial model for the lithosphere. This rheological model sketched in Fig. 5.13 and the following figures and will be the basis for our discussion in this book. Note that in these strength profiles (e.g. Figs. 5.13, 5.14 and 5.15) differential stress rather than deviatoric stress is usually plotted on the horizontal axis, because σ_d is a single scalar value that may be used to characterise the stress state. However, it is important to note throughout this chapter that σ_d corresponds only to the *maximum* deviatoric stress (s. discussion around eqs. 5.27 and 5.37).

The strength profiles in Fig. 5.13 consist of two different types of curves. The straight lines are for brittle fracture. They show increasing rock strength with increasing depth as the normal stresses in the crust increase with depth as shown in eq. 5.25 (Fig. 5.13b, s. sect. 5.1.2). The curved lines describe viscous deformation according to *Dorn's law*. The strength they describe decreases exponentially downwards, because temperature *increases* with depth roughly *linearly* (s. sect. 5.1.2) and viscosity for a given mineral decreases with temperature. It is important to note that these ductile shear strength curves are each for a given strain rate. A higher strain rate will yield a curve that has a higher strength at a given depth. Fig. 5.13b shows that, for a given strain rate, two different failure strengths may be associated with each depth. A rock at a given depth will always deform according to the deformation mechanism that requires *less* stress. Using this logic, we can draw

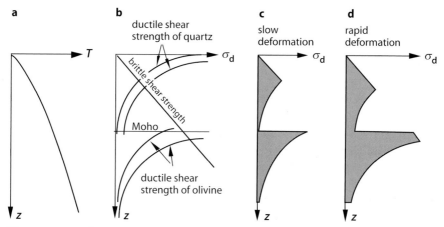

Figure 5.13. Schematic illustration of a *Brace-Goetze lithosphere*. **a** Temperature T as a function of depth z. This curve has the shape of a typical continental geotherm. **b**, **c** and **d** show shear strength as a function of depth. **b** Shear strength due to brittle failure (straight line) and viscous deformation (curved lines) for two different strain rates and the material constants for quartz and olivine. At any given depth, the curve with the higher strength corresponds to the higher strain rates. **c** and **d** Strength profiles constructed from **b** for low and high strain rates. Integrating the shaded area yields the vertically integrated strength. The units of this integrated strength are $\mathrm{N\,m^{-1}}$. This integrated strength may be interpreted as the force per meter length of orogen applied to the orogen in direction normal to the orogen (assuming the orogen is everywhere deforming). Note that the cartoon indicates that - at high strain rates (profile **d**)- the upper mantle will deform in a brittle fashion just below the Moho

strength profiles like those illustrated in Fig. 5.13c and 5.13d. We can see that the model predicts that, at shallow levels, rocks deform in a brittle fashion, as is familiar to us from the fact that rocks at near the surface break in brittle fashion or break (instead of flow) when we hit them with a hammer. At large depths and high temperatures, viscous deformation prevails as is familiar to us from the fact that high grade metamorphic rocks are often characterized by folding and other features indicating viscous (ductile) deformation. The depth at which brittle strength and viscous strength have the same magnitude is called the *brittle-ductile transition*. Note that the depth of this transition is strain rate dependent in this model.

In first approximation it is fair to assume that a rock will begin to deform when the rheologically weakest phase fails. As quartz is one of the softer minerals and most crustal rocks contain quartz, the ductile deformation of the crust may well be described with the rheological data for quartz. Rocks in the mantle part of the lithosphere are quartz absent and dominated by olivine. Therefore, Fig. 5.13 shows two pairs of curves for power law creep; one pair for the creep stresses of quartz at low and high strain rates, the other for the creep behavior of olivine at low and high stain rates. Together,

all these curves result in a strength profile for the continental lithosphere
that contains two strength maxima, one at mid crustal levels, the other in
the uppermost portions of the mantle part of the lithosphere. This extremely
simple model for the rheological stratification of the lithosphere is called a
Brace-Goetze lithosphere (after a suggestion by Molnar 1992). Before we begin
to describe the Brace-Goetze lithosphere quantitatively, we will now discuss
some qualitative features of this model.

Qualtitative Features of the Brace-Goetze Lithosphere. The model
of the Brace-Goetze lithosphere has a large number of features that are in
phenomenal correspondence with observations in nature. Before we determine
some of its characteristics quantitatively, we discuss some of its qualitative
features here.

• *Brittle failure in the mantle.* A comparison of Figs. 5.13c and d shows
an interesting qualitative difference between the two strength profiles. At
low strain rates the entire lithosphere below the brittle ductile transition
deforms viscously. However, at *large* strain rates, the ductile strength of the
upper mantle is larger than its brittle strength and the uppermost mantle
will fracture. Of course, the occurrence of brittle fracture in the upper mantle
depends on a large number of other factors as well. However, we want to note
that the brittle strength of the upper mantle is comparable to its viscous
strength at geologically realistic strain rates. Should it be true that the upper
most mantle deforms brittle under some circumstances, then this process

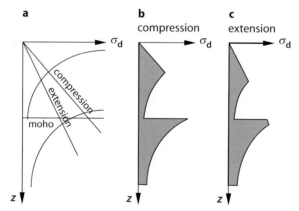

Figure 5.14. Schematic diagram showing the changes in mechanical strength of
a Brace-Goetze lithosphere when changing the deformation regime qualitatively,
i.e. from compression to extension. **a** The change from compression to extension
decreases the brittle failure strength (see. eq. 5.34 and eq. 5.35; Fig. 5.7), while
the viscous strength remains unaffected by this change, if the absolute value of the
strain rate remains constant. Potentially, this may be reflected in brittle failure of
the upper mantle. This is illustrated in the strength profiles in **b** and **c** that were
constructed from **a**: In **b** the entire mantle part of the lithosphere deforms viscously,
while in **c** the upper most mantle fails in brittle fashion

might have important consequences for the accumulation of underplates at the Moho (Huppert and Sparks 1988).

The transition from ductile to brittle failure in the upper most mantle may not only occur due to a change in strain rate (*increase* of viscous strength), but may also occur due to a *decrease* of the brittle strength. This may occur if there is a transition from compression to extension (sect. 5.1.2; Fig. 5.14, Sawyer 1985). Eq. 5.35 shows that the brittle strength of rocks is smaller in extension than it is in compression. It is therefore possible that brittle failure of the upper most mantle is caused by a qualitative change of the deformation regime.

- *Changes in the rheological stratification.* Changes in the strain rate of an orogen can also change the rheological stratification of the lithosphere. This is illustrated in Fig. 5.15 using a simple model lithosphere made up of three lithological layers. The figure shows that a change in the strain rate may change the rheological layering. At low strain rates there are three strength maxima, while at high strain rate there are only two (Fig. 5.15b,c). Such weak points may be the nucleus for the formation of a tectonic nappe boundary. Thus, it is possible that the thickness of nappes in a lithologically stratified crust is a function of the strain rate (Kuznir and Park 1986).

During viscous deformation it is not only changes in the strain rate that can change the strength of the lithosphere. Changing the geotherm may have the same influence. In the following we will encounter a series of examples where the shape of the geotherm is critical to the deformation mechanism. Other mechanisms that can cause changes in the strength of the lithosphere are, for example, strain hardening, or metamorphism.

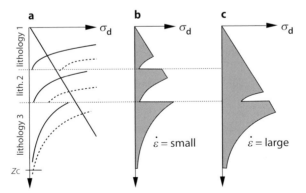

Figure 5.15. Schematic diagram showing how changes of the deformation rate can cause changes in the rheological stratification of the lithosphere. In the shown example the crust is assumed to consist of three rock strata with different viscous rheologies (e.g. the upper, middle and lower crust). During deformation at low strain rate the lithosphere has three strength maxima. If the strain rate increases, the central strength maximum disappears. In **a** the continuous lines are for low stain rate and the dashed curves for large strain rate

During metamorphism and deformation, both mineralogy and grain size change. It is therefore conceivable that a rock has a higher shear strength after metamorphism that it had before. For example, a garnet mica schist has a larger shear strength than its precursor: a clay. This is an interesting aspect which may be crucial in the consideration of postorogenic extension of mountain belts. In general it is thought that the stresses required for the late extension of an orogen are smaller than those required for its shortening. This is based on the fact that brittle deformation requires smaller stresses in tension than in compression. However, this is contrasted by the fact that the crust may have increased its strength by metamorphism by the time it wants to extend. It is therefore possible that *larger* rather than *smaller* stresses are required to extend an orogen, than to build it.

• *Rheology considering elastic criteria.* The Brace-Goetze lithosphere as we have discussed it in the last sections is based on the assumption that only viscous and brittle deformation mechanisms dominate its behavior. As such, it is in contrast to our assumption of the lithosphere as an elastic plate as we have done in sect. 4.2.2 when we have discussed flexural isostasy. Which model assumptions are made always depends on the question we are trying to answer (sect. 1.1).

Ranalli (1994) suggested to describe the rheology of the lithosphere using a coupled viscous, elastic and brittle approach. His model is illustrated in Fig. 5.16, but we need to be careful not to confuse it with the strength profiles in Figs. 5.13 or 5.14. Elastic deformation is instantaneous and does not reflect a strength envelop for a given strain rate as the viscous curves do. Thus, the model illustrated in Fig. 5.16 may be used to infer a stress state, but should not be interpreted as a failure envelope.

In a downward bent elastic lithosphere there is a stress neutral layer in the middle of the lithosphere. Above this point the lithosphere is under compression, below this point it is under extension (Fig. 4.20). These elastic stresses are shown in Fig. 5.16 by the straight line that goes from positive to negative stresses in the middle of the diagram. It may be seen that – in the upper most and lower most lithosphere – elastic stresses are extremely large but brittle and ductile stresses are small. However, in the central lithosphere, elastic stresses are smaller than brittle or viscous stresses and elastic stresses can therefore support internal and external loads. This model also shows that the elastic part of the lithosphere is significantly thinner than the thermally defined lithosphere (Fig. 5.16).

• *Quantitative description of a Brace-Goetze lithosphere.* In order to describe the Brace-Goetze lithosphere quantitatively we require quantitative information on 1. the depth dependence of temperature, i.e. a description of a geotherm; 2. the material constants (both 1 and 2 we need in order to calculate viscous stresses); 3. we need density and thickness of the crust and mantle part of the lithosphere in order to calculate vertical stresses and therefore the brittle strength. Table 5.3 lists typical numerical values for these parameters (Brace and Kohlstedt 1980). For the thermal structure of the lithosphere we

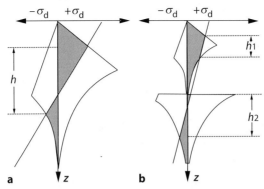

Figure 5.16. Rheology of the lithosphere considering elastic, viscous and brittle constitutive relationships. Note that this model illustrates a *stress state* rather than a *failure envelope* and should therefore not be confused with illustrations like those in Figs. 5.13 or 5.14. Compressive stresses are plotted towards the right, tensional stresses towards the left. **a** Strength profile for the crust (viscous stresses are only plotted for a single mineral phase, e.g. quartz). **b** Strength profile through the lithosphere considering the rheology of quartz and olivine. In both strength profiles elastic stresses are also plotted as the straight line with positive stresses in the upper crust and negative stresses in the lower parts (s. Fig. 4.20). The deformation mechanism that dominates at a given depth is given by the lowest stresses at a given depth. The elastic section of the crust in **a** is therefore restricted to the region h. In **b** the elastic portion of the lithosphere is in two regions h_1 and h_2, that are separated from each other (after Ranalli 1994)

will assume in the following that the radiogenic heat production decreases exponentially with depth according to eq. 3.67 with the characteristic drop off depth $h_r = 10$ km. We also assume that the thermal conductivity is $k = 2 \, \mathrm{J\,s^{-1}\,m^{-1}\,K^{-1}}$, and that the temperature at the base of the lithosphere is $T_l = 1\,280\,^{\circ}\mathrm{C}$. Assumption on thickness and density are also listed in Table 5.4.

We also make the assumption that the viscous behavior of olivine is described by eq. 5.41 below stresses of 200 MPa and by eq. 5.42 for stresses above 200 MPa. Brittle failure is described with Byerlee's law. That is, below 500 MPa brittle failure is assumed to occur without cohesion and an internal coefficient of friction of 0.8 and above 500 MPa the cohesion is 60 MPa and the internal coefficient of friction is 0.6. Some strength profiles calculated with these assumptions are dawn in Fig. 5.17.

Strength of the Lithosphere. When considering the distribution of stresses in the continental lithosphere, we have so far always only considered the stresses at a given depth. However, if we want to consider the deformation of entire continental plates, we need to know the mean stresses averaged over the entire lithosphere, or we need to know the total force that it needed to deform the entire lithosphere from top to base. Within the model of a Brace-Goetze lithosphere, this force is given by the vertically integrated stresses. This integrated strength is abbreviated with F_l and corresponds to the shaded region

Table 5.3. Rheological parameters of the continental lithosphere relevant for its viscous behavior. These data are largely after Sonder and England (1986). However, the activation energy was changed from Sonder and England (1986) ($Q_D = 5.4\cdot10^5$) to the value of $Q_D = 5.454 \cdot 10^5$ in order to allow a smooth transition between the stresses at 200 MPa. Instead of changing Q_D, Zhou and Sandiford (1992) changed for the same reason $\dot\epsilon_D$ from $5.7 \cdot 10^{11}$ to $3.05 \cdot 10^{11}$. If all other parameters remain constant, then the two changes have the consequence that stresses are about 40 MPa larger than those of Sonder and England (1986)

parameter	value/unit	definition
power law	*(eq. 5.41)*	
A_q	$5 \cdot 10^{-6}$ MPa^{-3} s^{-1}	pre exponent constant for quartz
Q_q	$1.9 \cdot 10^5$ J mol^{-1}	activation energy for quartz
n_q	3	power law exponent for quartz
A_o	$7 \cdot 10^4$ MPa^{-3} s^{-1}	pre exponent constant for olivine
Q_o	$5.2 \cdot 10^5$ J mol^{-1}	activation energy for olivine creep
n_o	3	power law exponent for olivine
Dorn's law	*(eq. 5.42)*	
Q_D	$5.4 \cdot 10^5$ J mol^{-1}	activation energy for olivine creep
$\dot\epsilon_D$	$5.7 \cdot 10^{11}$ s^{-1}	strain rate
σ_D	8 500 MPa	critical stress

Table 5.4. Rheological parameters of relevance for the brittle deformation of a Brace-Goetze lithosphere

parameter	value/unit	definition
$\mu_{(<500MPa)}$	0.8	coefficient of friction in the crust
$\mu_{(>500MPa)}$	0.6	coefficient of friction in the mantle
$\sigma_{0(<500MPa)}$	0	cohesion of the crust
$\sigma_{0(>500MPa)}$	60 MPa	cohesion of the mantle
λ	0.4 and 0.8	pore fluid/lithostatic pressure ratio
z_c	35 km	thickness of the crust
z_l	125 km	thickness of the lithosphere
ρ_c	2 750 kg m^{-3}	density of the crust
ρ_m	3 300 kg m^{-3}	density of the mantle lithosphere

in Figs. 5.13, 5.14 and 5.15). If we make the thin sheet approximation, then the integrated strength of the lithosphere may be calculated as:

$$F_l = \int_0^{z_l} (\sigma_1 - \sigma_3)\,dz \ . \tag{5.43}$$

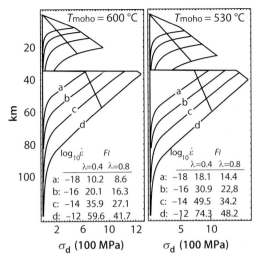

Figure 5.17. Strength profiles for the continental lithosphere as calculated with the model for a Brace-Goetze lithosphere and the data from Table 5.3 and 5.4). a, b, c and d are profiles for four different geologically relevant strain rates. The two diagrams show the strength profiles for two different Moho-temperatures that result from assumptions for the radiogenic surface heat production of $S_0 = 5 \cdot 10^{-6}$ W m^{-3} and $S_0 = 7 \cdot 10^{-6}$ W m^{-3}. In each diagram two linear curves for brittle failure for $\lambda = 0.4$ and $\lambda = 0.8$ are plotted. The stress curve with the higher stresses is for the lower value of λ. It was assumed that $\lambda_c = \lambda_l$. The vertically integrated stresses F_l are given in 10^{12} N m^{-1}

It has the units of force per meter or Pa m = N m^{-1}. F_l may be interpreted as the force acting in the direction normal to the orogen *per* meter length of orogen (i.e. in direction parallel to the orogen), that is required to deform the orogen with a given stain rate (Fig. 5.13, 5.14, 5.15). In the literature, the terms "strength" is often used very loosely. Strength (in Pa), *integrated strength* (in N m^{-1}), sometimes *stress* and occasionally even *force* are all often confused. We want to remember that *strength* has the units of stress (it is the stress that leads to brittle failure or viscous flow) and that *integrated strength* is a force per meter (which is equal to stress × meter). In the viscous regime strength is only defined for a given strain rate. This should be clear from eq. 5.41, where it is shown that the viscous stresses (strength) are strongly dependent on strain rate.

● *Strength as a function of Moho-temperature.* The strength of the lithosphere is very strongly dependent on the Moho-temperature. The details of the temperature distribution above and below the Moho are only a second order effect (Sonder and England 1986). Thus, for many mechanical questions on the scale of the lithosphere it is sufficient to characterize the geotherm by a single number: the Moho-temperature. Above and below the Moho it is sufficient to assume linear geotherms. However, in the following sections we continue to use curved geotherms characterized by exponentially decreasing radiogenic heat productions and so use the relationships derived in sect. 3.4.1. Thus, we determine the Moho-temperature indirectly by assuming the radiogenic surface heat production, the thermal conductivity and the surface heat flow. As a reminder to sect. 3.4.1, Fig. 5.18 shows the Moho-temperature and the surface heat flow as a function of thermal conductivity and heat production.

In order to quantify the integrated strength of the lithosphere we must integrate eq. 5.43. However, it is near impossible to integrate this equation analytically as the strength profile between $z = 0$ and $z = z_l$ is composed of several very different functions (Zhou and Sandiford 1992). The values for integrated strength quoted in the following sections were derived by numerical integration of eq. 5.43. Fig. 5.19 illustrates that the integrated strength rises dramatically with increasing strain rate (which corresponds to eq. 5.41). At geologically realistic strain rates the integrated strength ranges between 10^{12} and 10^{14} $\mathrm{N\,m^{-1}}$ (s. Fig. 5.17). These magnitudes correspond well with the magnitude of estimated plate tectonic driving forces (sect. 5.3).

5.2.2 Rheology of the Oceanic Lithosphere

The fundamental assumptions which we have made for the calculation of stresses and strength profiles for the continental lithosphere are also valid for the oceanic lithosphere. However, there are two important differences: 1. In contrast to the continental lithosphere, oceanic geotherms are time dependent and there is no radiogenic heat production in the oceanic lithosphere. As a consequence, different relationships must be used to calculate the temperature profile with depth and ultimately the rheology (sect. 3.5). 2. The rheology of oceanic lithosphere is dominated by olivine and not by quartz. As

Figure 5.18. Surface heat flow and Moho-temperature as a function of radiogenic heat production at the surface S_0. It is assumed that this heat production rate decreases exponentially with depth. The characteristic skin depth for this exponential drop off is assumed to be 10 km (s. eq. 3.67). Curves are labeled for conductivities in $\mathrm{J\,s^{-1}\,m^{-1}\,K^{-1}}$ (calculated with eq. 3.76, eq. 3.78 and the data from Table 5.4)

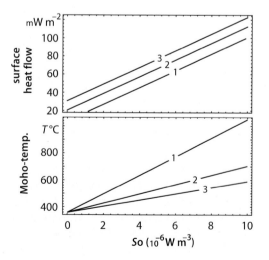

Figure 5.19. Vertically integrated strength of the continental lithosphere F_l as a function of strain rate. The curves are shown for three different geotherms characterized by three different surface heat productions (S_0 in $W\,m^{-3}$). The Moho-temperatures that correspond to these assumptions may be read from Fig. 5.18. Similar diagrams were first discussed by Sonder and England (1986). The integrated strengths shown here are relatively small, because the assumed temperature profiles are relatively high

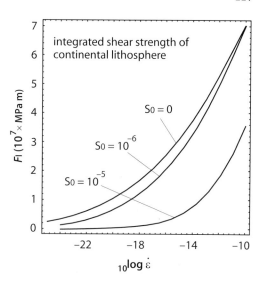

Figure 5.20. Strength profile through the oceanic lithosphere at ages of 10, 30 and 100 my. For each of these ages, stresses were calculated for three strain rates of $\dot{\varepsilon} = 10^{-16}$, 10^{-14} and 10^{-12} s^{-1}. For each age, the curve for the highest strain rate has the largest strength. The temperature profiles needed to calculate the stresses were calculated using eq. 3.81; for the rheological data the values of Table 5.3 and 5.4 were used

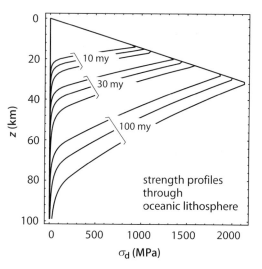

a consequence, there is only one maximum in the strength profile (Fig. 5.20). On the other hand, the strength profiles of oceanic lithosphere are highly dependent on its age (sect. 3.5.1). The depth dependent temperature profile of oceanic lithosphere may be calculated with eq. 3.81. A strength profile for the oceanic lithosphere may then be calculated using eqs. 5.34, 5.41 and 5.42 as well as the data from table 5.3 and 5.4. Fig. 5.20 shows some examples for such strength profiles through oceanic lithosphere.

Strength of the Oceanic Lithosphere. Eq. 5.43 may be used to calculate the integrated strength of the oceanic lithosphere just like we used it above to calculate the integrated strength of the continents. In fact, it is possible to

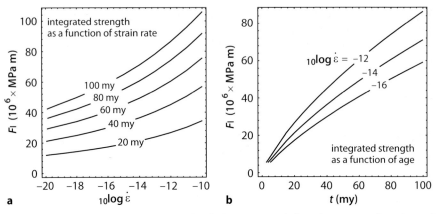

Figure 5.21. a Integrated strength of the oceanic lithosphere as a function of strain rate for 5 differently aged oceanic lithospheres. Curves were calculated with eq. 5.43 using the results of Fig. 5.20. **b** Integrated strength of oceanic lithosphere as a function of age for three different strain rates. **a** and **b** contain the identical information

calculate the integrated strength of the oceanic lithosphere with much higher accuracy than of the continental lithosphere, because oceanic geotherms are much better known than continental geotherms. Fig. 5.21a shows the calculated integrated strength of oceanic lithosphere as a function of different strain rates and age as calculated numerically with eq. 5.43.

Fig. 5.21a may be compared directly with Fig. 5.19. In Fig. 5.21b the same information is shown in a diagram with age on the horizontal axis. A comparison with Fig. 5.19 shows that only very young oceanic plates have a smaller integrated strength than continental lithosphere. This result from Figs. 5.21 and 5.19 corresponds to our observations: We know that most intra plate seismicity occurs in the continents and not in the oceans (Fig. 2.2). There is practically no deformation inside the oceanic plates of the earth. An oceanic plate acts - because of its high integrated strength - like a passive transmitter of stresses from the mid oceanic ridges to the continents (s. sect. 6).

• *Strength relationships between continental and oceanic lithosphere.* In the previous paragraph we have shown that oceanic lithosphere is significantly stronger than continental lithosphere, even though it is generally much thinner. We have come to this important conclusion by comparing the integrated strength of continents and oceans if they would deform under the same strain rate. However, in geodynamics it is often more meaningful to compare the deformation rate of continental and oceanic lithosphere under the same applied stresses, rather than the stresses under the same applied strain rates.

One of the most important plate tectonic driving forces is the potential energy of the mid oceanic ridges (sect. 5.3.2). The force that is exerted by these ridges onto the surrounding continents increases with the age of the oceanic lithosphere. We can now ask ourselves if it is the continents or the

Figure 5.22. The strain rate with which oceanic lithosphere will deform in response to ridge push. The strain rate is plotted against age of the oceanic lithosphere. Clearly, the older the oceanic lithosphere is, the larger the applied ridge push and the large the strain rate. Note that all strain rates are several orders of magnitude *below* anything that might be geologically relevant. In other words, oceanic lithosphere does practically *not* deform under the force exerted by ridge push force

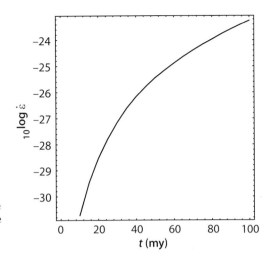

oceanic lithosphere that are deformed by this force. In order to answer this question, we have plotted the strain rate of oceanic lithosphere with which oceanic lithosphere would deform under its own ridge push against age of the oceanic lithosphere (s. sect. 6.2.2). Fig. 5.22 shows that these strain rates are geologically irrelevant. The vast majority of the plate divergence at the mid oceanic ridges is compensated by deformation inside the continents and not inside the oceanic lithosphere.

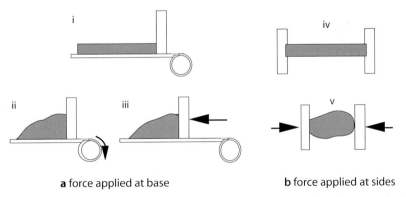

a force applied at base   **b** force applied at sides

Figure 5.23. Illustration of the two fundamental mechanisms for the transmission of plate tectonic driving forces. **a** illustrates transmission by *basal friction*. In *ii* basal friction is shown in a *Eulerian* reference frame ("conveyer belt" model). In *iii* basal friction is shown in a *Lagrangian* reference frame ("bulldozer" model). **b** illustrates transmission by lateral normal stresses ("side forcing")

5.3 Forces Applied to Lithospheric Plates

5.3.1 Transmission Mechanisms

Plate tectonic driving forces may be divided into two fundamental groups according to the way they are transmitted:

- transmission by shear stresses,
- transmission by normal stresses.

Because plate tectonic driving forces act horizontally, *shear stresses* must be applied to horizontal surfaces and *normal stresses* to vertical surfaces. If the transmission occurs by shear stresses, this is often called *basal traction*. If the transmission occurs by normal stresses, we speak of *end loading* or *side forcing*. Fig. 5.23 illustrates that different deformation geometries may arise, depending on the transmission mechanism. However, in reality it is often difficult to interpret which of the two mechanisms is responsible for a given plate motion. In fact, the scientific community remains divided between those who believe that plate tectonics is driven by *basal traction* and those who consider *lateral normal stresses* as the principal driving force.

Transmission of Stresses by Shear- or Normal Stresses. One model for the explanation of plate motions is that the friction between the base of the lithosphere and the convective motion in the asthenosphere is the principal driving mechanism (Ziegler 1992, 1993). The most important argument *for* this model comes from the reconstruction of past plate motions. These do not correspond very well with the global geometry of mid oceanic ridges and subduction zones. Thus, it is thought that these plate motions reflect the geometry of convection cells in the mantle instead. The most important argument *against* this model is implicit in Fig. 5.13. This figure shows that differential stresses at the base of the lithosphere are much too small to be able to transmit forces from the mantle into the lithosphere. It is therefore hard to imagine that this softest part of the lithosphere can transmit stresses large enough to build the mountain ranges of our planet (s. mechanical definition of the lithosphere in sect. 2.4.1). The tractions at the base of the lithosphere are not likely to be larger than 10^{-2} MPa (Richardson 1992).

The other - and by far more accepted - model for the explanation of plate motions is that *plate boundary* forces drive plate tectonics by lateral normal stresses (e. g. Forsyth and Uyenda 1975). These forces are predominantly caused by *potential energy variations*. Such variations occur inside the continents and along the boundaries of oceanic lithosphere and will be discussed on the following pages.

Despite the two different models for the origin of plate tectonic driving forces we should not forget that, ultimately, *all* plate tectonic forces find their origin in the thermal energy of earth. On the scale of an individual orogen it is often easier to determine the way forces are transmitted into the orogen. For example, the geometry of accretionary wedges shows clearly that they form by

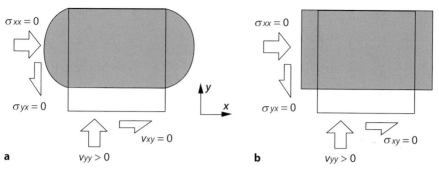

Figure 5.24. Illustration showing how apparently small differences in the boundary conditions can cause very different deformation geometries. In both examples a square body is subjected to two-dimensional deformation. In both **a** and **b** the tangential and normal components of the boundary conditions on the left and right boundaries are given by stresses, which are assumed to be zero ($\sigma_{xx} = \sigma_{yx} = 0$). Also, in both examples the normal component at the bottom boundary is given by a velocity with which the bottom boundary moves towards the top boundary ($v_{yy} > 0$). **a** and **b** differ only in the tangential component of the boundary condition along the bottom boundary. In **a** this is given by a velocity ($v_{xy} = 0$), in **b** it is given by a stress ($\sigma_{xy} = 0 =$ free slip)

basal traction with the subducting plate below (sect. 6.2.3). A nice example for deformation in response to lateral normal stresses is the deformation of intra continental mountain ranges like the Tien-Shan in central Asia.

Boundary Conditions of Deformation. In the last section we have discussed normal and shear *stresses* that cause the deformation of plates. However, the boundary conditions for the dynamic deformation of plates may be described more easily with velocity boundary conditions than with stresses. We discriminate between:

- Orogenic boundary conditions given by velocities,
- Orogenic boundary conditions given by stresses.

Both types of boundary conditions may have a *normal* and a *tangential* component. Thus, for a two-dimensional mechanical model with the two spatial coordinates x and y, we require a tangential and a normal boundary condition on each boundary. A total of four variables must be defined by the boundary conditions- Fig. 5.24 illustrates that the difference between velocity and stress boundary conditions may have a very profound influence on the deformation geometry. A boundary condition is called *free slip* if the shear stresses along this boundary are considered to be zero.

The reason why we require *four* boundary conditions for the description of two-dimensional deformation may be also seen from the stress balance equations (eq. 5.16, 6.16 and 6.17). If we integrate these equations with two variables, then there are *four* constants of integration. In order to determine these constants we need *four* independent pieces of information: the four boundary conditions.

If a medium is not everywhere a continuum, for example because it contains a brittle fracture, then all mechanical properties may have discontinuities. Such problems may not be solved using a single set of boundary conditions as in Fig. 5.24 and we must use internal boundary conditions or other special tricks to be able to solve such problems. For example, the medium can be subdivided into several continuous regions that are described separately.

Potential Energy. Practically all important plate tectonic driving forces find their origin in differences of the potential energy of different parts of the earth (Turcotte 1983). In this section we explain what we understand with the term *potential energy* in a plate tectonic context. We will return to this concept again in the sections 5.3.2, 5.3.3 and 6.2.2.

In sect. 4.2.1 we have shown that the vertical normal stress at a given depth in the crust z is given by the product of density, gravitational acceleration and the height, or thickness of the vertical rock column above it. This vertical normal stress is the vertically acting *force per area*. It may be calculated by integrating ρg between 0 and z, as we did in eq. 4.16. If the density over the thickness z remains constant, then this is simply $\rho g z$. This term has the units of Pa or $kg\,s^{-2}\,m^{-1}$ or $J\,m^{-3}$. We can see that *stress* has the same units as *energy per volume*. Thus, the vertical normal stress can also be interpreted as the potential energy of a cubic meter of rock at depth z. If we want to know the potential energy not of a single cubic meter, but that of a whole body, for example that of a mountain range, then we need to integrate this *potential energy per cubic meter* over the lateral and vertical extent of the range. Fortunately, it is usually sufficient to know the potential energy *per area*, i.e. that of a complete vertical column, but only for one square meter of area. Using this *potential energy per area* we can compare different regions on the globe, for example two neighboring lithospheric columns of different thickness and density distribution. In the following we will represent the potential energy *per area* with E_p. In order to determine E_p at depth z we simply need to integrate the vertical stresses in the lithospheric column of interest between the surface (which usually is $z=0$ in the reference frame we use) and the depth of interest z:

$$E_p = \int_0^z \sigma_{zz} dz = \int_0^z \int_0^z \rho_{(z)} g dz dz \quad . \tag{5.44}$$

Very often the "depth of interest" is the isostatic compensation depth. If the density is independent of depth, then eq. 5.44 may be simplified to give:

$$E_p = \int_0^z \sigma_{zz} dz = \int_0^z \rho g z dz = \frac{\rho g z^2}{2} \quad . \tag{5.45}$$

This integral corresponds to the gray shaded region in Fig. 5.25b. We want to remember that E_p has the units of energy *per* area and is, therefore, strictly speaking, no energy as such.

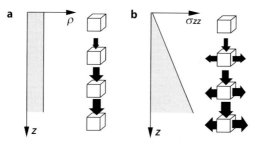

Figure 5.25. Density ρ and vertical normal stress σ_{zz} as a function of depth z. The value σ_{zz} is the vertically integrated density, times acceleration. Thus, the curve in **b** corresponds to the gray shaded region in **a**. The row of little unity cubes next to **a** illustrates how the vertical stress increases with depth. The column of cubes next to **b** illustrates that the horizontal force exerted by the column on its surroundings is given by the sum of all vertical stresses. This corresponds to the gray shaded area in **b**

Horizontal Forces Arising from Potential Energy Variations. In a static, non-deforming lithosphere the horizontal and vertical normal stresses have the same magnitude (see Fig. 5.25). It is true that:

$$\sigma_{zz} = \sigma_{xx} = \sigma_{yy} \; . \tag{5.46}$$

This is also stated in eq. 5.37, which says that there is no deviatoric stress if the strain rate is zero. The sum of all vertical stresses integrated over the thickness of a plate is the potential energy of the plate per area. Since horizontal and vertical stresses are the same, this potential energy per area is equivalent to the force exerted by the lithosphere onto its surroundings per meter length of orogen. If two neighboring vertical lithospheric columns have the same potential energy per unit area, then they also exert equally large horizontal forces onto each other and there is no "net force" between them. However, if they have different potential energies per area, then this potential energy difference between the two plates may be interpreted as the net force F_b that is exerted by one column onto the other in the horizontal direction and per meter length of orogen. This net force arising from potential energy differences is also called *horizontal buoyancy force* (somewhat cumbersome) or *gravitational stress* and it is important to remember that it has the units of force per meter length of orogen. This potential energy difference may be written as (s. Fig. 5.26):

$$\Delta E_{\mathrm{p}} = F_{\mathrm{b}} = \int_0^{z_{\mathrm{K}}} \int_0^{z_{\mathrm{K}}} \rho^{\mathrm{A}}(z) g \mathrm{d}z \mathrm{d}z - \int_0^{z_{\mathrm{K}}} \int_0^{z_{\mathrm{K}}} \rho^{\mathrm{B}}(z) g \mathrm{d}z \mathrm{d}z \; . \tag{5.47}$$

There, z_{K} could be any depth, but for many purposes it is useful to assume that it is the same isostatic compensation depth we used on p. 150. Below this depth there is no density differences between the vertical columns A and B (s. eq. 4.15). $\rho^{\mathrm{A}}(z)$ is the density of profile A as a function of depth z.

If density is a continuous function of depth, then eq. 5.47 may be usually integrated without too much trouble. However, in the lithosphere, the density distribution has a discontinuity at the Moho so that it may be necessary to split the integral in eq. 5.47, even for very simple assumptions on the density distribution in the lithosphere. For example, the potential energy of the two-layered column A in Fig. 5.26 for column A may be described by:

$$\int_0^{z_1}\int_0^{z_1}\rho^A(z)g\mathrm{d}z\mathrm{d}z = \int_0^{z_c}\sigma_{zz}\mathrm{d}z + \int_{z_c}^{z_1}\sigma_{zz}\mathrm{d}z \ ,$$

$$= \int_0^{z_c}\rho_1 gz\mathrm{d}z + \int_{z_c}^{z_1}\rho_1 gz_c + \rho_3 g(z-z_c)\mathrm{d}z \ ,$$

$$= \frac{\rho_1 gz^2}{2}\bigg|_0^{z_c} + \rho_1 gz_c z\bigg|_{z_c}^{z_1}$$

$$+ \frac{\rho_3 gz^2}{2}\bigg|_{z_c}^{z_1} - \rho_3 gz_c z\bigg|_{z_c}^{z_1} \ . \tag{5.48}$$

It should be easy to understand this equation graphically by plotting density and vertical stress as a function of depth as we did in Fig. 5.25, 5.27, 5.31 and 5.33.

The importance of the density *distribution* in the lithosphere for the potential energy may be illustrated nicely with an interesting example. Fig. 5.26 shows two columns in isostatic equilibrium. The two columns have the same isostatically supported surface elevation, because they are made up of sections of the same densities and thicknesses. However, they have different potential energies because in column B the dense part lies up high. Potential energy does not only depend on thickness and density, but also on the *distribution* of density with depth. Thus, there is a net buoyancy force between the two columns shown in Fig. 5.26. This net force is exerted by column B towards column A.

We can conclude that it is dangerous to infer lateral forces from topography on the surface of earth (England and Molnar 1991). Surface elevation is a linear function of thickness (eq. 4.17 and 4.18), while potential energy per unit area is a quadratic one (eq. 5.44 and 5.47). In fact, it is even possible, that topographically *lower* regions exert a gravitational stress on topographically *higher regions*, averaged over the thickness of the lithosphere (Stüwe and Barr 2000). Geoid anomalies on the other hand can be used to estimate the density distribution within the lithosphere. Coblentz et al. (1994) have used a combination of information on surface elevation and on geoid anomalies to estimate the potential energy of the lithosphere.

Force Balance Between Mountains and Foreland. In this section we estimate the forces exerted by a mountain range onto its foreland (Fig. 5.27). For this, we will follow the logic of Molnar and Lyon-Caen (1989) and also use their choice for the vertical axis of the cross section. We assume an origin at the Moho and measure the vertical direction positively upwards as illustrated in Fig. 5.27b. This choice for the vertical axis helps the intuitive

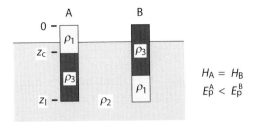

Figure 5.26. Schematic cartoon showing two columns in isostatic equilibrium ($\rho_1 < \rho_2 < \rho_3$). The surface of both columns has the same elevation above the liquid of density ρ_2, because both bodies consist of equally thick sections of the densities ρ_1 and ρ_3, i.e. they have the same weight. However, column B has a much higher potential energy per unit area than column A, because the distribution of density is different. In column B the high density part of the section lies higher. As a consequence, B exerts a net force towards A. The cartoon obviously represents no geologically realistic scenario for the lithosphere. However, it is useful to illustrate why mountain ranges need not exert a net force onto their lower lying surroundings even if they have a higher surface elevation. Note however, that this logic only applies if forces may be averaged over the thickness of the lithosphere. Problem 5.17 is related to this figure

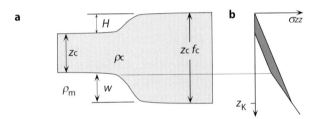

Figure 5.27. Cartoon contrasting the distribution of vertical stresses in mountain ranges relative to their foreland. **a** The thickness of the crustal root is w, the surface elevation relative to the reference lithosphere in the foreland H. In isostatic equilibrium it is true that: $H\rho_c = w(\rho_m - \rho_c) = w\Delta\rho$. In **b** the vertical stresses are drawn for the mountain range and the foreland. The dark shaded area between the two stress curves has the units of stress × meters or force per meter length of orogen exerted by the range onto the foreland. It corresponds to the potential energy difference between the mountain range and the foreland

understanding if the integration of eq. 5.47, as one of the integration limits is always zero (s. Molnar and Lyon-Caen 1989). However, note that the results are independent of which reference frame is chosen as we do not calculate *absolute* potential energies, but only potential energy differences between two neighboring columns. Thus, as long as we choose the same coordinate system for the two columns that are to be compared, it does not matter which reference frame we pick.

We begin by calculating the potential energy per unit area of the foreland. We can find this by integrating eq. 5.44. With foresight to the fact that there is a density difference between the foreland and the mountain range down to the compensation depth z_K (Fig. 5.27), we perform this integration between limits at the surface and depth z_K, resulting in:

$$E_p^{\text{foreland}} = \rho_c g z_c^2/2 + \rho_m g w^2/2 \quad . \tag{5.49}$$

Correspondingly, integration of eq. 5.44 between the same limits in the mountain belt gives the potential energy per unit area of the range:

$$E_p^{\text{range}} = \rho_c g(H + z_c)^2/2 + \rho_c g w^2/2 \quad . \tag{5.50}$$

Note that these relationships were derived according to the same principal as eq. 5.48. The potential energy difference per unit area is given by the difference of eq. 5.49 and eq. 5.50 (s. eq. 5.47). It is:

$$\Delta E_p = F_b = E_p^{\text{range}} - E_p^{\text{foreland}}$$

$$= \rho_c g H^2/2 + \rho_c g H z_c + \Delta \rho g w^2/2 \quad . \tag{5.51}$$

There: $\Delta\rho = (\rho_m - \rho_c)$. Eq. 5.51 may be significantly simplified because we assume that both, mountain range and foreland are in isostatic equilibrium. The isostasy condition states that: $\Delta\rho w = H\rho_c$. Using this we can simplify eq. 5.51 to:

$$\Delta E_p = F_b = \rho_c g H \left(H/2 + z_c + w/2\right) \quad . \tag{5.52}$$

The force F_b corresponds to the dark shaded region in Fig. 5.27b. It is the difference between the vertically integrated vertical stresses σ_{zz} of two vertical columns in the mountain range and in the foreland, respectively (Tapponier and Molnar 1976). For a 3 km high mountain range with a 30 km root, eq. 5.52 gives a force F_b of the order of 3–$4 \cdot 10^{12}$ N m^{-1}. We will see that this number is comparable with the forces applied to and exerted by mid ocean ridges.

Despite its simplicity, eq. 5.51 may be used to draw some very fundamental conclusions. For one, we can see that the third term is significantly larger than the first term. Thus, the potential energy difference between two mountain ranges of the same elevation becomes larger if the compensating root is thicker. For example, a 100 km thick root of a mountain range made up of low density mantle material contributes significantly more to the potential energy of a range than a 60 km thick root of crustal material. We can also see from eq. 5.51 that the potential energy of a mountain range grows with the square of both the surface elevation *and* the thickness of its root. The work that must be done to increase the surface elevation of a mountain range by one meter increases therefore as the mountain range gets higher (Molnar and Tapponier 1978; s. sect. 6.2.2). This is the reason why mountain ranges do not grow infinitely on this planet and have a limiting elevation. As potential energy variations are some of the most important driving forces in the lithosphere we will continue with more details in the following sections 5.3.2 and 5.3.3.

5.3.2 Forces in Oceanic Lithosphere

The forces exerted by oceanic lithosphere onto the continents around them are considered to be the fundamental driving mechanism for plate tectonic motion (McKenzie 1969b). There are two important driving forces in oceanic lithosphere: 1. the potential energy of the mid-oceanic ridges and 2. the tensional stresses that occur in association with subduction zones. The former is called *ridge push*, the latter are called *slab pull* and *trench suction*. In the following we discuss the nature of both types of forces and discuss their magnitude.

Ridge Push. Mid-oceanic ridges have a high topography and a high potential energy relative to the average oceanic lithosphere. This potential energy is one of the more important (and certainly best known) plate tectonic driving forces. While strictly speaking the mid-oceanic ridge applies a torque to the plate (s. p. 24), we will neglect here the curvature of the earth and continue using the term "ridge push". It is important to understand that ridge push finds its origin in the high potential energy of the ridge, rather than in the frictional stresses between an outward welling mantle plume and the oceanic plate as drawn in Fig. 5.28a.

The ridge push force per meter length of ridge (equivalent to the potential energy of the ridge per unit area) may be calculated with eq. 5.47, using similar assumptions to those we have made when designing a model to explain the water depth of the oceans (s. Fig. 4.17). The density of oceanic lithosphere must be expressed in terms of temperature (eq. 4.21) and temperature as a function of depth (eq. 3.81; s. Turcotte and Schubert 1982; Parsons and Richter 1980). Then - using the half space cooling model - it may be shown that the ridge push force is a function of the thermal profile through the oceanic lithosphere and therefore of age. Without reiteration the derivation of the ridge push force here, we simply state that it is given within this model by the equation:

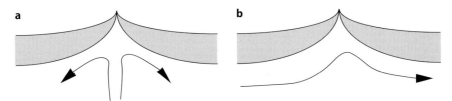

Figure 5.28. Cartoon showing two possible motions of the asthenosphere below mid oceanic ridges. **a** Asthenospheric material wells up below the mid ocean ridge in form of a *mantle plume*. During this process, adiabatic decompression of asthenosphere material will cause massive partial melting. It is thought that this situation pertains to regions where these melts are now present as large igneous provinces like the Karoo Basalts in southern Africa or the Deccan Traps in India and may be Iceland (Fig. 6.29). **b** shows the mantle motion that is thought to be representative for most mid oceanic ridges

Figure 5.29. The force exerted by mid-oceanic ridges onto the surrounding plate per meter length of ridge, shown as a function of age of the oceanic lithosphere. Calculated with eq. 5.53

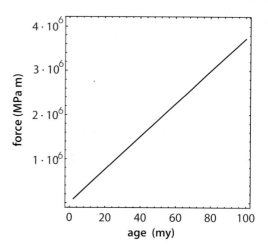

$$F_{\rm b} = g\rho_{\rm m}\alpha T_{\rm l}\kappa t \left(1 + \left(\frac{\rho_{\rm m}}{\rho_{\rm m} - \rho_{\rm w}}\right)\frac{2\alpha T_{\rm l}}{\pi}\right) \approx 1.19 \cdot 10^{-3} t \quad . \tag{5.53}$$

All parameters of this equation are explained in sect. 4.2.1. From eq. 5.53 we can see that the ridge push force is a *linear* function of age of the oceanic lithosphere (Fig. 5.29). As such it is different from water depth which - within this model is described by a square root function of age (Fig. 4.18). The numerical value of the proportionality constant between age and force in eq. 5.53 ($1.19 \cdot 10^{-3}$) is derived using the following constants: $T_{\rm l} = 1\,200°C$; $\rho_{\rm m} = 3\,200$ kg m^{-3}; $\rho_{\rm w} = 1\,000$ kg m^{-3}; $\alpha = 3 \cdot 10^{-5}$ K^{-1} and $\kappa = 10^{-6}$ m^2 s^{-1}. Fig. 5.29 shows that ridge push is about an order of magnitude *smaller* than the integrated strength of continents at normal orogenic strain rates (s. Fig. 5.19). Thus, we may conclude that ridge push alone is insufficient as the principal plate tectonic driving force.

• *Asthenospheric flow at mid-oceanic ridges.* In the past, ridge push has been interpreted to be related to frictional stresses of upwelling asthenosphere that "pushes" the ridge apart as illustrated in Fig. 5.28a. However, we now know that there are only very few places where mid-oceanic ridges coincide with diapirically upwelling mantle material. Rather, the asthenospheric flow at most mid-oceanic ridges is of the geometry shown in Fig. 5.28b. Among other arguments, this was recognized by McKenzie and Bickle (1988) using on geochemical arguments. These authors showed that partial melting that would occur due to adiabatic decompression of upwelling melt in a mantle plume would be enough to form a 15 km thick oceanic crust. In contrast, normal oceanic crust is measured to be only about 5–7 km thick. This thickness can be produced by adiabatic melting of only the upper most asthenospheric regions. Asthenospheric flow as sketched in Fig. 5.28b is sufficient to produce a 5–7 km thick oceanic crust. Thus, it is thought that the flow directions of asthenospheric convection have little to do with the position of the mid-oceanic ridges. There are only very few places where mid-oceanic ridges coincide with diapirically upwelling mantle material. One of these places is Iceland (Fig. 6.29). There, not only is the oceanic crust significantly thicker than 7 km, but also the mid oceanic ridge is uplifted by the stresses exerted to its base by the upwelling mantel material not unlike a jet of water shot from below onto a rubber sheet. Other regions where mantle plumes are thought to coincide with rifting margins are those of flood basalts (p. 298) (White and McKenzie 1989).

Slab Pull and Trench Suction. Old oceanic lithosphere is denser that the underlying asthenosphere and it has therefore a negative buoyancy and it wants to sink. However, because oceanic lithosphere is very strong and stiff, it cannot immediately do this as soon as it reaches this critical age where its density becomes large compared to that of the underlying asthenosphere. Rather, the oceanic plate "glides" along the surface of the asthenosphere until this gravitationally unstable configuration is brought out of balance and a subduction zone forms. Once the edge of such an old oceanic plate has begun to subduct, it drags the remainder of the plate behind it. This is what is called *slab pull*. Such subduction processes may cause, or may be caused by-, small scale convection in the upper mantle. This convection occurs predominantly in the wedge shaped region between the subducting and the upper plate. Once such a convection system is set up, it may actually drag both the upper plate and the subducting plate into the subduction zone (s. Fig. 3.26). This is what is called *trench suction*.

Slab pull is gravitationally induced simply by the dense oceanic lithosphere wanting to sink into the less dense upper mantle. In fact, the slab pull force is reinforced by the fact that the density of the down-pulling slab increases significantly once it has passed the olivine-spinel-transition at roughly 400 km depth. The magnitude of slab pull is roughly 10^{13} N m^{-1} (s. Turcotte and Schubert 1982). Thus, slab pull is about an order of magnitude larger than ridge push. However, it is likely that slab pull is being counteracted by frictional stresses of about the same magnitude between the sinking plate and the surrounding asthenospheric mantle. Thus, the net force exerted by subduction zones onto the foreland need not be very large. Estimates by Bott (1993) and Bott et al. (1989) suggest that both slab pull and trench suction may be of the magnitude of roughly $4 \cdot 10^{12}$ N m^{-1}. In general it may be said that the force balance in subduction zones is much less well understood than that around mid-oceanic ridges. Nevertheless, most authors agree that forces in and around subduction zones may be much larger than those exerted by the mid-oceanic ridges.

• *Roll back of subduction zones. Slab pull* and *trench suction* are predominantly forces acting downwards, while *ridge push* acts mainly in the horizontal direction. Slab pull and trench suction are not related to potential energy variations, but to *gravitational instabilities* similar to those that are responsible for convective motions. Because slab pull and trench suction act *downwards* it is possible that the kink in the subducting plate near the trench

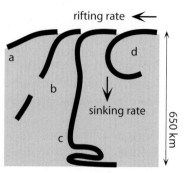

Figure 5.30. Four possibilities for the shape of subducted slabs at trenches. *a* Shallow subduction as probably occurs if subduction occurs in the same direction as the convective flow of the upper mantle (Doglioni 1993). *b* Steep subduction as it is often observed around the globe with slab break off at depth (von Blanckenburg and Davies 1995). *c* Vertically hanging plate with folding at the 650-km-discontinuity (Frottier et al. 1995, Houseman and Gubbins 1997). *d* Hypothetical (in nature not observed) shape of a subducted plate that would occur if *no* deformation of the plate would occur subsequently to subduction in the trench

shifts laterally. If the rifting rate at the mid-oceanic ridge exceeds the sinking rate of the subducting slab, then the trench will move *towards* the upper plate (Fig. 5.30). However, if the sinking rate is larger than the rifting rate, then the trench moves *away* from the upper plate towards the mid-oceanic ridge (e. g. Dewey 1988). This process is called *roll back* of a subduction zone. The most famous example for roll back is the Scotia arc west of South Georgia. The formation of extensional basins in front of subduction zones, in particular the formation of Fore-arc- and Back-arc-basins, is thought to be related to roll back (Royden 1993a). Roll back does not only depend on the relative rates of rifting and sinking of the oceanic plate, but also on how easily the asthenosphere may be displaced underneath the subducting plate.

• *Deformation of subducting plates.* The deformation of subducted plates in the upper mantle is not very well-understood. We know from bathymetric data that oceanic lithosphere is kinked at the trenches (Fig. 2.20). However, tomographic and seismic imaging indicates that subducted slabs are mainly planar slabs below the trench. This implies that the kinks of the subducted plates are unbent again at depth (Houseman and Gubbins 1997). Von Blanckenburg and Davies (1995) showed that slabs may also break off at depth (Fig. 5.30).

The long term evolution of subducting plates depends on the processes at the upper-lower mantle transition in about 650 km depth. Creager and Jordan (1984) showed that the processes there are of large importance to the possibility of recycling lithospheric material. In general it is thought that subducting plates cannot perforate this 650 km transition, predominantly because the density of the lower mantle is higher than that of the subducting slabs (e. g. Christensen and Yuen 1984) (Fig. 5.30c). This model has since been confirmed by tomographic imaging (e. g. van den Hilst et al. 1991). It appears that the 650 km discontinuity is a *graveyard* for subducted slabs. There is a range of recent analogue and numerical experiments that test details of the geometry of deformation of subducted slabs at this discontinuity (Frottier et al. 1995).

5.3.3 Forces in Continental Plates

Inside the continents, plate tectonic driving forces arise predominantly from lateral variations in the density structure, i. e. lateral variations in *potential energy*. When we discussed Fig. 5.27 we have already estimated the magnitude of these forces for a plate of constant density but variable thickness (eq. 5.51). In this section we want to refine our estimates by considering a more realistic lithosphere consisting of a dense mantle lithosphere and a much less dense crust (s. Fig. 2.15).

Fig. 5.31 illustrates two examples of potential energy differences between two lithospheric columns. Similar to Fig. 5.27 this potential energy difference is given by the shaded region between the two curves for vertical normal stress as a function of depth. This area corresponds to F_b in eq. 5.47 and may be interpreted as the net force exerted by one column onto the other

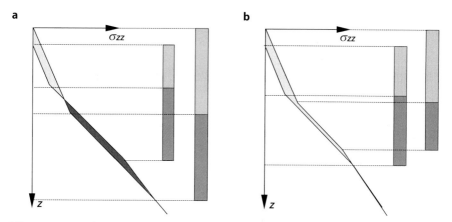

Figure 5.31. Illustration of vertical stresses and potential energy differences between two neighboring lithospheric columns. Vertical normal stress is plotted as a function of depth. The shaded region between the two curves are the potential energy difference per area between the two respective columns. In **a** this difference is positive in the upper part of the lithosphere (light shading) but negative in the lower part (dark shading). This means, that there is a net force acting from the right hand column towards the left hand column, while this net force is directed towards the right in the lower part. Because both shaded regions are roughly of the same area, there is practically no net force between the two columns, averaged over the thickness of the lithosphere. In **b** the entire right hand lithospheric column exerts a net force onto the left hand column

per meter length of orogen and averaged over the thickness of the lithosphere (horizontal buoyancy force). In Fig. 5.31b the vertical normal stresses in the right hand lithosphere column is larger than that of the left hand column at all depths. Thus, there is a net force from the right towards the left column at all depths. However, for the two columns shown in Fig. 5.31a the situation is different. In the upper part of the profile the vertical stresses in the column of lower surface elevation are smaller. Interestingly however, the vertical normal stresses are *smaller* for the column of *higher* surface topography in the lower part of the profile. This means that there is a net force exerted from the right hand column towards the left hand column in the *upper* part (light shaded region in the σ_{zz}-z-diagram), but that this force is directed in the opposite direction in the lower part (dark shaded region; see Problem 5.20).

The qualitative considerations of Fig. 5.31 may be quantified by integrating eq. 5.47 and using simple descriptions for density as a function of depth. If we assume a simple lithosphere of two layers (a crust and a mantle lithosphere) and assume a linear thermal profile in the lithosphere so that the density due to thermal expansion may be described with eq. 4.23, then the lateral buoyancy force is described by:

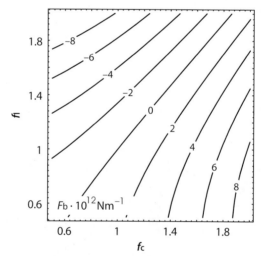

Figure 5.32. Diagram of lithospheric thickening strain f_l plotted against crustal thickening strain f_c and contoured for potential energy difference per area (equivalent to: "horizontal buoyancy force per meter" or: "lateral force"). The f_c-f_l plane was explained in detail in Fig. 4.5. The potential energy difference is always that between any point in f_c-f_l space and the reference lithosphere at f_c=f_l = 1. The diagram was calculated with eq. 5.54 and is contoured for F_b in 10^{12} N m^{-1}. Other assumptions are: $\rho_m = 3\,200$ kg m^{-3}; $\rho_c = 2\,750$ kg m^{-3}; $\alpha = 3 \cdot 10^{-5}$; $z_c = 35\,000$ m; $z_l = 125\,000$ m; $T_l = 1\,200\,°$C. The curvature of the contours arises because of the quadratic dependence of potential energy on thickness. As such, these contours for lateral force between two columns are fundamentally different from those for surface elevation (s. Fig. 4.16) (Sandiford and Powell 1990, Stüwe and Barr 2000)

$$\frac{F_b}{\rho_m g z_c^2} = \frac{\delta(1-\delta)}{2}(f_c^2 - 1) - \frac{\alpha T_l}{6(z_c/z_l)^2}\left(f_l^2 - 1 - 3\delta(f_c f_l - 1)\right)$$

$$+\frac{\alpha^2 T_l^2}{8(z_c/z_l)^2}(1 - f_l^2) \tag{5.54}$$

(Turcotte 1983; Sandiford and Powell 1990). All parameters in this equation are the same as those we used in eq. 4.29 to calculate the elevation of mountain belts in isostatic equilibrium (δ is the density ratio of crust and mantle lithosphere $\delta = (\rho_m - \rho_c)/\rho_m$, g is the gravitational acceleration, T_l the temperature at the base of the lithosphere and α is the coefficient of thermal expansion). Lateral forces calculated with eq. 5.54 are shown in Fig. 5.32. The shape of these curves hardly changes for more refined assumptions on the thermal structure of the lithosphere (Zhou and Sandiford 1992). Fig. 5.32 shows that the absolute values of the net lateral forces exerted by very *thin* or very *thick* continental lithosphere on its surroundings are of the order of 10^{12}–10^{13} N m^{-1}. Thus, they are comparable to the magnitude of forces arising from ridge push or slab pull. This is really no surprise, as the thickness variations within the continental lithosphere are themselves only caused by plate driving forces in the oceans. Fig. 5.32 can also be used to explain why there is no place on the earth where the crustal thickness is significantly thicker than double of normal ($f_c \gg 2$). Such regions can only form by forces in excess of $F_b > 10^{14}$ N m^{-1}, which is greater than any known plate tectonic driving force.

• *Potential energy excess created by external forces.* The potential energy of plates may not only be increased by internal deformation of the plates, but also by passively uplifting the entire plate, for example by the vertical stresses exerted from upwelling mantle plumes to the base of the lithosphere. McKenzie et al. (1974), McKenzie (1977b) as well as Houseman and England (1986b) showed that these forces are large enough to lift lithospheric plates by several hundreds of meters. On the abyssal planes of the oceans, such upwelling convection streams may even cause topography of the order of 1 km (Crough 1983; Watts 1976). Such topography has a higher potential energy than its surroundings and will - just like a mountain range in a region of thickened crust - exert a horizontal buoyancy force onto its surroundings. The geometry of this phenomenon is schematically drawn in Fig. 5.33. The horizontal force exerted by plate A onto plate B may also be calculated by integrating eq. 5.47, just as we did when calculating the potential energy difference between mountains and their foreland (Fig. 5.27). The magnitude of this force corresponds to the shaded region between the two curves for vertical stress as a function of depth in Fig. 5.33. The principal difference between this and the example discussed in Figs. 5.27 and 5.31 is that the vertical stresses of the two profiles do not converge, because the two columns in Fig. 5.33 are not in isostatic equilibrium. Integration according to the same principles we used in eq. 5.49 to eq. 5.52 gives:

$$F_{\rm b} = \Delta E_{\rm p} = \frac{\rho_{\rm m}gH^2}{2} + \rho_{\rm c}gHz_{\rm c} \quad . \tag{5.55}$$

Using the same numerical values for the physical parameters as we did in Fig. 5.32, we get for an uplifted elevation of $H = 1$ km a horizontal force of the order of $F_{\rm b} \approx 9 \cdot 10^{11} \ {\rm N\,m^{-1}}$. This results shows that convectional stresses in the mantle may have a significant influence onto the stress regime and therefore on the deformation of continents (s. sect. 6.1.4).

Tectonic Relevance of Momentum. In this section we discuss the nature and relevance of momentum in tectonic processes to show that momentum is practically negligible to most geological problems. The momentum of a body I is given by the product of its mass m and its velocity v (sect. 2.2.4):

$$I = mv \quad . \tag{5.56}$$

While the velocities of plate tectonic motions are very small, the mass of plates is very large and it is therefore not immediately obvious if momentum plays a role in the tectonic force balance.

Momentum is a physical quantity that is preserved: During the collision of two plates the momentum of the entire system remains constant. However, the momentum of one of the plates may be transferred to the other. This transfer of momentum occurs by a force. The magnitude of this force is given by the change of momentum ΔI per time Δt that occurs during the slowing of plate motion due to collision.

$$F = \frac{\Delta I}{\Delta t} \quad . \tag{5.57}$$

If a plate is slowed down due to collision very abruptly, then the force is large. If it slows over a large time period, the force is small. The slowing of a plate has also the consequence that its kinetic energy $E_{\rm k}$ decreases. Kinetic energy is given by the integrated momentum integrated over the change in velocity:

Figure 5.33. Illustration for the calculation of the potential energy change that is caused by actively lifting plate A with the thickness $z_{\rm c}$ and the density $\rho_{\rm c}$ by the amount H. Note that the two plates are *not* in isostatic equilibrium and that, therefore, the two curves for σ_{zz} do not meet at the depth $z_{\rm c} + H$. The shaded area corresponds to the net horizontal force exerted by plate A onto plate B

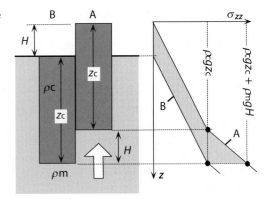

$$E_k = \int I dv = \int mv dv = \frac{mv^2}{2} \; . \tag{5.58}$$

Like *momentum*, the kinetic energy can change during the collision process.

Let us now estimate the force that arises as the consequence of the momentum change of the Indian plate during the India-Asia collision using some very simple assumptions. The area of the Indo-Australian plate is roughly $A = 5 \cdot 10^6$ km^2. If the mean plate thickness is $z_l = 100$ km and the mean density is $\rho = 3\,000$ kg m^{-3} then its mass is: $m = Az_l\rho = 1.5 \cdot 10^{21}$ kg. Let us further assume that the relative plate velocity between India and Asia is $v = 0.1\,\mathrm{m\,y}^{-1} \approx 3.2 \cdot 10^{-9}$ m s^{-1}. Thus, according to eq. 5.56 the momentum of the collision is: $I = mv = 4.7 \cdot 10^{12}$ kg m^{-1} s^{-1} and the kinetic energy of the Indian plate is: $E_k = 7.6 \cdot 10^3$ J. We also want to note that all assumptions made for this estimate are rather on the conservative side and that this value for the momentum can therefore be considered as a minimum. In the case of the India-Asia collision we happen to know that the relative rate of motion between the two plates has hardly changed since the Tertiary. Thus, the value of Δt in eq. 5.57 is very large and the force arising from the change of momentum is very small. Thus, the force balance of the India-Asia collision is not influenced by the effects of acceleration (or change in momentum). However, it is easy to show that even a dramatic slowing of the plate hardly changes the story. For example, assume that the Indian plate was brought to a complete halt within one million year. Then: $F = 4.7 \cdot 10^{12}/3.15 \cdot 10^{13} \approx 0.15$ N. Distributed over almost 5000 km of collision length this leaves only about: $3 \cdot 10^{-8}$ N m^{-1}. We can infer that plates would have to be brought to a halt within fractions of a second of a collision in order for momentum to have any influence on the orogenic force balance. In short, momentum is negligible in plate tectonics.

5.4 Problems

Problem 5.1. *Units of strain (p. 196):*
During orogenesis a continental crust has thickened from 30 km to 60 km. What is the stretch, the elongation and the vertical strain it has experienced? Use eq. 5.1.

Problem 5.2. *Difference between weight and mass (p. 196):*
What is the weight of 1 kg of rock at the surface? Give the result in SI units.

Problem 5.3. *Conversion of different energy forms (p. 199):*
A continent collides with another with a force of 10^{13} N and rams 100 km into the other continent. How much mechanical energy is released in the process? Discuss where this energy goes, i.e. into which other forms of energy it may be transformed.

Problem 5.4. *Formulations of stress state (p. 202):*
A continent is under extension because an attached subducting plate pulls
it apart. The tensional stress has the absolute value A. Another continent is
under extension because it collapses under the weight of a mountain range
on its surface. Let us assume that the vertical normal stress exerted by the
mountain range onto the plate has also the magnitude A. Are the states
of stress of the two continents the same? For simplicity, consider the two
continental plates to be represented by little cubes that have no weight of
their own: one that is pulled on its side and the other that is loaded from
above. Consider the problem only in two dimensions.

Problem 5.5. *Stress balance, pressure and deviatoric stress (p. 196– 205):*
Fig. 5.34 shows a rock that lies atop the plane $z = 0$. a) How large are the
vertical and horizontal normal stresses inside the rock at a given depth z?
b) How large is the pressure at this depth? c) How large are the components
of the deviatoric stress tensor? For your answers, use the stress balance equations (eq. 5.16 and corresponding relationships in the other spatial directions;
s. also eq. 6.16 and 6.17) and ignore elastic effects. Note that the coordinate
system for this problem (Fig. 5.34) implies that the upper surface of the rock
is at negative z. This is different from many other examples in this book
where the origin of the vertical axis is often located at the highest point of
the surface (see Fig. 4.1).

Figure 5.34. Illustration to Problem 5.5

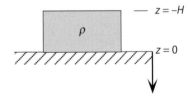

Problem 5.6. *Stress balance, pressure and deviatoric stress (p. 196– 205):*
The rock from Problem 5.5 has melted. The shape it has in Fig. 5.34 is
only maintained because we have put a box around it. a) How large are the
vertical and horizontal normal stresses inside the rock in a given depth z?
b) How large is the pressure at this depth? c) How large are the components
of the deviatoric stress tensor? For your answers, use the stress balance equations (eq. 5.16 and corresponding relationships in the other spatial directions;
s. also eq. 6.16 and 6.17).

Problem 5.7. *Stress balance and deviatoric stress (p. 196– 205):*
The rock from Problem 5.5 is warm and soft. It deforms under its own weight
according to the constitutive relationship of a Newtonian fluid with viscosity
η (eq. 5.37). We also assume that there is no friction between the rock and
the surface it lies on at $z = 0$. Find the relationship that describes the strain

rate with which the rock flows apart? Note that the horizontal strain rate
will depend on z.

Problem 5.8. *Stress balance and deviatoric stress (p. 196– 205):*
Assume that the rock from Problem 5.7 cannot flow freely apart but pushes
onto a fixed side wall (e. g. an indenter). How large is the force per meter
that the rock exerts onto the side wall?

Figure 5.35. Illustration for
Problem 5.9

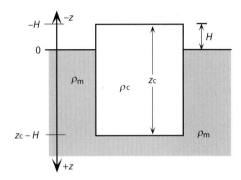

Problem 5.9. *Stress balance and deviatoric stress (p. 196– 205):*
Fig. 5.35 shows a rock of density ρ_c, floating in a fluid with the density ρ_m.
Write relationships describing the principal stresses, the pressure and the
principal components of the deviatoric stress tensor in the rock and in the
fluid, both as a function of depth.

Problem 5.10. *Understanding elastic deformation (p. 206):*
Granite has a Young's modulus of roughly 50 GPa. What is the elastic strain
of a granite to which a uniaxial stress of 50 MPa is applied? Use eq. 5.20 and
eq. 5.21.

Problem 5.11. *Understanding elastic deformation (p. 206):*
What is the elastic change in thickness of the lithosphere (assuming that
it is perfectly elastic) that arises solely as a function of its own weight?
Assume the pre-elastic thickness $z_1 = 100$ km as well as $E = 60$ GPa, and
$\rho = 3\,000$ kg m^{-3}.

Problem 5.12. *Fault plane solutions (p. 211):*
Draw qualitative fault plane solutions for the following faults: a) a north
- south striking vertical dip-slip fault where the eastern block is displaced
downwards; b) a horizontal fault where the top wall is displaced to the west;
c) a north - south striking vertical fault at which the eastern side was displaced with roughly 45° towards the north and down; d) a thrust that dips
a bit steeper than 45° to the east along which the upper plate was thrust
obliquely to the north.

Problem 5.13. *Power law rheology (p. 216):*
Use the material constants for quartz and olivine from Table 5.3 (the activation energies Q, the pre exponent constants A and power law exponent n) and calculate the shear stresses supported at strain rates of $\dot{\epsilon} = 10^{-14}\mathrm{s}^{-1}$ and a temperature of 500° C. Investigate how the stresses change if the power law exponents were actually 2 or 4 instead of 3. Compare your result with Fig. 5.12.

Problem 5.14. *Power law rheology (p. 216):*
The data from experiments that are used to derive the rheological material constants for viscous deformation (those required in eq. 5.41 are usually represented as lines in diagrams where $\log(\sigma_d)$ is plotted against $\log(\dot{\epsilon})$, or in diagrams where $\log(\dot{\epsilon})$ is plotted against $1/T$. Why ?

Problem 5.15. *Lithospheric strength (p. 228):*
Fig. 2.4 shows that seismicity along continental plate margins is distributed over much larger regions than seismicity along oceanic plate margins. This indicates that continental lithosphere is softer than oceanic lithosphere, although it has a larger thickness. Why is this so?

Problem 5.16. *Understanding potential energy (p. 232):*
Calculate the potential energy per unit area of the rock shown in Fig. 5.34.

Problem 5.17. *Understanding potential energy (p. 232):*
Estimate the potential energy (per unit area) of the two columns in Fig. 5.26 above the base of the lithosphere algebraically and graphically. For your graphic estimate, use the scheme illustrated in Fig. 5.31. For the calculation use the scheme explained in eq. 5.48 and $z_c = 30$ km; $z_1 = 100$ km; $\rho_1 = 2\,700\,\mathrm{kg\,m}^{-3}$, $\rho_3 = 3\,200\,\mathrm{kg\,m}^{-3}$ and $g = 10\,\mathrm{m\,s}^{-2}$.

Problem 5.18. *Understanding gravitational stress (p. 236):*
What is the lateral buoyancy force exerted by a mountain range onto its foreland if the mountain range is characterized by a crust that is twice the thickness from the foreland ($f_c = 2$), but the same thickness of the entire lithosphere ($f_l = 1$). Use eq. 5.52 and $z_c = 30$ km, $\rho_c = 2\,700\,\mathrm{kg\,m}^{-3}$, $\rho_m = 3\,300\,\mathrm{kg\,m}^{-3}$ and $g = 10\,\mathrm{m\,s}^{-2}$. In order to use eq. 5.52 we also need to know the surface elevation of the range H and the thickness of the root w, both at $f_c = 2$ and $f_l = 1$. Recalling eq. 4.29 the elevation is $H{=}5\,454$ m. Consequently, the thickness of the root is $w = 24\,546$m. (See also Problem 6.10.)

Problem 5.19. *Understanding gravitational stress (p. 243):*
Refine your estimate from Problem 5.18 using eq. 5.54 with $T_l = 1\,200°C$, $z_1 = 100$ km and $\alpha = 3 \times 10^{-5°}C^{-1}$. (Note that the surface elevation need not be known when using eq. 5.54). Compare your result with the result from Problem 5.19 and the graphically presented result in Fig. 5.32.

Problem 5.20. *Understanding gravitational stresses (p. 241)*:
Fig. 5.36 shows a schematic lithosphere: a) Normal thick lithosphere of thickness z_l with crust of thickness z_c; b) after homogeneous thickening to double thickness $(2z_l, 2z_c)$; c) after doubling the entire lithosphere by over thrusting. How large are the net horizontal forces that the columns exert on each other? Estimate the result graphically using the scheme used in Fig. 5.31. Calculate the result using eq. 5.44 to eq. 5.48) and use the parameter values given in the caption of Fig. 5.36.

Figure 5.36. Illustration to Problem 5.20. $\rho_c = 2\,700\,\mathrm{kg\,m^{-3}}$, $\rho_0 = 3\,300\,\mathrm{kg\,m^{-3}}$, $\rho_m = 3\,200\,\mathrm{kg\,m^{-3}}$, $g = 10\,\mathrm{m\,s^{-2}}$. Note that the surface elevation of the columns in b and c is the same as both columns have the same weight

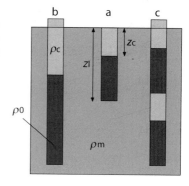

Problem 5.21. *Understanding momentum (p. 244)*:
A 100 km thick continental plate that has a mean density of $3\,000\,\mathrm{kg\,m^{-3}}$ and is $1\,000 \times 1\,000$ km large collides with a much larger continent at a velocity of $0.03\,\mathrm{m\,y^{-1}}$. The collision stops the plate. How large is the kinetic energy of the plate? How high can a mountain belt be built if all kinetic energy of the collision process is transformed into potential energy? Assume that all other forces that may apply to the plates may be neglected.

6. Dynamic Processes

This chapter is the first of two chapters in which we integrate the information of the previous three chapters. The first two thirds of this chapter are dedicated to the description of continents in extension and in collision, respectively. In the last third of this chapter we touch upon a range of interesting and currently very topical geodynamic problems.

6.1 Continents in Extension

Under certain stress states continents may extend. In the process, they usually decrease in their thickness. In general, extensional processes are divided into *active* and *passive* processes (e.g. Allan and Allen 1990; Ruppel 1995). An extensional process is considered to be *active* if the extension occurs as the consequence of forces *inherent* to the extending area, for example gravitational collapse of a region of high potential energy. Examples are mid ocean ridges, high continental plateaus or regions actively uplifted by mantle plumes (Keen 1980). We describe an extension process to be *passive* if the forces causing extension are applied *outside* the extending area, for example the force of a subducting plate that pulls at the passive margin of an adjacent continent (e. g. Le Pichon 1983). However, we should remember that the stress state of an extending plate is identical for both active and or passive extension (s. Fig. 5.3 and Problem 5.4).

• *Uplift or subsidence.* It is not trivial that extension of the lithosphere must lead to subsidence of the surface. The density of continental crust is *lower* than that of the underlying asthenosphere and the density of the mantle lithosphere is *higher* (s. Fig. 2.15). Whether or not extension leads to subsidence depends therefore on the partitioning of the extensional strain between the crust and the mantle part of the lithosphere. It also depends on the initial thickness ratio of the two prior to onset of extension (s. Fig. 4.16). McKenzie (1978) showed that surface subsidence during extension will only occur during homogeneous lithospheric extension if the crustal thickness at the onset was more than 14 km (Fig. 6.1), as part of a thermally equilibrated continental lithosphere. If the crust was thinner than this, then lithospheric extension will lead to surface uplift.

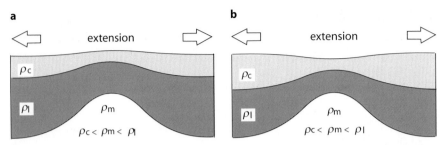

Figure 6.1. The influence of the initial thickness ratio of crust (light shaded area) and mantle part of the lithosphere (dark shaded areas) on the qualitative nature of the vertical motion of the surface during homogeneous lithospheric extension. The variables ρ_c, ρ_l and ρ_m are the density of the crust, mantle lithosphere and asthenosphere, respectively. In **a** extension causes surface uplift, because the mantle lithosphere constitutes a large proportion of the lithosphere by thickness. In **b** extension causes subsidence because the crust forms a larger proportion of the lithosphere

Subsidence and the development of sedimentary basins must not have been caused by extension: Several types of sedimentary basins form in collisional environments during crustal thickening. However, the processes of *continental extension* and *sedimentary basin formation* are so intimately related, that this first section of this chapter is concerned with the origin, nature and processes of sedimentary basin formation and we include in it the processes for some basins that form in collisional environments as well.

The temporal evolution of the subsidence of sedimentary basins is one of the most important sources of information for the geodynamic interpretation of continental extension processes. In many ways, the processes and methods of interpretation of the subsidence of basins is analogous to the processes and interpretation of surface uplift during collisional processes (e. g. in sect. 4.1.1, 4.2.1). However, they differ fundamentally in one respect: the thickness of a basin fill is usually substantially larger than the amount of subsidence that could be accredited to tectonic processes. Earth scientists first explained this phenomenon in connection with the COST1- and COST2-drill holes on the continental shelf of the eastern USA: the density of sediments is substantially larger than that of water or air. Thus, in isostatically compensated basins, sedimentary loading of tectonically formed basins will cause additional subsidence which in turn makes room for additional sediment loads.

Thus, we discriminate between *tectonic* subsidence and *total* subsidence. *Total* subsidence is the total amount of vertical change of the former surface. The *rate* of downward motion of the former surface is called the subsidence *rate*. As such this definition is not quite analogous to the definition of uplift rate (s. sect. 4.1). The *tectonic* subsidence is only the component of total subsidence that is caused by tectonic mechanisms. In order to interpret the tectonic processes that lead to sedimentary basin formation it is necessary to know the total subsidence. It is then necessary to subtract the influence of sedimentary loading from the total subsidence to determine the tectonic subsidence (Sclater and Christie 1980; Steckler and Watts 1978, 1981). This mental exercise is referred to as *back stripping* and will be discussed in sect. 6.1.3.

6.1.1 Basin Subsidence Mechanisms

The subsidence of just about all sedimentary basins is caused by one or more of the following three processes. Each of these processes has been discussed in previous sections of this book:

– isostatic subsidence,
– flexural subsidence,
– thermal subsidence.

The three mechanisms may be intimately related. In the following we discuss briefly the nature of these three mechanisms and then go on to explain different types of sedimentary basins using these terms. Good summaries of models for the development of sedimentary basins are published by Angevine et al. (1990) and Allen and Allen (1990) and we rely here – in part – on their work.

Isostatic subsidence is caused by physical changes in the thickness of the lithosphere. For example, if physical stretching of the lithosphere causes thinning, then isostatic compensation will generally lead to subsidence (sect. 4.2.1; Fig. 6.2).

Flexural subsidence relies on elastic bending of the lithosphere (s. sect. 4.2.2). If the lithosphere is loaded, it bends and a basin forms near the load (Fig. 4.19). For very strong plates, such basins are wide and shallow, while for less competent plates such basins are narrow and deep. However, the basin volume is independent of the rigidity of the plate (s. Fig. 4.22).

Thermal subsidence occurs because the density structure of the lithosphere is thermally changed by cooling (sect. 4.2.1). Thus, thermal subsidence is also a type of isostatic subsidence, except that the thickness change is caused thermally and not mechanically. As cooling of the lithosphere occurs only in thermally *de*stabilized lithosphere, thermal subsidence can only occur in lithosphere that has previously been heated. Everything else being equal,

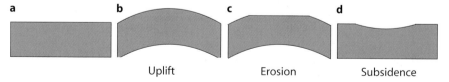

a	b	c	d
	Uplift	Erosion	Subsidence

Figure 6.2. Sketch illustrating one of the first models for the development of sedimentary basins (Sleep 1971). A continent is uplifted by external processes, for example by upwelling asthenospheric material of a mantle plume. Erosion thins the uplifted region. When the uplifting process terminates, the continent returns to its original position and a shallow sedimentary basin is formed

the amount of thermal subsidence during cooling is exactly as large as the amount of thermal uplift during the heating phase. Thus, no sedimentary basin can form as the consequence of thermal processes on their own. Erosion or extension must have separated the thermal uplift and thermal subsidence in order to form a sedimentary basin (Fig. 6.2; sect. 6.1.4; McKenzie 1978).

6.1.2 Basin Types

Different types of sedimentary basins were classified by Dickinson (1976) according to the three subsidence mechanisms discussed above (s. also Buck 1991).

• *Passive margins and rift basins.* Rift basins form as the consequence of continental extension and ultimately rifting (Fig. 2.21a). The extensional process during the formation of rift basins may be symmetrical (Keen et al. 1989) or asymmetrical (Wernicke 1985; Lister et al. 1986, 1991) about the rift axis. The subsidence associated with the isostatic compensation of the rifting is usually followed by a later phase of thermal subsidence during which the mechanically rifted mantle lithosphere thickens by cooling. Thus, the subsidence of rift basins may usually be divided into a *rift phase* and a *sag phase*, both of which are characterized by specific sedimentary environments. During the rift phase, sedimentation is rapid, highly energetic and associated with the development of half grabens and other tectonic structures. During the sag phase, sedimentation is slow and static. Both phases are best developed if the rifting has not gone to completion. Examples for rift basins that *have* gone to completion in their successive stages of development are to be found in the Rhein Graben, the East African Rift system, the Red Sea and the Atlantic coast.

• *Transform basins.* Transform- or pull-apart basins also form due to continental extension. The most important difference between these and rift basins is that they are smaller as their extensional phase terminated much earlier. Transform basins never get to a rifting stage. They are bound on at least two sides by strike slip faults and they are usually rectangular or diamond shaped. Because of their limited size, heat conduction processes do not only occur in the vertical direction, but also in the lateral direction. As a consequence, thermal thinning of the mantle lithosphere is limited in transform basins. Therefore, transform basins usually lack the sag phase that is so typical for rift basins (Pitman and Andrews 1985). Subsidence of transform basins is usually short-lived and is largely a linear function of time. This is especially because the deformation history of most brittle structures that control their shape is very short-lived, before other structures in the same orogen partition the strain. Examples of transform basins are the Death Valley in California, the Vienna basin in Austria as well as a large number of small intramontane basins within the European Alps.

• *Foreland basins.* Foreland basins form during the collision of two continental plates and are the continental analogue to fore-arc and back-arc basins (see next paragraph). The principle subsidence mechanism is elastic flexure of the plate in response to the loading by external and internal loads (Beaumont 1981; Jordan 1981; Karner and Watts 1983). According to their location relative to the lower plate, foreland basins may be divided into two groups (Fig. 2.20). *Peripheral* foreland basins form near subduction zones in collisional environments as a consequence of loading of the lower plate by the upper plate. *Retroarc* foreland basins form on the upper plate in the hinterland of a subduction zone. Good examples of peripheral foreland basins are the molasse basins near the Alps or the Himalayas. Examples of retro arc basins are those that form east of the Andes. Their formation is not very well-explained. However, the subsidence rate in peripheral foreland basins may be used to determine the rate of loading of the plate and therefore ultimately the collision rate of two plates.

• *Fore-arc basins.* Fore-arc-basins have their name because they form in front of an island arc. There is a range of models that have been used to explain their origin, but none of them is really well-constrained or completely satisfactory. Some of the models include: 1. Subduction of an oceanic plate underneath another leads to a doubling of the plate thickness beneath the accretionary wedge. Since the density of oceanic lithosphere may be higher than that of the underlying asthenosphere, doubling the plate thickness leads to subsidence and formation of a fore-arc basin. 2. Subduction of a cold plate underneath a hot plate may cause cooling of the upper plate and thus lead to thermal subsidence and basin formation. 3. Loading of the plate from above by an island arc and loading from below by the buoyancy of the accretionary wedge may lead to elastic back-bending of the plate.

Basins on oceanic lithosphere that form behind a subduction zone are called *back-arc* basins. Their formation is usually interpreted as the consequence of upwelling asthenospheric material in the mantle wedge. However, they also have been thought to be connected with potential energy differences (Stüwe and Barr 2000).

• *Intracontinental basins.* Some large sedimentary basins form intracontinentally, for example the Michigan basin in the USA. The amount of tectonic subsidence in these basins is rarely more than 2 km. Their round shape and slow subsidence rates indicate thermal subsidence as a subsidence mechanism. However, the origin of these basins remains largely unconstrained.

6.1.3 Subsidence Analysis

The large variety of sedimentary basins discussed in the last section illustrates the necessity to find a data set that can be collected in the field and that can be used to constrain the nature of the basin forming process. Such a data set exists in the subsidence history of a basin as recorded by the sedimentary basin fill. In order to constrain the evolution of tectonic subsidence from the stratigraphic record, three steps are necessary:

– Documentation of the stratigraphic section,
– Consideration of compaction of sediments,
– Consideration of the water depth.

If we are also interested in the tectonic component of the subsidence (as we usually are), then a fourth step is necessary:

– Consideration of the sedimentary loading: back stripping.

When mapping the stratigraphy of the basin fill with the intention of using it for subsidence analysis, the following data must be collected or assumed for each layer: 1. thickness, 2. lithology, 3. age and 4. water depth at deposition. Porosity of the sediments and information on the thermal evolution are additional data that can be extremely helpful. On the following pages we show how the subsidence history may be extracted from this data.

Compaction. Because of their porosity, sedimentary strata are compacted by overlying layers after their deposition. Thus, the thickness of each layer in a sedimentary sequence was larger at the time of its deposition than it is when measured in the field. In order to consider the influence of sediment compaction on the thickness and density of the stratigraphic column, the porosity must be known. Empirical studies show that the porosity of rocks decreases exponentially with depth. In general we can describe this with the relationship:

$$\phi = \phi_0 e^{-cz} \quad . \tag{6.1}$$

There, ϕ is the porosity of the rock at depth z, ϕ_0 is the porosity at the surface and c is a rock specific compaction constant (Fig. 6.3). As we will need to keep track of porosities, thicknesses and densities of a whole sequence of layers, we need to define a few more variables for the following calculations. We define L as the measured thickness of a general sedimentary layer in the sequence, L_i as the thickness of a specific layer i in the column,

Figure 6.3. The decrease of porosity of a range of rock types with depth. Calculated with eq. 6.1 and using the following material constants. Sandstone: $\phi_0 = 0.4$, $c = 3 \cdot 10^{-4}$ m^{-1}; limestone: $\phi_0 = 0.5$, $c = 7 \cdot 10^{-4}$ m^{-1}; slate: $\phi_0 = 0.5$, $c = 5 \cdot 10^{-4}$ m^{-1}. The grain density ρ_g of these three rock types is: sandstone: $\rho_g = 2\,650$ kg m^{-3}; shale: $\rho_g = 2\,720$ kg m^{-3}; limestone: $\rho_g = 2\,710$ kg m^{-3} (data from Sclater and Christie 1980; s. also Bond et al. 1983)

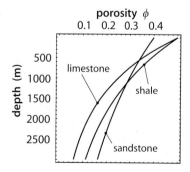

L_0 as the thickness of a general layer at the time of sedimentation, L^* as the decompacted thickness of the column after removing a single layer, and L_i^* as the decompacted thickness of a specific layer. In order to derive the original thickness of a layer L_0 from that measured for this layer in the field, L, we must solve an integral. Assuming that the thickness was only changed by changing the porosity (and not by metamorphism, diagenesis, cementation or dissolution), it must be true that:

$$\int_0^{L_0} (1 - \phi)\mathrm{d}z = \int_{z_L}^{z_L+L} (1 - \phi)\mathrm{d}z \quad . \tag{6.2}$$

This equation states that the rock volume without the pore space $(1-\phi)$ remains a constant, regardless of the fact whether the layer is at depth $z = z_L$ or at $z = 0$. It may be said that eq. 6.2 is a one-dimensional volumetric balance. This equation forms the basis of our following considerations. Clearly it is important to check the applicability of this equation before embarking on a compaction analysis. If the sediment is cemented or partially dissolved, then it is easily possible that the processes did *not* occur at constant volume and the compaction analysis becomes more complicated. It may then be necessary to study the cementation material petrographically to see if it was derived internally or externally, and so on. It comes back to the universal fact that field and laboratory data determine how simple a model is allowed or how complicated it must be designed.

By substituting eq. 6.1 into eq. 6.2 and integrating we get:

$$L_0 + \frac{\phi_0}{c}e^{(-cz_0)}(e^{(-cL_0)} - 1) = L + \frac{\phi_0}{c}e^{(-cz_L)}(e^{(-cL)} - 1) \quad . \tag{6.3}$$

Sadly, it is impossible to solve eq. 6.3 for L_0 - the original depositional thickness of a layer of which we measured the thickness L at depth z. When using eq. 6.3, L_0 can only be determined numerically by iteration (sect. A.5.2). However, for most cases this is not necessary. It is usually sufficient to use the following approximation:

$$L_0 = \frac{(1 - \phi)L}{1 - \phi_0} \quad . \tag{6.4}$$

Using eq. 6.4 it is now possible to calculate the original thickness of a layer at the end of its sedimentation L_0 from field data on the porosity ϕ and present thickness L. For this, the porosity ϕ_0 is calculated using eq. 6.1. The original thickness of the layer is needed for the further steps in the subsidence analysis. Obviously, eq. 6.4 can also be used to determine the thickness of a layer at any other stage of the decompaction process L^*. We just need to use the porosity at the respective stage (and depth) of the analysis ϕ^* instead of the original porosity ϕ_0. ϕ^* can also be calculated with eq. 6.1 (Fig. 6.3).

If we have additional information on the porosity or thermal evolution of our rocks, it is possible to refine the compaction analysis. Using eq. 6.1 and eq. 6.4 it is possible to calculate subsidence curves for the basin floor. The method is also illustrated on Fig. 6.4.

The subsidence evolutions that may be calculated with the method described above give us the depth evolution of the basin floor underneath the surface of the basin fill as a function of time. However, it is important to note that the surface of the basin fill need not remain at a constant depth below (or above) sea level. If, for example the water depth in a sedimentary basin changes over time, then the water depth must be added to the subsidence curve to obtain the subsidence evolution relative to a fixed reference level. In marine basins, the water depth at the time of deposition can usually be constrained by lithology, sedimentary structures and fossil record. If the sea level itself changes during deposition, the interpretation becomes more difficult. Sometimes it is possible to document sea level changes by comparing synchronous stages in the sedimentation record of two independent basins subjected to the same sea level change. Basins that develop in terrestrial environments are much harder to interpret, as it is difficult to document the changes of surface elevation through time.

Backstripping. In the previous sections we have shown how to determine the evolution of total subsidence (i. e. the depth evolution of the basin floor) as a function of time. This subsidence history is the sum of *tectonic subsidence* and subsidence caused by the *sedimentary loading*. The process of determining the *tectonic* subsidence from the *total* subsidence is called *back stripping*

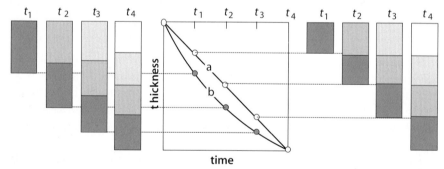

Figure 6.4. Illustration showing the influence of compaction on a sedimentary section. The columns on the left and right of the central diagram show cartoons of stratigraphic columns as they evolve through times t_1, t_2, t_3 and t_4. On the left these columns are for sediments that compact during successive deposition, on the right these columns are for sediments that do not compact. For both sets of columns the oldest layer is shaded the darkest, the youngest the lightest. At present (t_4) the stratigraphy of both profiles is identical. However, these identical columns at time t_4 were arrived at by different sedimentation and subsidence histories. On the right sedimentation rates were constant through time and none of the units were compacted (curve a on the central diagram). On the left, sedimentation rates were rapid at time t_1 and then decreased through time (curve a). The similar thickness of all units at time t_4 arises here only because compaction has balanced the variations in sedimentation rate for the columns on the left

(Watts and Ryan 1976; Steckler and Watts 1978; Sclater and Christie 1980). The inverse of back stripping – *back stacking* – is also possible, using thermochronological data from sedimentary basins (Brown 1991).

Back stripping is a mental exercise in which sedimentary layers are removed successively from a basin. During each step of removal, the hypothetical depth of the basin floor *without* being loaded is calculated. Depending how careful we need to do this (as required by the large scale geological situation) we discriminate between:

– Back stripping assuming hydrostatic isostatic compensation,
– Back stripping assuming flexural isostatic compensation.

Back stripping assuming hydrostatic isostasy is straight forward as all we need to do is apply eq. 4.17. We will illustrate this with a simple example of a marine (i. e. water covered) basin that was created by a single tectonic process and is now filled with a single sedimentary layer of thickness L and density ρ_L (Fig. 6.5). We assume some tectonic process that created a basin of the depth z_T at the start. On Fig. 6.5 we can see that this tectonic amount of subsidence (z_T) can be written as the sum of the water depth at present w and the basin depth change due to sedimentary loading: $z_T = w + z_s$. In order to derive the tectonic amount of subsidence from our present observation after the basin was filled with sediments we need to calculate z_s: the depth of the basin floor prior to the sedimentary fill below sea level. The calculation of z_s follows the same principle as the relationships we have used for the calculation of isostatic equilibrium (sect. 4.2.1; Problem 6.4), (Steckler and Watts 1978):

$$z_s = L \left(\frac{\rho_m - \rho_L}{\rho_m - \rho_w} \right) \quad , \tag{6.5}$$

if we disregard any change in sea level during the period or:

$$z_T = L \left(\frac{\rho_m - \rho_L}{\rho_m - \rho_w} \right) + w - \Delta SL \frac{\rho_m}{\rho_m - \rho_w} \quad , \tag{6.6}$$

if we formulate it in terms of z_T and also consider any potential sea level change.

Remember that the densities ρ_m, ρ_L and ρ_w are those of the asthenospheric mantle, that of the sediment layer L and that of water, w is the water depth and ΔSL is the change in sea level during the deposition of the unit of thickness L (Fig. 6.5).

The thickness of the crust or even that of the lithosphere are not needed for eq. 6.5, as we assume that no thickness change of the crust occurred during the sedimentation, i.e. we assume that the tectonic amount of subsidence occurred before the onset of sedimentation. If the basin was not filled by water at any stage of the evolution, then ρ_w must be substituted by $\rho_{air} = 0$ for the corresponding period. Eq. 6.5 is analogous to eq. 4.20, but it is solved for another variable. The density of the porous sediment layer ρ_L, that occurs in eq. 6.5 may be determined from the grain density ρ_g and the pore fluid

Figure 6.5. Cartoon illustrating the origin of eq. 6.5 (s. also Problem 6.4). The right hand column shows a crustal profile in isostatic equilibrium *after* tectonic subsidence but *before* fill of the depression by sediments. The left hand profile shows the total subsidence of the column after the sedimentation of a layer with the thickness L. ρ_c, ρ_m and ρ_w are the densities of crust, asthenospheric mantle and water, respectively and z_c is the thickness of the crust. z_s is the change of surface after removing the sedimentary basin fill and compensating isostatically. As ρ_c and z_c are the same in both columns, they cancel out in eq. 6.5

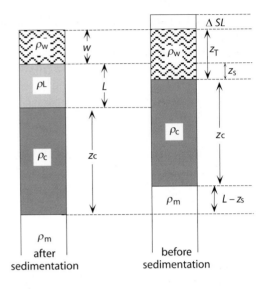

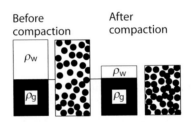

Figure 6.6. Cartoon illustrating eq. 6.7. Compaction of sediments decreases the water filled pore volume, but not the grain volume (unless diagenesis accompanies compaction). This is schematically shown by two profiles of water volume and grain volume for both the situation before compaction and after compaction. The variable ρ_w is the density of the pore fluid and ρ_g is the grain density

density ρ_w (which - in most cases - is the density of water: $\rho_w = 1\,000$ kg m^{-3}) from the relationship:

$$\rho_L = \phi \rho_w + (1 - \phi)\rho_g \quad , \tag{6.7}$$

if the porosity ϕ is known (Fig. 6.6). The amount of tectonic subsidence of basins that are filled by more than one sedimentary layer may also be determined with eq. 6.5. All we need to do is use the mean values for L and ρ_L from all layers in the sequence. In fact, using eq. 6.5 we can determine the complete evolution of tectonic subsidence by stepwise removal of the respective top layer and using the mean densities and thicknesses of the remaining sedimentary column to calculate the amount of tectonic subsidence during each time step. For this, we substitute L and ρ_L by L^* and ρ_{L^*}. The value of z_T is then the tectonic amount of subsidence during sedimentation of the top most layer; L^* and ρ_{L^*} are the decompacted thickness and density of the remaining sediment column after removal of layer i. The thickness of the total sediment pile at the start of sedimentation of layer i is:

$$L^* = \sum_{j=1}^{i} L_j^* \ . \tag{6.8}$$

The density of the sedimentary column underneath layer i is given by the mean density of all remaining layers. This is given by the sum of all densities, multiplied by the respective thickness and divided by L^*:

$$\rho_{L^*} = \frac{\sum_{j=1}^{i} L_j \left(\phi_j \rho_w + (1 - \phi_j) \rho_g \right)}{L^*} \ . \tag{6.9}$$

Now we can use eqs. 6.8 and 6.9 to determine the tectonic subsidence history of a basin by stepwise reconstruction. If we do this by hand, then we need to iteratively apply eq. 6.1, eq. 6.5 and eq. 6.7. An example is shown in Fig. 6.7 (s. also Problem 6.6).

6.1.4 Models of Continental Extension

We have shown in the last section how a careful analysis of the sedimentary basin fill may be used to constrain the tectonic subsidence history of a basin. Based on analyses of this kind it was recognized that the evolution of subsidence of many sedimentary basins follows very process-specific patterns in time. Some of these characteristic patterns and simple models that have been used to explain them are discussed on the next pages.

The McKenzie- and its Follow up Models. Subsidence analysis has shown that the tectonic subsidence rate of many sedimentary basins is rapid at first and then decreases abruptly to continue at a much slower rate for a much longer time. The model of McKenzie (1978) was one of the first and certainly most famous models that has been successfully used to explain this pattern. Despite its simplicity it remains the basis of a large range of more refined models. Like other models of its time (e.g. Le Pichon et al. 1982) the model is one-dimensional and describes the subsidence of the surface as a function of lithospheric extension. Within the model, typical subsidence histories are divided into two phases: a *rift phase* and a *sag phase*. The model holds well in examples where the phase of physical extension of the lithosphere (rift phase) is short compared to the duration of the subsequent thermal equilibration (sag phase). In fact, within the original McKenzie (1978) model, the stretching phase (rift phase) is assumed to have occurred instantaneously (Fig. 6.8). The amount of subsidence during this instantaneous rifting phase H_{rift} may be calculated with the relationship:

$$H_{\text{rift}} = z_{\text{c}} \left(\frac{\rho_0 - \rho_{\text{c}}}{\rho_0 - \rho_{\text{w}}} \right) \left(1 - \frac{1}{\delta} \right) - z_{\text{l}} \left(\frac{\rho_0 \alpha T_{\text{l}}}{2(\rho_0 - \rho_{\text{w}})} \right) \left(1 - \frac{1}{\beta} \right) \ . \qquad (6.10)$$

Equation 6.10 is largely analogous to eq. 4.29, which we have previously used to describe the changes in surface elevation as a function of changed thicknesses of crust and mantle part of the lithosphere. Therefore, we do not derive the details of eq. 6.10 here and the reader is referred to sect. 4.2.1. In eq. 6.10, z_{c} and z_{l} are the thickness of crust and lithosphere. ρ_{c}, ρ_{w} and ρ_0 are the densities of crust, water and mantle at $0°C$. α is the coefficient of thermal expansion, T_{l} is the temperature at the base of the lithosphere, δ the stretching parameter of the crust and β that of the mantle part of the lithosphere (where stretching parameter is defined as the ratio of starting thickness to stretched thickness). As such eq. 6.10 is already a refined McKenzie model: it allows to explore the consequence of different amounts of stretching in the crust and in the mantle lithosphere (s. e. g. Royden and Keen 1980). In the original McKenzie model $\delta = \beta$. In other words, the original McKenzie model assumed homogeneous stretching of the entire lithosphere. The stretching parameter of the crust δ is the inverse of the thickening strain f_{c} which we have

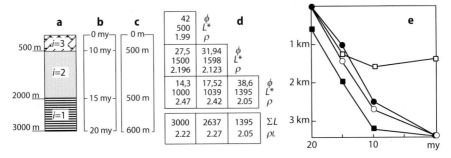

Figure 6.7. Example for a subsidence analysis with consideration of compaction. **a, b** and **c** show the field data of a simple basin filled by three sedimentary strata. **a** Lithology and thickness of strata (slate: *horizontal lines*; sandstone: *shaded*; limestone: *brick signature*); **b** age of the unconformities; **c** water depth as derived from fossil record. No sea level changes occurred during the sedimentation process. Strata are numbered from base to top. The center of each layer is the reference point used for the following calculations. For the values of porosity and density we use the data from Fig. 6.3. Porosity and density of the top layer (with a mean depth of 250 m) may be derived from eq. 6.1 and eq. 6.7. These data are written in the 1st column of Table **d**. The second column lists the porosity of the second layer (with a mean depth of 750 m = mean depth of second layer minus the thickness of the first layer) and so on. The thickness of the second layer is given by eq. 6.4 using the porosities of the 1. and 2. column as well as the thickness from the first column. The sum of the thicknesses and densities in the bottom two rows were calculated with eq. 6.8 and eq. 6.9 respectively. Their mean values are written in the third column. In **e** the result of the subsidence analysis are illustrated graphically. *Black dots* show the measured field data, *white dots* the calculated decompacted thicknesses and *black squares* show the decompacted thicknesses plus water depth. The curve given by the white squares was calculated with eq. 6.5 and shows the tectonic component of subsidence

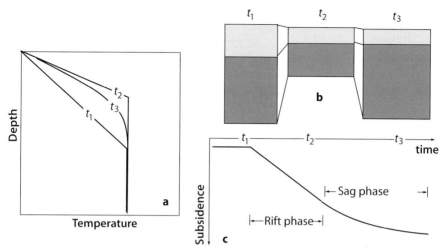

Figure 6.8. Subsidence of sedimentary basins according to the model of McKenzie (1978) and that of Jarvis and McKenzie (1980). **a** Geotherms at three different times: prior to onset of extension (t_1), at the end of a rapid stretching phase (t_2) and during subsequent thermal equilibration of the lithosphere (t_3). **b** Thickness of the crust (light shaded region) and that of the mantle part of the lithosphere (dark shaded region) at these three different time steps. **c** Schematic subsidence curve corresponding to the cartoons in **a** and **b**

used in several other sections of this book: $\delta = 1/f_c$ (s. sect. 4.0.1). However, β does *not* correspond to $1/f_l$, but rather to $1/f_{ml}$ (s. Problem 4.2). Because of this eq. 6.10 looks somewhat different from eq. 4.29, although both may be used to answer similar questions (Fig. 4.16). When we discussed eq. 4.29 we saw that the initial ratio of crust to mantle lithosphere determined if homogeneous thickening of the whole lithosphere leads to surface uplift or subsidence. The same is true for eq. 6.10 (s. Problem 6.7).

The subsidence described by eq. 6.10 (Fig. 6.9a) is followed by a phase of thermal equilibration: the *sag phase*. During this subsequent phase, active extension has stopped and thickening of the mantle part of the lithosphere occurs due to cooling. The surface subsides as a consequence of cooling (Fig. 6.8). This is largely analogous to the surface subsidence associated with cooling of the oceanic lithosphere (Fig. 3.22). Thus, we can describe this subsidence with a similar model to the one we have discussed in sect. 3.5.1. The principle difference to the cooling of oceanic lithosphere arises because McKenzie assumed the bottom boundary not at infinity, but at depth z_l. Because of this boundary condition at *finite* depth, the heat conduction equation may only be solved with the aid of Fourier series (s. sect. A.4). As a consequence, the solution presented below contains trigonometric functions instead of an error function (s. sect. A.4). The solution is:

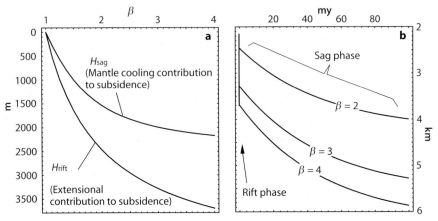

Figure 6.9. a Subsidence of rift basins at the end of the rift phase (H_{rift}) and at the end of the sag phase (H_{sag}) both as a function of the amount of initial stretching as given by β. It is assumed that the lithosphere stretched homogeneously, i.e. $\delta = \beta$ (s. eq. 6.10). The total amount of subsidence is given by the sum of the two curves. For $\beta < 1$ the curve go to negative values, i.e. thickening leads to uplift. **b** Subsidence during the rift and subsequent sag phase as a function of time for two different values of β (describing different amounts of rifting at the onset). All curves were calculated using eq. 6.10, eq. 6.11 and the following values for the variables: $z_1 = 125$ km; $z_c = 35$ km; $T_1 = 1\,280\,°C$; $\rho_0 = 3\,300$ kg m^{-3}; $\rho_c = 2\,750$ kg m^{-3}; $\rho_w = 1\,000$ kg m^{-3}; $\alpha = 3 \cdot 10^{-5}\,°C^{-1}$ and $\kappa = 10^{-6}$ m^2 s^{-1}.

$$H_{sag} = \left(\frac{4\rho_0 \alpha T_1 z_1}{\pi^2 (\rho_0 - \rho_w)} \right) \left(\frac{\beta}{\pi} \sin(\pi/\beta) \right) \left(1 - e^{-t/\tau} \right) \ . \qquad (6.11)$$

This relationship describes the subsidence of rift basins during the sag phase as a function of time t. The parameter τ is the time scale of thermal equilibration and is given by: $\tau = z_1^2/(\pi^2 \kappa)$ (which differs from the time constant discussed on p. 58 by the factor π^2, but is in principle equivalent). While we will not derive the origin of this equation in any detail, we note that it should be possible, at least in principle, to follow it from the information given in previous chapters of this book (s. e. g. eq. 4.63). In contrast to eq. 4.63, eq. 6.11 contains no infinite summations, because it is an approximation in which all terms for $n > 1$ were omitted. Aside from this, eq. 6.11 (and also eq. 4.63) describe a similar model to that of eq. 4.39, which we have used to calculate the water depth in the oceans. Examples of subsidence during the sag phase are shown in Fig. 6.9b. For very large times, the last term of eq. 6.11 becomes 1 and the simplified equation may be used to estimate the total amount of subsidence during the entire sag phase.

● *Finite extension.* In contrast to the assumptions of the McKenzie model, many sedimentary basins show evidence that the duration of physical stretching or rifting of a continent was comparable to the thermal time constant of the lithosphere, i.e. rifting and sagging phase may be compared on the same time scale. In order to account for finite duration of the rifting phase Jarvis and McKenzie (1980) and Cochran (1983) expanded the McKenzie model. Jarvis and McKenzie (1980) suggested as a rule of thumb that the duration of stretching, t, must only be considered in their model for basin development if the following relationship holds:

$$t < \frac{60\mathrm{my}}{\beta^2} \ , \text{if } \beta < 2 \quad \text{or} \quad t < 60\mathrm{my}\left(1 - \frac{1}{\beta}\right)^2 \ , \text{if } \beta > 2 \ . \tag{6.12}$$

If the duration is shorter than time t, then it is sufficient to assume that stretching was instantaneous and occurred prior to any thermal equilibration.

• *Two-dimensional models for continental extension.* Most rift basins are one-dimensional in their geometry, i. e. they are long compared to their width. If however, lithospheric stretching occurs two-dimensionally so that basins extend into two direction, then the modeling of subsidence during the sag phase must account for two-dimensional heat conduction. Some of the first two-dimensional models describing this were designed by Buck et al. (1988), Issler et al. (1989) and Wees et al. (1992). Those models assume that extension occurs symmetrically about the rifting axis.

• *Heterogeneous extension.* One of the first models describing asymmetric extension of the lithosphere was designed by Oxburgh (1982). Oxburgh assumed that both the crust and the mantle part of the lithosphere extend homogeneously, but that the location of maximum crustal extension is laterally displaced from the location of maximum extension of the mantle part of the lithosphere. Wernicke (1985) and Lister et al. (1986) were among the first who explored the consequences of simple shear geometries on lithospheric extension processes by assuming low-angle normal faults that transect the entire lithosphere. These models were based on observations in the Basin and Range province of the western US. Lister and Etheridge (1989) and Lister et al. (1991) applied this model to the east coast of Australia in order to explain the simultaneous uplift of the Australian Great Dividing Range and the subsidence of continental lithosphere at the west coast of New Zealand (Fig. 6.10).

An elegant analytical solution describing some aspects of the models of Wernicke and Lister was published by Voorhoeve and Houseman (1988). Some geometric considerations of extension models based on heterogeneous stretching geometries may be found in Buck (1988).

Dynamic Extension Models. In all previous paragraphs we have considered the evolution of extension and sedimentary basin formation purely on the basis of kinematic and thermal assumptions. We will now consider *dynamic* models, that is, models for continental extension that are based on mechanical assumptions. One of the many questions that can only be explained with dynamical models relates to the causes for terminating extension: we observe that some rift basins extend until rifting occurs and a passive continental margin forms (sect. 2.4.4), while in others the extension was limited. Examples for the former are the central African rift or the Red Sea, examples for the latter include the Michigan basin in the US, the Cooper basin in Australia of the Pannonian basin in Europe. There is two fundamentally different processes that may be responsible for the termination of extension:

1. *External reasons.* The forces or velocities rate with which a plate is pulled decrease. In this case the termination of extension has nothing to do with the plate itself and is only a function of processes in the surrounding plates. Extension is controlled by the boundary conditions.
2. *Internal reasons.* Extension may terminate because the rheology of the plate changes to become stronger. In this case no changes in the boundary conditions need to occur; the plate boundary forces may remain constant. The extension process is subject to a force balance (s. sect. 6.2.2) and is called *self limiting.*

Analogous to the terms *active* and *passive extension*, these two processes might be called *active* and *passive termination* to extension. Which of these two mechanisms caused the termination of a particular basin is a question which often can only be answered using dynamical models. Some of the first dynamic models for continental extension were those of Bassi (1991), Bassi et al. (1993) and Cloetingh et al. (1995). These models showed that the extension geometry is strongly dependent on the tectonic processes in a plate that occurred *prior* to the onset of extension. For example, they showed that continental extension onsets most easily in regions of thickened continental

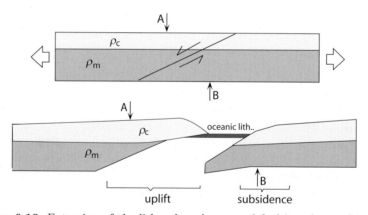

Figure 6.10. Extension of the lithosphere by normal faulting along a low angle normal fault that transects the entire lithosphere. The upper cartoon shows the situation at the onset of extension, the lower cartoon after full separation of the two plates and the development of passive continental margins. Note that at location A, extension only decreased the thickness of the mantle part of the lithosphere, while at location B, extension only decreased the thickness of the crust. Lister et al. (1986) and Wernicke (1985) interpreted that the uplift of the left hand plate at location A (e. g. eastern Australia) and the subsidence of the right hand plate at location B (i. e. western New Zealand) are only caused by the changed thickness ratio of crust and mantle lithosphere at these two locations

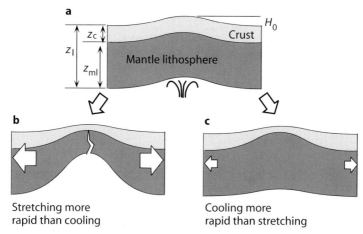

Figure 6.11. Possible evolution of a passive extension process initiated by uplifting of the entire lithosphere for the amount H_0 by upwelling asthenospheric material over mantle plumes. **a** Starting geometry at the onset of extension. **b** If the extensional forces are large enough, then the lithosphere is thinned and thermally weakened until a rift zone develops. **c** If the extensional forces are small, thickening (and strengthening) of the mantle lithosphere by cooling is more rapid than physical stretching. The extension process is self limiting. Note that the mantle lithosphere in **c** is thickest in the region of maximum crustal thinning

crust, as this is the region where the plate is the weakest (s. sect. 5.2.1, 6.3.6; Houseman and England 1986b). As another example, it was shown by Buck (1991) that the width of continental rift basins depends on the geothermal gradient and ultimately on the rheology. One of the more elegant models explaining the temporal limitation of extension processes is that of Houseman and England (1986b). In the following section we discuss this model in some more detail.

• *The model of Houseman and England.* The one-dimensional model for continental extension by Houseman and England (1986b) is a coupled thermal and mechanical model in which extension processes are investigated as a function of both thermal and rheological development. For this, the authors assumed a model geotherm and a model rheology for the lithosphere. Extension is assumed to be driven by upwelling asthenosphere of a mantle plume which initially lifted the entire continental plate by the amount H_0 (Fig. 6.11a). As the extension is triggered only by the increased potential energy due to this uplift, the extensional process is passive. Fig. 6.12 illustrates the evolution of extension in response to this uplift (s. also Fig. 5.33). According to the calculations of Houseman and England (1986) this extension can ultimately lead to three different scenarios:

1. If H_0 is smaller than about 100 m, then the extensional forces caused by the excess potential energy are too small to be reflected in any appreciable strain rate.

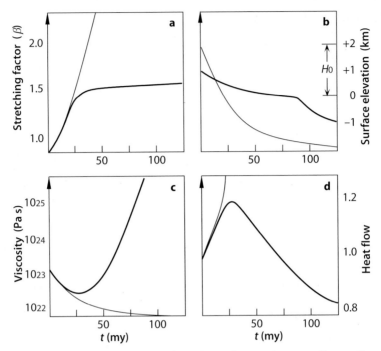

Figure 6.12. Two model evolutions of continental extension according to the model of Houseman and England (1986b). *Thick lines* are for conditions that cause a self limiting extension process. *Thin lines* are for conditions that ultimately lead to rifting. **a** the evolution of the stretching factor; **b** evolution of surface elevation; **c** evolution of viscosity; **d** evolution of surface heat flow (normalized to that at the onset of extension). The scaling of the vertical axes of all four plots depends strongly on the model assumptions. The shoulder in the evolution of surface elevation as seen for the thick line in **b** corresponds to the transition from rift phase to sag phase

2. If H_0 is of the order of several hundreds of meters, then the extension is self limiting. The cooling of the mantle part of the lithosphere is rapid enough so that thermal thickening of the lithosphere is more rapid than mechanical thinning (as in Fig. 6.8 between t_2 and t_3). Thus, the mantle part of the lithosphere thickens, while the crust thins. This leads ultimately to a termination of the extension process as the rheology of the mantle lithosphere is much stronger than that of the crust (thick lines in Fig. 6.12; Fig. 6.11c).

3. If H_0 is of the order of one or even several kilometers, then the extension rates are rapid enough so that physical extension outweighs thermal thickening (as in Fig. 6.8 between t_1 and t_2). The geotherm steepens, the lithosphere is thermally weakened and the excess potential energy leads to an acceleration of the extension process. Extension leads to rifting and ultimately to the development of a new passive margin (thin lines in Fig. 6.12; Fig. 6.11b).

The basic principles of the force balance underlying the scenarios discussed above will be explained in sect. 5.3.1 and 6.2.2. Models equivalent to the model of Houseman and England (1986b) but with application to continental *collision* will be discussed in the context of Fig. 6.21 (Sonder and England 1986; Molnar and Lyon-Caen 1989).

6.2 Continents in Collision

When two plates of continental lithosphere move toward each other, they will eventually collide. In contrast to the collision between two oceanic plates (during which one of the plate generally dives below the other and *no* intense deformation occurs in either of the plates), continental collision leads to intense deformation and interfingering of both plates (Fig. 2.19). This fundamental difference between the collision of continental and oceanic plates occurs because of three reasons:

 − because of their different thickness
 − because of their different strength
 − because of their density difference.

Continental lithosphere is much *thicker*, much *weaker* much *less dense* than oceanic lithosphere. The lower denisity and larger thickness of continental plates make it more difficult to subduct them underneath each other. However, importantly, the weakness of continental lithosphere allows internal pervasive deformation including nappe stacking, folding and more, while the much larger strength of oceanic lithosphere does not allow such processes. This internal deformation leads to mountain building, metamorphism and a series of other tectonic events which we can observe today in active orogens and which are preserved for us in the metamorphic rocks of ancient orogens.

Although only a relatively small number of all plate boundaries around the globe are formed by collisional orogens, such orogens are among the best studied tectonic features of our planet. This is certainly in part because continental collision processes form some of the most eye catching features on the globe: spectacular mountain ranges like the Himalayas or the European Alps. In this section we deal with aspects of the thermal and dynamics evolution of such orogens.

6.2.1 Thermal Evolution of Collisional Orogens

It is widely observed that collision of two continental plates leads to the heating of rocks at depth. In present day orogens this is evidenced by increased heat flow and in ancient orogens it is documented by the preservation of metamorphic parageneses that form only at elevated geothermal conditions.

However, heating and collisional deformation phases do not occur simultaneously and it is often observed that metamorphism occurred later than the collisional deformation phases that were responsible for the crustal thickening. 340 Many aspects of this typical relationship between deformation and metamorphism may be explained by a simple comparison of the duration of three processes: 1. The duration of crustal thickening processes; 2. The duration of thermal equilibration of the crust and 3. the time scale of exhumation processes. A comparison of these three time scales was the basis of the model by England and Richardson (1978) which was one of the first elegant models unifying deformative and metamorphic processes in collisional orogens into one model. This model will form the basis of the following section. Precursors of the "England and Richardson" model were already published by Oxburgh and Turcotte (1974) as well as Bickle et al. (1975) and the model was quantified later by England and Thompson (1984).

Fundamentals of the Thermal Evolution. England and Richardson (1978) recognized that the following relationships are fundamental to the thermal characteristics of collisional orogens (s. also Ridley 1989):

– Thickening of the crust is substantially more rapid than thermal equilibration of the crust. The former generally occurs at rates of the order of $\dot{\epsilon} \approx 10^{-14}$ s^{-1} (i. e. doubling the crustal thickness in less than 10 my), the latter takes of the order of several tens of my.
– The time scale of thermal equilibration is comparable to that of many exhumation processes: both often take several tens of my.

The time scale of deformation and the time scale of exhumation are functions of the geological boundary conditions of the orogen in question and may vary. Their absolute and relative magnitudes are known to us from field observations. In contrast, the time scale of thermal equilibration is independent of the geological processes. It is given directly by the laws of heat conduction. These state that this time scale increases quadratically with the thickness of the crust (s. sect. 3.1.4).

• *Thermal evolution during thickening.* The fact that continental deformation is typically much more rapid than thermal equilibration on the length scale of the crust has two immediate consequences: 1. Thickening leads to burial of rocks without appreciable heating of these rocks. This is indicated in Fig. 6.13a by the small vertical arrow in the T-z-diagram. 2. From the point of view of an Eulerian observer the crust cools. This is indicated in Fig. 6.13a by the small horizontal arrow in the T-z-diagram. Most of the interplay of heating and cooling mechanisms that governs the thermal evolution of collisional orogens occurs *after* the thickening. In order to understand this interplay it is useful to discuss heating and cooling mechanisms separately.

• *Heating mechanisms following thickening.* There are two mechanisms that lead to heating:

1. Prior to thickening the stable gotherm formed an equilibrium between surface heat flow *out* of the crust, Moho heat flow from the mantle *into* the crust and radiogenic internal heat production *in* the crust. This equilibrium was perturbed by the thickening process. The geotherm was cooled. Given that the mantle heat flow does not change, heating will occur to reestablish the equilibrium geotherm.

2. Thickening of the crust increases the total amount of radiogenic elements in the crust (coarse dotted part in Fig. 6.13). This increased radiogenic heat production has the consequence that the new equilibrium geotherm has a steeper gradient than before and heating will occur.

• *Cooling mechanisms following thickening.* Crustal thickening leads not only to heating, but also to mountain building. This is followed by erosion and extension processes thinning the crust and leading to exhumation of metamorphic rocks. In analogy to the last paragraph, there is two cooling processes that interact with the heating processes:

1. Denudation of the upper crust removes the heat producing elements from the upper crust. This has the consequence that the thermal equilibrium heat flow decreases in the whole crust.

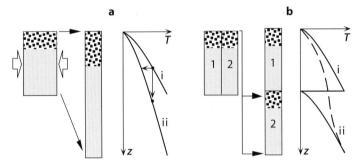

Figure 6.13. Cartoon illustrating the thermal changes that occur in the crust as the consequence of rapid crustal thickening. **a** and **b** show two end member scenarios of crustal thickening geometries: homogeneous crustal thickening in **a** and overthrusting of the whole crust (block 1 over block 2) in **b**. In each **a** and **b** a cartoon of a crustal column is shown on the left (the radioactive heat producing part of the crust is shown with the coarse dots) and a corresponding temperature T depth z diagram is shown on the right. The T-z diagrams show geotherms before thickening (labeled i) and after thickening but before thermal equilibration (labeled ii). The dotted line in **b** is the geotherm during early subsequent thermal equilibration. It shows that the "saw tooth geotherm" is rapidly equilibrated to a T-z profile not unlike ii in **a** (rapidly in comparison to the overall duration of subsequent thermal equilibration). Thus, the overall thermal evolution of thickened crust is robust towards the initial thickening geometry. The vertical arrow between ii and ii in **a** indicates the T-z path of a rock during thickening. The horizontal arrow shows that thickening leads to cooling at a constant depth

Figure 6.14. Geotherms (*thin lines*) and *T*-*z*-paths (*thick lines*) of rocks in typical collisional orogens as predicted by the model of England and Richardson (1978). t_0 labels the starting geotherm at the onset of collision, t_1 the geotherm immediately after thickening, t_2, t_3 and t_4 geotherms at later time steps. The Moho heat flow (thick drawn lower tip of geotherms) remains constant throughout the evolution. The black and white dots that lie on each geotherm label two different rocks at the same time. Note that the rocks of the shallower *T*-*z*-path (black dots) experience their thermal peak at time t_3 and cool thereafter, while the rocks at larger depths (white dots) reach their metamorphic thermal peak only at time t_4

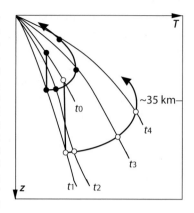

2. All rocks must cool to the surface temperature by the time they reach the surface. The closer rocks get exhumed to the surface, the stronger the cooling influence of this surface becomes.

The thermal evolution of rocks in a given orogen depends on the interplay of the heating and cooling mechanisms listed in the last paragraphs. Usually, heating mechanisms outweigh cooling mechanisms in the early phase following the crustal thickening. During the later evolution, in particular after the onset of denudation at the surface, the heating mechanisms wane and the influence of cooling mechanisms increases. Eventually, cooling processes win over the heating processes and rocks begin to cool (s. Fig. 6.14). The interplay of these various heating and cooling mechanisms has the consequence that the *T*-*z*-paths of rocks in collisional orogens have smooth curvature in a *T*-*z* diagram. They follow this curvature in a clockwise sense if the depth axis is drawn upwards and the temperature axis is drawn to the right (s. sect. 7.3.1). Depending on the relative importance of heating and cooling mechanism, such paths are more or less tight in a *T*-*z* diagram (Fig. 6.14).

The Model of England and Thompson. The qualitative comparison of time scales we discussed in the last section was first performed by England and Richardson (1978). Based on their work, England and Thompson (1984) designed a quantitative numerical model that is still among the best explaining the shape and nature of metamorphic *P*-*T* paths in collisional orogens. Thus, we will discuss the assumptions and results of their model in some more detail here.

• *Thermal assumptions.* The thermal assumptions made by England and Thompson (1984) to describe the evolution of collisional orogens are those necessary to calculate an initial geotherm with eq. 3.24 (s. sect. 3.4.1. Following boundary conditions are assumed:

– The temperature at the surface of earth T_s is constant.
– The mantle heat flow at the Moho q_m is also constant.

These boundary conditions are assumed to remain the same throughout the thermal evolution. It is also assumed that the radiogenic heat production is of the constant value S_{rad} down to the depth $z_{rad} = 15$ km and is zero below that (sect. 3.4.1). With those assumptions, integration of eq. 3.24 gives a description of a stable geotherm as given by eq. 3.65.

The numerical values needed in this equation are those for q_m and S_{rad}. England and Thompson chose them so that the surface heat flow q_s (which is one of the few thermal parameters that may be measured directly) is between 0.045 and 0.075 $W\,m^{-2}$, which is reasonable for continental shield regions (Table 6.1).

• *Heat flow relationships.* The relationship between surface heat flow, mantle heat flow and radioactive heat production can be illustrated clearly by interpreting the surface heat flow q_s as the sum of the mantle heat flow q_m and the heat flow caused in addition by radiogenic heat production q_{rad}:

$$q_s = q_m + q_{rad} \quad . \tag{6.13}$$

In this equation, the radiogenic heat flow is given by: $q_{rad} = S_{rad} z_{rad}$, as we explained when we discussed Fig. 3.15 (see also eq. 3.61). England and Thompson (1984) assumed that the radiogenically caused heat flow is comparable to the mantle heat flow ($q_{rad} \approx q_m$; Table 6.1) and that the mantle heat flow remains unchanged, regardless of the thickness of the crust. Thickening of the crust only doubles the radiogenic heat flow (because z_{rad} is doubled). We can write:

$$q_s = q_m + 2q_{rad} \quad . \tag{6.14}$$

Thus, the surface heat flow in thermal equilibrium after thickening is expected to be of the order of 1.5 times as high as before if $q_{rad} = q_m$ (Eq. 6.13).

However, if the mantle part of the lithosphere thickens together with the crust, then this halves the heat flow through the Moho (as the mantle lithosphere is thermally defined). We can then write:

Table 6.1. Three different simple but geologically realistic assumptions about the distribution of heat sources in the crust (after England and Thompson 1984). The mantle heat flow q_m and the radiogenic heat production q_{rad} are chosen in a way so that they give three different values for surface heat flow covering a realistic range of measured surface heat flows. The thickness of the heat producing layer in the crust is always assumed to be $z_{rad} = 15$ km. Remember that we can write: $q_{rad} = S_{rad} z_{rad}$. England and Thompson have coupled these three heat source distributions with three different thermal conductivities of 1.5, 2.25 and 3.0 $W\,m^{-1}\,K^{-1}$ to obtain a total of nine groups of T-z-paths

q_s ($W\,m^{-2}$)	S_{rad} ($\cdot 10^{-6}\ W\,m^{-3}$)	q_{rad} ($W\,m^{-2}$)	q_m ($W\,m^{-2}$)
0.045	1.666	0.025	0.020
0.060	2.000	0.030	0.030
0.075	2.333	0.035	0.040

$$q_s = \frac{q_m}{2} + 2q_{rad} \ . \tag{6.15}$$

Thus, if $q_{rad} = q_m$ and the entire lithosphere thickens to double thickness, the surface heat flow in thermal equilibrium and after thickening would be only 1.25 times as large as the value given by Eq. 6.13 (according to eq. 6.15). If $q_{rad} = q_m/2$, then thickening or thinning of the lithosphere as a whole does not change the surface heat flow at all.

• *Kinematic assumptions.* England and Thompson (1984) made the following extremely simple assumptions about the kinematic evolution of collisional orogens (s. Fig. 6.15):

– Thickening of the crust occurs homogeneously and instantaneously at the start of an orogenic cycle (t_0 and t_1 in Fig. 6.15 occur at the same time (s. p. 270). This assumption was made because many observations of plate tectonic velocities show that the rate of continental deformation is indeed about an one or two orders of magnitude more rapid than the lifetime of an orogen (at rates of centimeters per year, crustal thickening events last only few millions of years, while orogenic cycles last of the order of tens to one hundred million years).
– Following initial thickening, there is no vertical motion in the crust for about 20 my (Fig. 6.15). This assumption was made based on the observation that erosion and extension (causing the most important vertical motions in orogens) do not commence with the onset of thickening, but generally only *after* substantial topography is developed.
– After 20 my erosion sets in. Denudation of the thickened crust is linear in time and lasts for several times of my. While this assumption may be also an extreme simplification, it is the most simple assumption that can be made for the orogen to return to its normal thickens after some tens of my.

These three kinematic assumptions are illustrated in Fig. 6.15. The z-t-paths in this figure correspond to homogeneous thickening. Fig. 6.13 shows the thickening geometry and geotherms during the initial thickening process for two end member scenarios of thickening explored by England and Thompson (1984): homogeneous thickening and stacking of the entire crust. The dashed curve in Fig. 6.13b is drawn for a short time (some few my) after thickening by thrusting. It shows that the subsequent T-z-paths of rocks are quite robust towards the geometry of initial thickening. This justifies the assumption of the most simple of all thickening geometries.

• *Model results and application.* Fig. 6.14 illustrates the thermal evolution of a rock in a collisional orogen according to the model of England and Richardson (1978). According to this figure and the model, the involved processes occur in the following order:

1. At the onset of the evolution (time t_1) crustal thickening causes burial of rocks to great depths. In metamorphic rocks this is likely to be documented by an early high pressure metamorphic event.

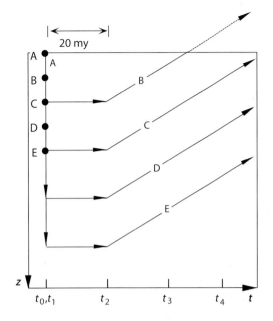

Figure 6.15. Kinematic assumptions of the continental collision model of England and Thompson (1984). The depth-time-(z-t)-paths A–E correspond to 5 different rocks that originally (before thickening) were located at the depths indicated by the black dots. The labeled times t_0 to t_4 correspond to those of Fig. 6.14

2. Some tens of my later rocks reach their thermal metamorphic peak at mid-crustal depths. In Fig. 6.14 it may be seen that, depending on crustal level of the rocks, this occurs at times t_3 or t_4. In metamorphic rocks this stage may be documented by the metamorphic peak paragenesis.
3. Following the thermal peak, T-z-paths are first characterized by isothermal exhumation and later by cooling and finally exhumation to the surface (s. a. sect. 7.3, 7.3.1).

This evolution is shifted in time for different crustal levels (s. sect. 7.4.1). In Fig. 6.14 cooling of the upper crust commences at time t_3, while the lower crust heats at least until time t_4. Overall, there is a positive correlation between metamorphic grade and the time of metamorphism: The higher the metamorphic grade of a rock, the later its peak metamorphism occurred. For contact metamorphic rocks this relationship is exactly the opposite (sect. 3.6.2). We can conclude that the relationship between metamorphic grade and time of peak metamorphism is an important tool for the interpretation of heat sources of metamorphism (s. sect. 7.4.2).

The space, time and grade relationships predicted by the model discussed above (and shown in Fig. 6.14) are documented in the eastern Alps. There, Cretaceous high pressure parageneses are overprinted by mid Tertiary amphibolite facies parageneses. In fact, in the early seventies these very observations were made by Oxburgh and his students and formed later the basis of the model discussed in this section (e. g. Oxburgh and Turcotte 1974).

• *An example of some calculated T-z-paths.* Fig. 6.16a shows some depth-time paths and Fig. 6.16b shows some temperature-depth paths as they were calculated with the model of England and Richardson as discussed above. We can see that these paths pass through conditions typical of Barrovian metamorphism. Metamorphic peak temperatures between 600 °C and 800 °C are measured at metamorphic peak depths of 20–40 km in many regional metamorphic terrains. However, the temperatures that the model predicts for the lower crust are unrealistically high.

For a comparison of these paths with paths that rocks describe if metamorphism and deformation occurred simultaneously, Fig. 6.16c and 6.16d show T-t- and T-z-paths calculated with the kinematic assumptions discussed in sect. 4.1.2. We can see that there are important differences between the T-z-paths shown in b and d.

– If deformation, exhumation and thermal development occur on similar time scales, then metamorphic rocks reach their maximum depth and peak metamorphic temperature roughly at the same time (Fig. 6.16d). If deformation of the orogen is short, compared to the thermal development, then metamorphic depth and temperature peak are clearly separated by a hiatus (Fig. 6.16b).
– The pro- and retrograde sections of the T-z-paths are distinctly different if deformation precedes thermal development (Fig. 6.16b), but are much more similar if deformation and metamorphism occur simultaneously (Fig. 6.16d).

These differences may form useful criteria for the tectonic interpretation of a metamorphic terrain for which the T-z- and T-t-path is roughly known, but any other geodynamic information is unknown.

• *Problems of the model.* The England and Richardson model has been often applied implicitly and uncritically to the description of PT paths in collisional orogens. Thus, it seems appropriate to dedicate a few paragraphs to its limitations.

The most important and fundamental limitation of the model is made by the assumption that the mantle heat flow through the Moho remains constant during orogenesis. Considering the thermal definition of the mantle part of the lithosphere, this assumption implicitly determines that the thickness of the mantle part of the lithosphere remains also constant during orogenesis. (s. sect. 3.4). The evolution of the thickness of orogens implied by this assumption is illustrated in Fig. 6.17 and 6.18 (s. also Fig. 3.18). In summary, the following evolution is implied for the lithosphere as a whole:

1. Initial thickening only affects the crust. The thickness of the mantle part of the lithosphere remains unchanged.
2. The subsequent heating of the crust by the increased radiogenic heat production causes the mantle part of the lithosphere to thin continuously during orogenesis.

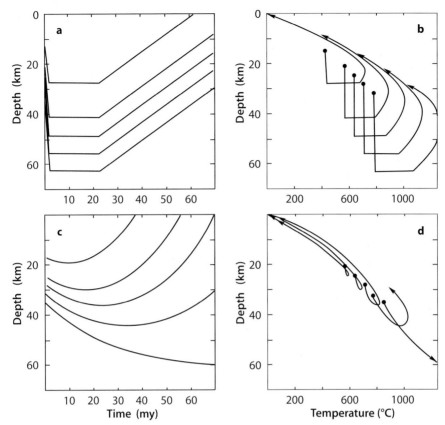

Figure 6.16. Kinematic and thermal evolution of rocks in collisional orogens: **a** and **b** show depth-time (z-t) and depth-temperature (z-T) curves using the kinematic assumptions of England and Thompson (1984); **c** and **d** show corresponding paths assuming deformation, exhumation and thermal evolution occur simultaneously (as described in terms of a simple model in sect. 4.1.2 (eq. 4.13) (Stüwe and Barr 1998)

This implicit evolution of the thickness of the mantle part of the lithosphere places a severe limitation on the applicability of the model.

Another problem is that the implicitly determined evolution of the thickness of the mantle part of the lithosphere leads ultimately to unrealistically high temperatures in the lower crust (Fig. 6.16). Moreover, the model has some quite peculiar implications for the evolution of surface elevation and the dynamic state of orogens as a whole (Fig. 3.18 and Fig. 4.16). For example, England (1987) showed that orogens with geotherms as predicted by the England and Thompson (1984) model, would be mechanically extremely unstable (s. sect. 6.2.2). The origin of metamorphic T-z-paths of the shape like those shown in Fig. 6.16b are therefore still subject to some debate (England 1987; Ridley 1989).

Interestingly, the evolution of the thickness of crust and mantle part of the lithosphere that is described by the model of England and Richardson (Fig. 6.18) has a lot of similarities with an orogenic evolution characterized by early homogeneous thickening of the whole lithosphere and later delamination of the mantle part of the lithosphere (Fig. 6.18b, s. also sect. 6.3.2). The latter is mechanically plausible for several orogens (Houseman et al. 1981).

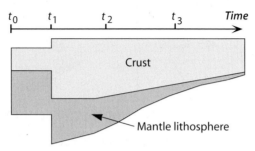

Figure 6.17. The thickness evolution of the lithosphere according to the models of England and Richardson (1978) and England and Thompson (1984). The thickness evolution of the *crust* is explicitly prescribed by the model. The thickness evolution of the *mantle* part of the lithosphere is determined implicitly by the thermal development. The labeled times t_0 to t_3 correspond to the time steps labeled in Fig. 6.14 and 6.15. The evolution illustrated here may also be read from Fig. 6.18. The mechanical, topographic and thermal consequences that the *implicit* thickness evolution of the mantle part of the lithosphere implies are rarely considered by those using the model for the interpretation of a metamorphic depth-temperature path

Figure 6.18. Schematic illustration of two possible thickness evolutions of collisional orogens in the f_c-f_l plane. *a* Evolution of orogens according to the model of England and Richardson (1978) or England and Thompson (1984) as in Fig. 6.17. *b* Evolution of orogens during homogeneous lithospheric thickening followed by convective removal of the mantle part of the lithosphere. f_c and f_l are the vertical thickening strains of the crust and the lithosphere. For a more detailed explanation of this diagram see sect. 4.0.2 and Fig. 4.5

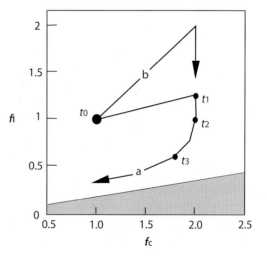

6.2.2 Mechanical Description of Colliding Continents

So far we have considered the collision of continents only in terms of their *thermal* (sect. 6.2.1) and *kinematic* (sect. 4.1.2) evolution. If we want to consider their *mechanical* evolution, we need to use a constitutive relationship that relates motion to forces or vice versa (sect. 5.3.1). In *thermo mechanical* descriptions, thermal processes and mechanical evolution are considered simultaneously (e. g. Sonder and England 1986; Sahagian and Holland 1993).

The differential equations describing a force balance are the basis for the mechanical description of orogenic processes (sect. 5.1.1). If we limit ourselves to a two-dimensional description using the spatial coordinates x and z these are (s. eq. 5.16) :

$$\frac{\partial \sigma_{zz}(x, z)}{\partial z} + \frac{\partial \sigma_{xz}(x, z)}{\partial x} = \rho(x, z)g \quad , \tag{6.16}$$

$$\frac{\partial \sigma_{xx}(x, z)}{\partial x} + \frac{\partial \sigma_{xz}(x, z)}{\partial z} = 0 \quad . \tag{6.17}$$

These equations must be integrated using geologically meaningful initial and boundary conditions. If the mechanical problem we are interested in is a three-dimensional one, then all three equations of mechanical equilibrium must be integrated. Because this is often very difficult, a series of authors have discussed meaningful simplifications of three-dimensional problems to two-dimensional ones (e.g. England and Jackson 1989; England and McKenzie 1982). England et al. (1985) used for their study analytical solutions of simplified mechanical equilibria to describe collisional orogenesis.

While some of the simplifications are elegant and allow very transparent descriptions of orogenic processes, many processes are just too complicated to justify the use of these simple solutions. Thus, the trend of recent years has been away from analytical descriptions and towards sophisticated numerical solutions of eq. 6.16 and 6.17 and a corresponding formulation in the y direction. Some of the major advances in our understanding of collisional orogenesis were made using numerical models (England and Houseman 1986, 1988; Houseman and England 1986a).

Because a two-dimensional problem is only fully described by both equation 6.16 and 6.17 and both these equations contain two derivatives we need a total of four boundary conditions to solve them. In general, these boundary conditions are defined by the knowledge of the stress state. Then, these equations may be solved and – using an appropriate constitutive relationship (which relates stress and strain rate) – they may be used to calculate a velocity and strain rate field. Because the constitutive relationship links stress and strain state, it is often possible to formulate boundary conditions in terms of a velocity field instead of a stress field. In modeling studies it is therefore common to discern between "*constant velocity*" and "*constant stress*" type boundary conditions.

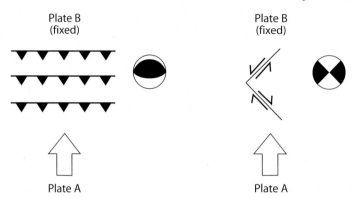

Figure 6.19. Illustration of the fact that the *same* boundary conditions may be reflected in two completely different stress states and strain fields (after Jackson and McKenzie 1988; England and Jackson 1989). (For an interpretation of the fault plane solutions see p. 211.) The processes in the left diagram may best be illustrated on a vertical section. The processes in the right hand diagram are best described with a plane strain model

• *Boundary conditions to orogens.* Orogens may be subject to either "constant stress" or "constant velocity" type boundary conditions and it is important to discriminate between the two (s. Fig. 5.24). The India-Asia collision is an example for an orogensis subject to a "constant velocity" type boundary condition: The convergence rate between the two plates has been largely constant over the last 30 million years, even though the potential energy of the Himalayas has increased dramatically during this time and is now opposing the driving force of the Indian plate.

It is also important to realize that the *same* boundary conditions may lead to very different strain distributions. Fig. 6.19 illustrates an example. In both cartoons of this figure plate A moves towards plate B. However, in the example on the left, the convergence is compensated by overthrusting. In the example on the right, the convergence leads to orogen parallel extension. Although the overall velocity across the boundaries is the same, the strains are different. Clearly, the two geometries are different, because the boundary conditions in the *vertical* direction are different.

Force Balances in Orogens. The forces that keep orogens in mechanical equilibrium may be divided (very loosely and not very precisely) into three groups:

1. *Driving forces*: Driving forces are forces applied from the outside to an orogen, for example ridge push or slab pull. In the following we abbreviate these forces with F_d. Some of these forces were already discussed in sect. 5.3.2.

2. *Internal forces*: These are the forces internal to the lithosphere which "resist" the driving forces and are limited by the inherent strength of the rocks int he lithosphere. As we are often interested in the whole lithosphere we

often use the vertically *integrated* strength of the lithosphere, which has the units of force/meter. We represent this in the following with F_l.

3. *Potential energy:* Forces resulting from the potential energy difference of an orogen relative to its surroundings are also called *gravitational stresses* or: *horizontal buoyancy forces*. We denote those in the following with F_b.

This division is not completely sound, as many of the plate tectonic driving forces themselves are also caused by potential energy differences and many of the other forces are also coupled. However, it helps us to understand the balance of forces in orogens. Note also that all three forces are usually not given in the units of force alone (N), but that they are discussed in terms of force *per* meter (Nm^{-1}) and that the unit of "force per meter" is equivalent to the units of "potential energy per area" or the units of "stress × distance". We begin by considering aspects of the potential energy of mountain belts.

• *Potential energy.* In sect. 5.3.1 we showed that the potential energy of orogens grows with the square of the surface elevation *and* with the square of the thickness of the orogenic root. Thus, it takes significantly more energy to increase the surface elevation of a high mountain range by one meter than it takes to increase the elevation of a low range by the same amount (Molnar and Tapponier 1978). As a consequence, the height of a mountain range and the thickness of an orogenic root are limited, if the driving force is a constant. This limiting elevation is reached when the potential energy of the range per square meter area is exactly as large as the tectonic driving force per meter length of orogen. Then, a steady state equilibrium of the forces is reached.

In order to understand how this equilibrium is reached, consider Fig. 6.20a, which illustrates a very simple model orogen. The left of this diagram shows normal thick crust of the thickness z_c and the density ρ_c. On the right, this diagram shows an elevated mountain range in isostatic equilibrium of the elevation H. The diagram is equivalent to Fig. 5.27. The difference in potential energy between the two mountain range and the foreland per square meter of area is given by eq. 5.51 and 5.52. Let us also recall that ΔE_p is a potential energy *per* area and has the units of Jm^{-2} and may also be interpreted as the mean net horizontal force exerted by the mountain range onto the foreland per meter length of orogen. By analogy, the potential energy *per meter length of orogen* may also be interpreted as the product of the potential energy per area times the width of the mountain range l. From eq. 5.52 we can derive directly that:

$$\Delta E_{p,m^{-1}} = \rho_c g H l \left(H/2 + z_c + w/2 \right) \quad . \tag{6.18}$$

The subscripts are used to emphasize that we are dealing with the units of potential energy difference *per* meter, while the ΔE_p, we used in eq. 5.51 and eq. 5.52 has the units of potential energy difference per area. Further growth of the mountain range may now proceed either in the vertical direction (Fig. 6.20b) *or* in the horizontal direction (Fig. 6.20c). If the crust inside the orogen is doubled in *thickness*, then the potential energy of the range per meter grows to the following value:

$$\Delta E_{p,m-1}^{\text{high}} = 2\rho_c g H l \left(H + z_c + w\right) \quad . \tag{6.19}$$

If the growth of the mountain range is by doubling its width (at constant thickness, as shown in Fig. 6.20c), then the potential energy per meter growth to the following value:

$$\Delta E_{p,m-1}^{\text{wide}} = 2\rho_c g H l \left(H/2 + z_c + w/2\right) \quad . \tag{6.20}$$

The difference of the potential energy increases between the two deformation styles is given by the difference between eq. 6.19 and eq. 6.20:

$$\Delta E_{p,m-1}^{\text{high}} - \Delta E_{p,m-1}^{\text{wide}} = \rho_c g H l \left(H + w\right) = \left(\frac{\rho_c \rho_m}{\rho_m - \rho_c}\right) g l H^2 \quad . \tag{6.21}$$

The last simplification in the equation above was performed using the isostasy condition $\Delta \rho w = H \rho_c$ that we also used for the step from eq. 6.19 to eq. 6.20. Eq. 6.21 shows us that it takes significantly less energy to thicken the crust in the foreland of a mountain belt than it takes to increase the thickness of the crust in the mountain range itself. Because of this, it is not necessary that convergence between two plates stops when the gravitational extensional force that acts from the orogen towards the foreland has reached the same magnitude as the tectonic driving force acting towards the orogen. It is just that the convergence cannot be compensated anymore by *vertical*

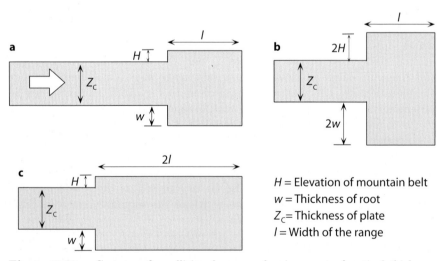

H = Elevation of mountain belt
w = Thickness of root
Z_c = Thickness of plate
l = Width of the range

Figure 6.20. a Cartoon of a collisional orogen showing crust of normal thickness on the left and a mountain range on the right. Further displacement of the crust from left to right is compensated in **b** by further thickening and in **c** by lateral growth of the range. The difference in deformation style between *b* and *c* causes a significant difference of the potential energy of the mountain ranges in **c** and **c**. This may be seen by comparing the equations eq. 6.18 to 6.21 (s. also Fig. 5.27; after Molnar and Lyon-Caen 1988)

growth of the range, but must be compensated by *lateral* growth of the range towards the fore- or hinterland. Thus, active deformation in the range itself will come to a halt, the zone of active deformation propagates into the fore- and hinterland and a plateau will form in the center. In the process, the transition zone between the region where the largest principle stress is oriented horizontally and the region where it is oriented vertically will shift also into the foreland.

Note that - despite these dramatic changes of the deformation and stress fields in the orogen - nothing has changed in the overall kinematics or stresses of the collision zone as a whole (Molnar and Lyon-Caen 1988) (s. p. 285). This should be considered as a warning to structural geologists who are tempted to infer the overall kinematics of an ancient orogen from field observations on the kinematics of the orogen from a sub-area of that orogen.

- *Evolution of orogens in the equilibrium of forces.* The force balance we have discussed in the last paragraphs may be summarized in the following equation:

$$F_{eff} = F_d - F_b \quad . \tag{6.22}$$

There, F_d is the tectonic driving force *per* meter length of orogen, F_b is the gravitational stress *times* the thickness of the lithosphere. F_b is also called *horizontal buoyancy force*, or: *extensional force* or: *potential energy per area*. The difference between the driving force and the horizontal buoyancy force is the effective driving force applied to a continent F_{eff}. Equation 6.22 is often referred to as the "orogenic force balance". Note that – although this equation is called a "force balance" – it *really* balances parameters that have the units of force per meter or stress × meter. Eq. 6.22 is often also written as:

$$F_{eff} = F_d - F_b = F_l \quad . \tag{6.23}$$

There, F_l is the vertically integrated strength of the lithosphere in Nm^{-1} and corresponds to the area under the failure envelope discussed in Figs. 5.13, 5.14 and several others. Note that F_l can only equal the left hand side of the equation if the orogen is deforming (i.e. at the point of failure). When $F_{eff} < F_l$, there is no deformation. However, we assume that active orogens are always on the point of failure so that $F_{eff} = F_l$ (sect. 5.2.1, eq. 5.43). The bulk of the lithosphere is dominated by viscous deformation mechanisms where deviatoric stress and strain rate are proportional. Thus, an orogen will always deform with a strain rate that is just large enough so that the vertically integrated deviatoric stresses balance exactly the effective driving force (per meter). If the strain rate would be *lower* than this, the integrated strength of the lithosphere would be smaller than the effective driving force (per meter) and the deformation rate would increase. Conversely, if the strain rate would be *larger* than the effective driving force, then the strength would be too large for any deformation to occur. Note also that, within eq. 6.23,

Figure 6.21. Schematic illustration of the temporal evolution of some important geodynamic parameters during collisional orogenesis. Surface elevation and crustal thickness converge to a steady state when the magnitude of the horizontal buoyancy force approaches the tectonic driving force. Then, the convergent strain rate goes towards zero

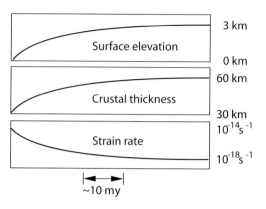

the integrated failure strength of the lithosphere is zero if it is at failure (i.e. deforming) when the effective driving force is zero. Conversely, if the driving force is zero and the integrated strength is greater than zero, there will be no deformation. Because of the balance described by eq. 6.22 it is possible to solve this equation for strain rate of an orogen, if the constitutive relationship is known that determines the relationship between stress and strain rate.

If we can simplify an orogen enough so that we can characterize it with a one-dimensional description and use vertically averaged values for stress and strain rate, then eq. 6.22 may be used for a full description of the mechanical evolution of a collisional orogen (e. g. Sonder and England 1986; Sandiford et al. 1991; Stüwe et al. 1993). If the tectonic driving force is assumed to remain constant through time, then all important physical parameters of model orogens in such descriptions evolve to asymptotic limits over time, as the equilibrium where $F_b = F_d$ and $F_{eff} = F_l = 0$ is slowly reached (Fig. 6.21). Thus, collisional orogens are self limiting. This is important to note, as extensional orogens are not necessarily self limiting (sect. 6.1.4).

The limiting values for a range of geodynamical parameters (e.g. surface elevation, crustal thickness) depend not only in the driving force F_d, but also on the integrated strength of the lithosphere. This may be illustrated if we reformulate eq. 6.23 to:

$$F_d = F_l + F_b \quad . \tag{6.24}$$

Remember that F_b is a direct function of the surface elevation and the thickness of the lithosphere. Thus, if the strength (F_l) of the lithosphere becomes smaller, for example because the lithosphere becomes warmer or the material changes due to metamorphism, then it is possible that the potential energy of the range per unit area (F_b) may increases even at constant driving force - the mountain belt grows (Houseman and England 1986a). Because of this process, sudden changes of the thermal structure of the lithosphere may cause sudden deformation events without changes in the driving force.

• *The mean strength of the lithosphere.* Differences in surface elevation of the continental lithosphere can only be created if the lithosphere has a finite

strength (Fig. 2.2). That is: if the horizontal and vertical principle stresses
are of different magnitude (Artyushkov 1973, McKenzie 1972, Molnar and
Lyon-Caen 1988). If the were no stress differences, then the surface of a plate
subjected to lateral forces from the outside would lift everywhere by the same
amount; like water between two converging sides of an aquarium. There would
be no mountain ranges and the surface of the continents would look rather
boring. Conversely, it is possible to use the thickness and surface elevation
of a mountain belt to estimate the mean strength of the lithosphere (Molnar
and Lyon-Caen 1989).

Consider a mountain range which collapses under its own weight and to
which there is no externally applied force. We could then reformulate eq. 6.24
to:

$$F_b = -F_l \quad . \tag{6.25}$$

If we not worry about the sign difference between the left hand and right
hand side of this equation, we can see that the vertically integrated strength
of the lithosphere is just as large as the horizontal buoyancy force of the
range. Correspongingly, at the start of the growth of a new mountain belt
$F_b = 0$ and the strength of the lithosphere will balance the driving force
$F_d = F_l$.

The left hand side of eq. 6.25 is the potential energy difference between
mountains and foreland per unit area and was evaluated in eq. 5.52 or, somewhat more precisely, with eq. 5.54 (s. also Fig. 5.27). The right hand side of
eq. 6.25 is the integrated strength of the lithosphere (s. eq. 5.43, Fig. 5.19
and Fig. 5.21). It is the product of the mean differential stress of the extending mountain range and its thickness. Thus, the elevation contrast between
mountain belts and their foreland (which is reflected in the buoyancy force
F_b) may directly be used to provide an upper bound on the mean strength
of the lithosphere.

According to the estimates of Molnar and Lyon-Caen (1988), the surface
elevation contrast between the Tibetan plateau and the Indian foreland indicates a mean strength of the Asian lithosphere of $\sigma_d = 69$ MPa. For the Alti-
plano in the Andes similar estimates indicate a mean strength of $\sigma_d = 52$ MPa.
This mean strength is estimated purely on the basis of topography differences and is quite a sound estimate. If we acknowledge that some parts of
the lithosphere will be significantly softer than this value (e. g. the uppermost
and lowermost parts of the crust as shown in Fig. 5.13), then there *must* be
other parts of the lithosphere that are significantly stronger than this value
to maintain the mean value given by these estimates.

Mechanics on Vertical Sections. Many continental orogens are long compared to their width. In such orogens many parameters do not change very
much in the direction parallel to the orogen and it is often possible to neglect
this direction altogether when describing the orogen: We can characterize

them with a description on a vertical cross section. In this section we introduce modeling on vertical cross sections by expanding on the discussion of the last section. We begin by discussing the changes of the stress state along an orogenic cross section and then introduce an elegant two-dimensional model that has been applied in recent years to the description of many orogens.

• *Changes in the stress field in collisional orogens.* In the discussion of eq. 6.21 we have shown that the stress field in an orogen may change over time, even if the far field plate boundary stresses remain constant. Here we illustrate this in some more detail by looking at the changes of the stress state across a mountain belt. In this discussion we follow the logic of Dalmayrac and Molnar (1981) as well as Molnar and Lyon-Caen (1988).

If the shear stresses at the base of the lithosphere are negligible, then the *horizontal* forces in a simple orogen (simplified as shown in Fig. 6.22) are constant, regardless of thickness of the plate or surface elevation (Artyushkov 1973; Dalmayrac and Molnar 1981) (s. Problem 5.8). In other words, the product of the mean horizontal stress σ_{xx} and the thickness of the plate remains a constant. Thus, if the stresses are a similar function of depth in different parts of the orogen, then the horizontal stress σ_{xx} is constant at any one depth across the orogen. This also implies that mountain ranges and plateaus transmit horizontal forces from the foreland to the hinterland of the orogen without changing their magnitude. On Fig. 6.22 this is indicated by the horizontal white arrows that are of the same size everywhere across the orogen.

This logic does *not* apply to the vertical stresses. Vertical stresses are the largest in regions where the overlying rock column is the thickest and the smallest where it is the thinnest (s. Fig. 5.27). As a consequence, the stress distribution in an orogen may be like that shown in Fig. 6.22. In the foreland (on the left in this figure) the *vertical* stress is *smaller* than the horizontal stress. The region is thickening, for example by thrusting. In the mountain belt (strictly: in the region of high potential energy, s. sect. 5.3.3), the largest principle stress is the vertical stress. The region is extending. In short: although the horizontal stress on Fig. 6.22 is everywhere the same, there is thickening in parts of the Figure and extension in others.

The lateral qualitative change in the deformation regime is *not* caused by changes in the horizontal- but changes in the vertical stress. This also explains why the observation of extension in mountainous regions must not occur because the surrounding plates are moving apart. The Tibetan plateau is an example for such a situation: although the plateau is extending laterally, there is thrust tectonics in the surrounding regions (s. p. 288 and p. 292).

• *Changes in the stress field during the aging of plates.* If the driving force in an orogen remains constant, the qualitative nature of the orientation of the principal stresses may be changed during orogenesis because of: 1. Increasing the potential energy of the mountain range or 2. *decreasing* the mean potential energy of the entire plate (Coblentz and Sandiford 1994; Sandiford and

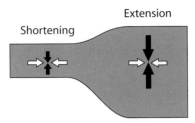

Figure 6.22. Distribution of horizontal and vertical stresses in a simple collisional orogen. If the topographic gradients at the surface and the base of the lithosphere are small, then the horizontal and vertical stresses σ_{xx} and σ_{zz} are parallel to the principal stresses. The horizontal stresses are constant across the orogen. However, the vertical stresses at a constant crustal level are higher in the orogen and smaller in the foreland. Thus, the largest principal stress in the foreland is given by σ_{xx}, while it is given by σ_{zz} in the orogen

Coblentz 1994). The former was discussed in the last section, the latter is the subject of this paragraph.

The geoid anomalies at continental margins tell us that the potential energy of old oceanic lithosphere is very low, while that of the mid oceanic ridges and that of large parts of the continental lithosphere is relatively high. Thus, the mean potential energy of a lithospheric plate is largely governed by the relative proportions of oceanic and continental lithosphere within this plate. If a plate is largely surrounded by mid ocean ridges (as both the African and the Antarctic plate are) then the proportion of oceanic lithosphere in the plate will grow over time and the mean potential energy of the plate will sink. Correspondingly, the part of the *continental* lithosphere that has a higher potential energy than the mean value, will also grow over time. Thus, some areas that have long been under compression because of the high potential energy of the surrounding mid ocean ridges, may go into extension only because the plate is aging (the proportion of oceanic to continental lithosphere rises as the area of oceanic lithosphere rises). Sandiford and Coblentz (1994) have suggested that all continental plates will ultimately go into extension as a function of their age and suggested that features like the central African rift system may be largely caused by the aging of the African plate.

• *An elegant model for collisional orogens.* All considerations of the last paragraphs are based on the thin sheet model assumption. That is, we have assumed thickening processes to be homogeneous and, in fact, we have largely compared one-dimensional profiles with each other, rather than making fully two-dimensional considerations. Using this simplification we were able to understand some important aspects of the mechanics of continental orogens, but we have limited ourselves to the understanding of symmetrical orogens and plateaus. However, many active orogens like the European Alps have a fundamentally asymmetric geometry in cross section (e. g. Pfiffner et al. 1997). In order to describe such orogens, we have to depart from the thin sheet model and apply more appropriate boundary conditions. One of the most elegant models for asymmetric orogenesis was developed originally by Willet et al. (1993) for the description of doubly vergent orogens. The model was later expanded by Beaumont et al. (1996) to describe oceanic and continental orogens with the same set of boundary conditions (Fig. 6.23). Because this model has since been applied with great success to a large range of problems related to continental orogenesis, we explain these boundary conditions in the next paragraph is some detail.

The asymmetric geometry of many collisional orogens occurs because crust and mantle part of the lithosphere behave differently during collision. The mantle lithospheres of the two plates generally remain internally largely undeformed and move one over the other, while the two crusts collide head on and pervasively deform. This collision geometry is described within the simple model of Beaumont et al. (1996) by a lateral discontinuity in a boundary condition imposed at the Moho. In Fig. 6.23b the Moho is given by the lower edge of the shaded rectangle. This rectangle is subjected to stress boundary conditions at the sides and at the top (where the stress is zero). The lower boundary is subjected to kinematic boundary conditions.

In Fig. 6.23b the velocities to the left of point S are given by $v_T = v_P$ and $v_z = 0$. v_P is the velocity with which the plate moves from left to right, v_T is the tangential velocity of the model boundary and v_z is the normal velocity of the model boundary. To the right of point S it is assumed that: $v_T = v_z = 0$. On the other three boundaries of the plate the tangential and normal stresses are assumed to be zero, so that these boundaries are allowed to move freely in response to the velocities applied at the base. The lower boundary conditions implies that it is assumed that the mantle lithosphere of the plate PB subducts at point S towards the right underneath plate RB.

Beaumont et al. (1996) expanded this model to describe subduction of the entire lower plate within the same model (Fig. 6.23a). For this modification they assumed a vertical load L (simulating a downwards pulling plate) to apply at point S'. To the left of point S' and to the right of point S, the same boundary conditions as in Fig. 6.23b apply. Between S' and S the velocities are assumed to be the same as at point S'. These boundary conditions imply that the entire crust is subducted between these two points. As the load L in Fig. 6.23a becomes zero, the model becomes that of Fig. 6.23b.

Model runs performed with these boundary conditions reproduce a series of structures commonly observed in collisional orogens. In particular, these are the conjugate shear zones that develop dynamically during the model runs and are schematically indicated in Fig. 6.23b. In Fig. 6.24 these shear zones are labeled with "Thrusting" and "Back thrusting". Note that the crustal scale shear zones give the orogen quite a symmetric appearance, despite the fundamental asymmetry of the basal boundary conditions. This apparent symmetry warns us to be careful with the interpretation of the direction of subduction of plates in an orogen, if this is solely interpreted on the basis of observations at the surface

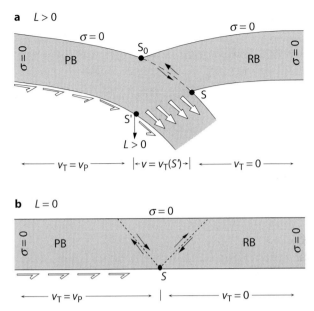

Figure 6.23. The boundary conditions of the two-dimensional model for the mechanical description of collisional orogenesis by Beaumont. **a** The subduction-collision model of Beaumont et al. (1996). The white arrows symbolize the velocity boundary condition along the lower margin of the plate. On the other three boundaries it is assumed that both normal and shear stresses are zero. The geometry of the initial condition shown here is the consequence of the magnitude of the downwards acting load L applied at point S'. PB and RB are the lower and upper plate (PR for pro beam and RB for retro beam). The shear zone between S and S_o is *no* model assumption but develops dynamically during model runs with these boundary conditions. **b** If the load L is assumed to be zero, the model shown in **a** recovers the collisional model of Willet et al. (1993) shown here. The dashed lines with the shear sense indicators are not part of the model assumptions. They illustrate regions of high shear as they develop dynamically during model simulations with these boundary conditions (s. Fig. 6.24)

Mechanics in the Plane. Many collisional orogens have features that may only be described by considering stresses and strains in plan view, for example processes like lateral extension or lateral extrusion. In order to explain such observations it is necessary to use kinematic and mechanic models that describe processes in *two* horizontal coordinates. If a full three-dimensional description is to be avoided, then the depth dependent parameters must be substituted by an appropriate single mean value (e. g. temperature, stress etc.).

• *Simplifications in the third dimension.* The two most common assumptions that are made to reduce a three-dimensional problem to a two-dimensional model are the *plane strain* assumption and the *thin sheet* (plane stress) assumption (sect. 1.2.1). The two assumptions are profoundly different (s. Fig. 6.19).

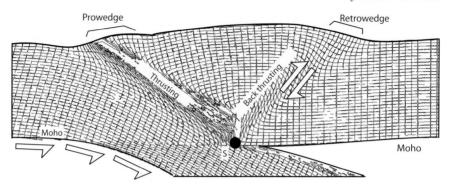

Figure 6.24. Typical result of a model simulation of collisional orogens using the boundary conditions of Beaumont et al. (1996). One of the most important results is the development of two conjugate crustal scale shear zones

Which of the two assumptions is more appropriate for the description of orogens has been a subject of debate between the schools of Tapponier on the one side (e. g. Molnar and Tapponier 1975, 1978) and that of England, Houseman and McKenzie on the other side (e. g. England and McKenzie 1982; Houseman and England 1986a; England and Houseman 1986; Molnar and Lyon-Caen 1988).

In a two dimensional model using either of these two assumptions it is not possible to include geotherms or other depth dependent parameters. It is therefore necessary to design models with which these depth dependent parameters may be averaged. For example, for many collisional orogens it is appropriate to assume that they deform in response to applied stresses according to the simple viscous constitutive relationship:

$$\dot{\epsilon} = B^{-n}(\sigma_1 - \sigma_3)^n \quad . \tag{6.26}$$

This equation is a simplification of the viscous relationships explained from eq. 5.37 to eq. 5.40. The constant B summarizes all temperature dependent terms of the power law (eq. 5.41). A comparison of eq. 5.41 with eq. 6.26 shows that: $B = A^{(-1/n)}e^{Q/nRT}$. England and McKenzie (1982) showed that B depends largely on the ratio Q/Tm, where Tm is the temperature of the lithospheric section with the largest strength. If this is in the upper mantle (as in Fig. 5.13), then Tm is the Moho-temperature. However, eq. 6.26 also holds if the rheological stratification of the lithosphere is different from that shown in Fig. 5.13. In short, B depends on temperature, but is independent of the thermal gradient in the lithosphere and independent of the depth of the strongest part of the lithosphere.

Using the simplification of eq. 6.26 the lithosphere may be considered as a simple medium deforming according to a power law relationship between stress and strain rate without the need to consider depth dependence of rheology or temperature. This is the basis of many dynamic models for the description of continental deformation, for example those of England and McKenzie (1982) or Vilotte et al. (1982). In these models, the nature of deformation may often be characterized by a single value: the Argand number.

● *The Argand number.* The Argand number Ar is a measure for the ease with which the lithosphere deforms in response to gravitational stresses. It tells us if an orogen is likely to flow apart at the same rate it is being built, or if significant amounts of potential energy may be stored within it before it would collapse slowly under the influence of gravitational stress.

The Argand number is defined by the dimensionless ratio of the additional pressure $P_{(L)}$, that arises because of the thickness difference L between two plates and the stress $\sigma_{(\dot\epsilon_0)}$, that is necessary to deform a plate with a significant rate $\dot\epsilon_0$. The significant strain rate is defined as $\dot\epsilon_0 = \frac{u_o}{L}$, where u_o is the collision velocity between the two plates (England and McKenzie 1982). We can write this as:

$$Ar = \frac{P_{(L)}}{\sigma_{(\dot\epsilon_0)}} \quad . \tag{6.27}$$

The Argand number is an important parameter for determining the deformation of an orogen. In fact, it is possible to prescribe the deformation geometry of a model orogen by using the Argand number as an input parameter. In effect, the Argand number incorporates the viscosity, density and thickness of a plate into a single number. Rheology or constitutive relationships need not be specified (s. p. 310, 308). In other words: the influence of different constitutive relationships or different temperatures on the evolution of an orogen may be indirectly tested by investigating the evolution of a model orogen as a function of Ar.

If we want to calculate the Argand number, we express $P_{(L)}$ in eq. 6.27 in terms of the crustal and mantle densities ρ_c and ρ_m and in terms of the thickness difference between two lithospheric plate (similar to how we did this in eq. 4.16 or eq. 5.55). Also, we describe $\sigma_{(\dot\epsilon_0)}$ in terms of the parameters of a viscous constitutive relationship eq. 6.26. Ar may then be written as:

$$Ar = \frac{\rho_c g L (1 - \rho_c/\rho_m)}{B(u_o/L)^{1/n}} \quad . \tag{6.28}$$

In this form, Ar may be used as an input parameter for mechanical modeling of orogens without having to explicitly consider the rheology, the material constants or the temperature profile of the lithosphere. The additional pressure rises linearly with the thickness of the orogen and the stress $\sigma_{(\dot\epsilon_0)}$ increases with the effective viscosity of the plate.

We can see that – if the effective viscosity of a plate is large, then the Argand number is small. Then, the flow properties of a mountain belt will depend largely on the orogenic boundary conditions. The belt will only begin to extend once its potential energy is very large. In contrast, if the Argand number is large (say between 10 and 20), then the effective viscosity of the range is very small and the forces caused by potential energy differences are large. The crust will quickly flow in response to applied forces. No significant thickness variations between foreland and orogen will ever develop during orogenesis. England and McKenzie (1982) have shown that orogens characterized by an Argand number of 30 show practically no thickness variation and their deformation is already nearly plane strain.

- *The Deborah number.* A number related to the Argand number is the Deborah number. In contrast to the Argand number (which is defined in terms of stresses), it is defined as the ratio of two time scales, namely the time scale of viscoelastic stress relaxation (usually given in terms of the ratio of viscosity to shear modulus) and the characteristic time scale of deformation (given in 1/strain rate) (Reiner 1964, 1969). Akin to the Argand number it can be used as a measure of the fluid-like behavior of continents. When applied to the evolution of continents, the Deborah number may be interpreted as the duration for which an orogenic driving force is applied to a plate, relative to the duration an orogen takes to flow apart (England 1996). If the Deborah number is significantly larger than 1 then the deformation of the orogen will be largely confined to the size of the collision front. In contrast, if the Deborah number is smaller than 1, no high mountain range will develop and the deformation of the orogen will dissipate rapidly and far into the foreland. The Deborah number has also been employed on a much smaller scale to characterize outcrop scale folding (e.g. Mancktelow 1999; Schmalholz and Podladchikov 1999).

- *Orogen parallel extension.* Collision of continents causes displacement of rocks in all three spatial directions. The *vertical* displacement that occurs

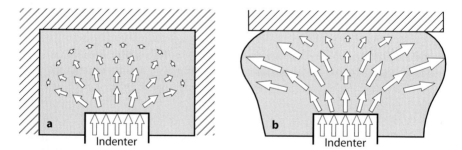

Figure 6.25. Different deformation regimes that occur during collision of an indenter with a much larger continental plate (gray shaded region). The arrows are velocity vectors of rocks. In **a** the plate is infinite or bound at all sides. the velocity vectors get shorter with increasing distance from the indenter indicating that the entire plate is in a compressive regime, even though individual rocks are displaced in orogen parallel direction. In **b** the side boundaries of the plate are free. The plate flow to the sides so that individual rocks are not only displaced laterally, but there is actually lateral extension as well. The absence of confined model boundaries is one of four mechanisms that can account for orogen parallel extension during convergent plate motion

during thickening was discussed already in the chapters on isostasy. In this section we want to consider the displacement in the direction parallel to the orogen in some more detail. In this context we emphasize at the outset that lateral *displacement* of rocks must not be confused with lateral *extension*. Sadly, this confusion is common in the literature.

Fig. 6.25a shows the collision of a plate with a rigid indenter that deforms the plate in front of it. The rocks in front of the indenter are displaced both in the direction of indentation and perpendicular to that direction. The amount of displacement decreases with distance from the indenter as the deformation there dissipates more. In Fig. 6.25a this is illustrated by a decreasing length of the displacement vectors. However, despite the orogen parallel *displacement* of rocks, all points of the indented plate are under *compression*. There is no lateral extension. This conclusion from Fig. 6.25a is in contrast to many observations in active collisional orogens where lateral extension *does* occur. For example, in the eastern European Alps, extensional tectonics is observed along the east and west margins of the Tauern window (Selverstone 1988; Genser and Neubauer 1989; Ratschbacher et al. 1991). This fact is not readily explained with the simple model of Fig. 6.25a.

According to England and Houseman (1989) orogen parallel extension in convergent orogens must find its nature in one of the following four processes:

1. unconstrained boundaries,
2. decrease in the convergence rate between two plates,
3. changes in the rheology of the plate,
4. external addition of potential energy to the plate.

The first of these four processes is illustrate in Fig. 6.25b. There – in contrast to Fig. 6.25a – the gray shaded region is not bound on the sides. The second, third and fourth process may be illustrated with an analysis of eq. 6.24. A decrease in the convergence rate is reflected in this equation by a decrease in F_d. If F_l remains constant, the horizontal buoyancy force must decrease and extension sets in. This process is generally known as "post orogenic collapse". Changing the rheology of the plate (e. g. by heating, recrystallization, metamorphism etc.) is reflected in eq. 6.24 by changes in F_l. In order to maintain the force balance, strengthening of the plate must be accompanied by a decrease in the deformation rate or a decrease in the horizontal buoyancy force F_b. Both has extension as a consequence. The external addition of potential energy, for example by delamination of the mantle part of the lithosphere, has a similar influence on eq. 6.24. It may also cause the transition from compression to extension.

• *Lateral extrusion.* Lateral *extrusion* of material in orogens is a not very well defined term describing orogen parallel displacement of rocks. Generally, lateral extrusion processes are understood to be a plane strain deformation on the lines of the right hand diagram in Fig. 6.19. Individual rocks may extend of compress during this process. Tapponier et al. (1982) have described lateral extrusion processes in some detail and its dependence on boundary conditions was recently discussed by Jones (1997). (s. p. 307).

6.2.3 Accretionary Wedges

During the collision of plates, wedge shaped packages of rock form often in the foreland of the orogen. These wedge-shaped packages are characterized by the angle between the surface of the earth and a detachment surface at the base of the plate. Such wedges form both on land and under water during both the collision of continental and oceanic plates. Wedges that form in connection with the subduction of oceanic lithosphere are usually below sea level and are called *accretionary wedges*. In continental orogens, such wedges form above sea level and are called *fold and thrust belts* (McClay 1992). One of the fundamental characteristics of both fold and thrust belts and accretionary wedges is their constant overall shape during growth and in different examples throughout the world. Most accretionary wedges have a surface slope which dips at about $1°$. In fold and thrust belts, this angle is typically of the order of $3°$. In the following section we use the term "wedge" or "orogenic wedge" for both accretionary wedges and fold and thrust belts.

The wedge shape of all orogenic wedges stems from the fact that an inclined plane moves material towards a fixed back stop. This inclined plane is usually referred to as the basal detachment and the transmission of forces into the wedge is by friction along this detachment (s. sect. 6.2.2). In accretionary wedges marine sediments that lie on top of the subducting oceanic lithosphere are moved towards the upper plate (s. Fig. 2.20; 5.23). The upper plate is the back stop or indenter, depending on whether the process is seen in a Lagrangian or Eulerian reference frame (Fig. 6.27). A good example of such a wedge is the accretionary wedge that forms between the Pacific plate and the North American continent in Alaska. The best known example of a fold and thrust belt is Taiwan, which formed as a consequence of the subduction of the Eurasian continental margin underneath the island arc of Luzon (Suppe 1981, 1987). Using Taiwan as an example, a number of models have been developed in the past 20 years that may be used to explain orogenic wedges around the world (Davis et al. 1983; Dahlen et al. 1984; Dahlen 1984; Barr and Dahlen 1989; Dahlen and Barr 1989; Platt 1990; Platt 1993a).

Modeling orogenic wedges is a typical two-dimensional problem. The most important parameters necessary for the geometrical description of the wedge are the angle of inclination of the basal detachment β and that of the surface α (both relative to the horizontal Fig. 6.26). Models describing accretionary wedges may be divided into:

1. Models, describing the geometry and the state of stress.
2. Models, describing the kinematics and thermal development.

Both types of models will be discussed briefly in the next sections. In the context of doing so we will show that indenters are not necessary for the formation of an orogenic wedge.

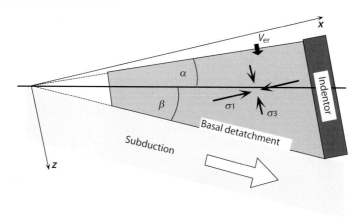

Figure 6.26. Geometrical parameters of accretionary wedges and fold and thrust belts. The angles α and β are the inclination angle of the surface and that of the basal detachment both measured relative to the horizontal. In models that are based on Mohr-Coulomb criteria, the coordinate system is often chosen so that the axes and the indenter are parallel to the direction of the principle stresses as shown here

1. Geometry and State of Stress. The origin of the typical shape of orogenic wedges may be understood by comparing orogenic wedges with the familiar analog of a snow plow. If a snow plow starts moving through a fresh layer of snow, we know that there are two possibilities of how the snow deforms: 1. If the internal strength of the snow is higher than the friction with the road (e.g. if the snow is icy and the road is warm), then the snow will be moved as an undeformed slab in front of the plow. 2. If the internal strength is smaller than the friction with the road (as usually is the case) then the snow deforms internally and its surface will become inclined. According to the Mohr-Coulomb criterion, the strength of the snow will increase as the thickness of the snow wedge increases (and the normal stresses within it do, s. eq. 5.25). This will continue until a critical taper between surface and basal detachment (in our example: the road) is reached where the strength of the snow is exactly as large as the basal traction (on Fig. 6.26 this is the angle $\alpha + \beta$) (Davis et al. 1983). In this state, the wedge can now move along the basal detachment and during progressive deformation the wedge stays constant in shape. However, while constant in *shape*, it still may grow in size as more snow is scraped up always retaining a balance between (increasing) internal strength and (increasing) basal friction. The wedge has a self similar geometry.

Dahlen (1984) showed that many orogenic wedges may be described very well with the assumption that they consist of cohesion-free material deforming according to Mohr-Coulomb criterion (sect. 5.1.2). In this case the orientation of the principal stresses are constant everywhere within the wedge (Fig. 6.26). Summarizing this we can write:

$$\alpha + \beta = \text{constant} \quad . \tag{6.29}$$

The constant in this equation depends on two parameters: the strength of the wedge material and the strength of the basal detachment. Both parameters are functions of the coefficient of internal friction μ and the fluid pressure (eq. 5.29). High stresses on the basal detachment *increase* the critical taper, high internal strength decreases it. Similarly, high fluid pressure inside the wedge decreases the strength of the wedge and increases the critical taper, while high fluid pressure along the basal detachment decreases the friction along this surface and thus decreases the critical taper.

The model discussed in the last paragraph was a great advance in our understanding of orogenic wedges (Dahlen 1984). Before the consideration of Mohr-Coulomb rheologies, wedges were generally described with models that do not consider the depth dependence of stresses (Chapple 1978; Stockmal 1983).

• *Differences between limited and unlimited wedge size.* In the last paragraph we have shown that an orogenic wedge will grow infinitely while maintaining a self similar shape. Depending on the position of the observer this growth may be interpreted in two different ways. An observer on the indenter (e. g. a snow plow driver) will see that the wedge grows at the toe and increases in thickness. This is illustrated in Figs. 6.27a and Fig. 6.28a and this situation corresponds to most accretionary wedges. However, if seen from the point of view of the wedge toe on the subducting plate, the wedge grows at its wide end and the region of most intense deformation moves progressively towards the foreland (Fig. 6.27b and Fig. 6.28b). This is observed in Fold-and-thrust belts where the deforming area progressively propagates forward into the orogen.

There are two possibilities to limit not only the *shape* but also the *size* of a wedge:

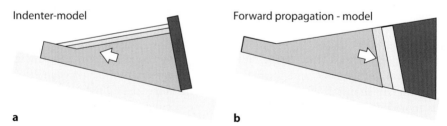

a   b

Figure 6.27. Cartoon illustrating the growth of accretionary wedges and fold and thrust belts. The arrows and light shaded wedge sections indicate the growth direction. The black regions show undeformed parts of the foreland. **a** From the point of view of the indenter, the wedge grows at its toe. **b** From the point of view of the basal detachment, the wedge growth appears to be caused by forward propagation of the deformation into the foreland. The indenter model is more appropriate to accretionary wedges. The forward propagation model more appropriate to fold-and-thrust belts

– In the snow plow model, the size of the wedge is limited by the height of the plow. During progressive deformation, the wedge will eventually become as high as the plow itself and the snow will be removed from the system over the top of the plow blade. Koons (1990) showed that exactly this process may be relevant to the geometry of deformation in New Zealand.
– The surface of the wedge may erode rapidly enough so that the erosive removal of material from the system balances the input of material at the wedge toe.

The second possibility appears to be realized in many accretionary wedges around the world. It forms the basis of the kinematic models we discuss in the next section.

2. Kinematics and Thermal Structure of Orogenic Wedges. If there is no erosion and the height of the indenter is infinitely high, then wedges would grow infinitely. The velocity field of rocks in the wedge relative to the indenter would look like a radial vector field starting at the toe of the wedge (coordinate origin in Fig. 6.26). There is no exhumation, However, if the wedge formation is accompanied by erosion at the surface, then the rock trajectories follow curved paths. If the total volume of the eroded material is exactly as large as the amount of material that is put into the wedge at its toe by the subducting plate, then the wedge does not grow and the rock trajectories are in a steady state (Fig. 6.28c). For such wedges with a constant shape and size, there is a range of models that describe their internal kinematics.

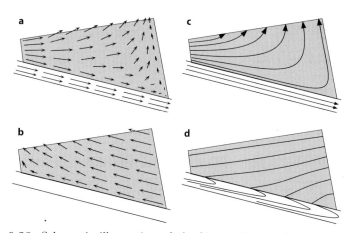

Figure 6.28. Schematic illustration of the kinematic and thermal structure in accretionary wedges (after Dahlen and Barr 1989; Barr and Dahlen 1989). The velocity vectors penetrate the surface because it is assumed that there is a steady state equilibrium between material input at the wedge toe (left edges) and material output by erosion at the surface. **a** Velocity vectors relative to the indenter (upper plate); **b** Velocity field relative to the subducting plate; **c** Rock trajectories in the accretionary wedge; **d** Isotherms

The most simple models are based on methods used in fluid dynamics for the description of the flow behavior of fluids in corner (*corner flow models*; Cowan and Silling 1978; Cloos 1984; Cloos and Shreve 1988). In further developments, Barr and Dahlen (1989) as well as Dahlen and Barr (1989) succeeded in developing analytical models that may be used to describe the internal kinematics of wedges that are subject to the Mohr-Coulomb-criterion. For this, they chose a coordinate system that is parallel to the principal stresses in the wedge (Fig. 6.26).

The thermal structure of wedges has also been described in models designed by Royden (1993b); Platt (1993a) as well as Barr and Dahlen (1989); (Fig. 6.28d). Similar models have been designed by Bird and Piper (1980), Beaumont et al. (1992) and Willet (1992).

6.3 Selected Geodynamic Processes

In this section we discuss, very superficially, a random selection of some important and currently actively debated geodynamic problems. There is no direct connection between the individual sections.

6.3.1 Flood Basalts and Mantle Plumes

Around the globe there are a large number of regions where enormous quantities of basalts have erupted (Fig. 6.29). These regions are known by the name of "*large igneous provinces*" or just "LIP" and are the second largest accumulations of mostly mafic igneous rocks on earth (after the rocks formed at spreading centers). LIPs include three types of basaltic provinces:

- continental flood basalts,
- rifted continental margin volcanic sequences,
- oceanic plateaus.

In particular, LIPs are a typical feature of Phanerozoic geology. The formation of LIPs has been discussed by a large number of authors (see recent summary edited by: Mahoney and Coffin 1997). In general, it is accepted that most of such provinces are caused by plume activity in the mantle. Initially mantle plumes were considered to be part of the general convection system of earth (Morgan 1971). However, within our modern understanding of plumes, they are known to be secondary features unrelated to the plate scale convection in the mantle (Sleep 1992). Today, authors discern between two types of plumes:

- Plumes that initiate from the core - mantle boundary that have narrow stems and large heads (Fig. 6.30). These plumes will be associated with "passive" rifting that follows active uplift of the overlying lithosphere by the plume (s. p. 237).

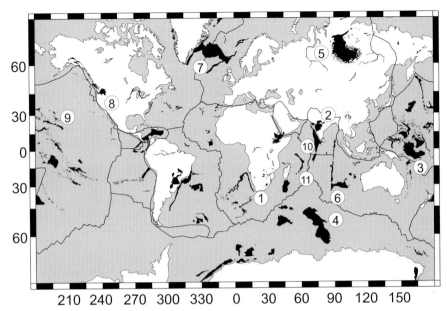

Figure 6.29. Large igneous provinces on the earth. The most important labeled provinces are: 1=Karoo Basalts; 2=Deccan Traps; 3=Otong-Java plateau (largest province with $2 \times 10^6 \mathrm{km}^2$); 4=Kerguelen plateau; 5=Siberian traps; 6=Broken Ridge plateau; 7=Iceland; 8=Columbia river basalts; 9=Hawaii; 10=Chagos Kaccadive ridge; 11=Maskarene plateau. Data are from Coffin and Eldholm (1993a,b)

– Plumes that form due to adiabatic upwelling of the asthenosphere in response to lithospheric extension (White and McKenzie 1989). Such plumes are themselves the consequence of "active" rifting but may provide positive feedbacks for an accelerating extension process.

For a discussion of active versus passive rifting see also p. 251 and 237). The island chain of Hawaii was the first place where mantle plume activity was suggested to be responsible for the appearance of the volcanic chain (Wilson 1963), although Hawaii is now known to be one of the smaller provinces on the earth (Fig. 6.29). Some of the largest LIPs form in places where rifted continental margins coincide with plumes (see also discussion of Fig. 5.28 and White and McKenzie 1989). There, LIPs may occur on the surface or below sea level, depending on the relationship between rifting rate and magma supply (Saunders et al. 1996). If the spreading rate is low in comparison to the rate of magma supply, then the plateaus become subareal as in Iceland. If the spreading rate is low the igneous province remain submarine.

Geometry. In the nineties people have begun to explore the geometry and volumes of LIPs mainly using the facilities of the international ocean drilling program ODP. For example, it is now known that plumes account at present

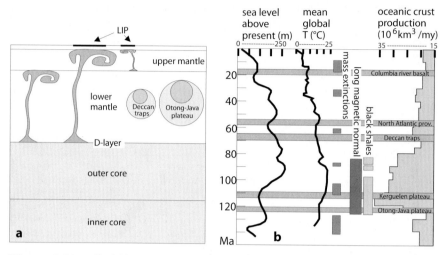

Figure 6.30. a Scaled cross section through earth showing the size and geometry of mantle plumes. The circles for the Otong-Java plateau and the Deccan Traps show minimum and maximum estimates for the volumes of melt extracted from the mantle during their activity (after Coffin and Eldholm 1994). See also Fig. 4.19. **b** The occurrence of the last 5 large igneous provinces in the last 150 my in relationship to other episodic events during this time. Shown are the time of the long magnetic normal in the Cretaceous, some important mass extinctions, important black shale deposition events as well as sea level, global temperature and oceanic crust production curves (after Larson 1991, as well as Coffin and Eldholm 1994)

for abut 5-10% of the mass and energy flux from the mantle to the crust and that this value may have been much larger in the past (Coffin and Eldholm 1994). Thicknesses of LIPS are between 20 and 40 km and appear to have formed in relatively short lived pulses of increase global production. Many oceanic plateaus have refractory depleted keels that are more buoyant than normal oceanic crust. As a consequence, oceanic plateaus may be preserved much longer in the plate tectonic cycle than normal oceanic crust. When they collide with active margins, they choke the subduction zone and may cause backsteppping of the zone. Thicknesses and areal extent are well enough known to estimate magma production rates. For example, it is now known that the magma production rate for the formation of the Otong-Java plateau may have exceeded the entire production rate of mid ocean ridge around the world at present (Coffin and Eldholm 1994).

Plumes or no Plumes. While mantle plumes have been generally accepted as the cause for the formation of most LIPs, it is worth mentioning that they may not be the only cause (Coffin 1997). Several authors have discussed alternative mechanisms for the formation of large igneous provinces (King 1996; Seth 1999). Three alternative ideas have been discussed:

- 1. Linear chains of volcanoes may indicate propagating rifts where the volcanic chain delineates the stress field, rather than the displacement field. This idea has been discussed by Turcotte and Oxburgh (1973) and Jackson and Shaw (1975) and the origin of the Deccan Traps as well as the Laccadives-Reunion hotspot has been suggested as an example by Seth (1999). Within this model volcanic chains are piezometers rather than speedometers.
- LIPs may be leaky transform faults (Smoot 1997).
- 3. Linear chains of volcanoes are produced by magma surge channels (Meyerhoff 1995).

However, it should be said that none of these models have found wide acceptance and have been applied only to some individual igneous provinces.

Large Igneous Provinces and Mass Extinctions. Large igneous provinces have formed episodically during the entire Phanerozoic. This episodicity has been brought in connection with a range of other episodically occurring events (Fig. 6.30b). For example, super plume activity and oceanic crust production have been brought in connection with cause for cessation of the magnetic field reversal of earth in the mid Cretaceous and in the Permian (Larson 1991). An even larger scale of correlation was discussed by Yale and Carpenter (1998). Correlations with mass extinction events have been attempted by a number of authors. However, in a recent summary Wignall (2001) considers most of these correlations as unduly optimistic. On the other hand, Wignall (2001) does recognize correlations with some mass extinction events, namely with the Karoo basalts, the Siberian traps, the central Atlantic volcanism and the Emeishan flood basalts.

6.3.2 Delamination of the Mantle Lithosphere

The mantle part of the lithosphere is colder and thus may be denser than the underlying asthenosphere (Fig. 2.15, eq. 4.24). Thus, the mantle part of the lithosphere can have a *negative* buoyancy (in contrast to the crust) and it is conceivable that it falls down into the asthenosphere. If this happens, the overlying orogen experiences dramatic changes to its potential energy, surface elevation and thermal structure. There are several mechanisms why and how in detail such sinking of the mantle lithosphere into the asthenosphere may occur. In particular, there are two models regarding how this may happen:

1. Delamination of the entire mantle part of the lithosphere from the crust along the Moho.
2. Convective removal of only the mechanically unstable thick root of the mantle lithosphere which, during progressive thickening, successively protrudes into the asthenosphere (Fig. 6.31).

- *Delamination.* The first mechanism was initially suggested by Bird (1979) to explain the uplift of the Colorado plateau in the western US. However, the model implies that the asthenosphere comes in direct contact with the crust and it is therefore expected that abundant crustal melts occur in connection with the surface uplift. This is not observed in Colorado. However, mantle xenolith studies in several regions around the world show that the uppermost mantle lithosphere is significantly younger than the overlying crust, indicating that this process does happen under some conditions. Jull and Kelemen (2001) have even suggested that the lowermost (eclogitized) crust may delaminate together with the mantle lithosphere.

- *Convective removal.* The second mechanism was suggested by Houseman et al. (1981) and has since been confirmed by many field observations (e.g. Platt and England 1994). The model relies on the following argument: The uppermost part of the mantle lithosphere is so viscous that its sinking rate is geologically irrelevant, despite its high density. The viscosity of the lowest part of the mantle lithosphere, on the other hand, approximates that of the asthenosphere. This part, where heat is still being transported mainly by conduction (and therefore part of the thermally defined lithosphere), but which has a negative buoyancy and a viscosity comparable to that of the asthenosphere, is also called the *thermal boundary layer* (Parsons and McKenzie 1978) (Fig. 2.16). This part of the mantle lithosphere may be removed from the rest of the mantle lithosphere by convective processes in the surrounding asthenosphere and it may ultimately sink (Houseman et al. 1981; Fleitout and Froidevaux 1982; Houseman and Molnar 1997; Molnar et al. 1998).

Temporal Evolution of Convective Removal Processes. The evolution of the convective removal of the mantle part of the lithosphere may be divided into three temporal stages:

- The first stage is the development of a lithospheric root during collision. It is necessary to develop a substantial root of the mantle lithosphere in order for the negative buoyancy forces to be large enough to cause stain rates that overcome geologically relevant values. Only then convective removal can set it.
- The second stage is the *beginning* of the removal process. This process is initially very slow and may take of the order of 1–10 my.
- The third stage is the *completion* of the removal. This occurs very rapidly once the sinking velocity has reached its maximum.

At the end of the detachment of the thickened root, the mantle part of the lithosphere is of similar thickness, or even thinner than at the start of the orogenic process (Fig. 6.31b)

The numerical experiments of Houseman et al. (1981) showed that convective removal may occur much more rapidly than orogenic evolution as a whole. Removal may be completed within a total of 1–10 my, while orogenic cycles last on the order of several tens of my. Thus, convective removal of

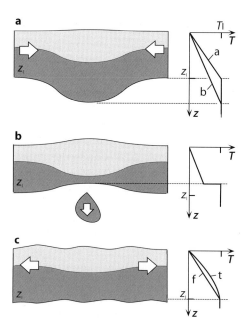

Figure 6.31. Schematic illustration of the Houseman et al. (1981) model for the convective removal of the mantle lithosphere in thickening orogens. **a** Thickening of the lithosphere during collisional orogenesis (crust is shaded light, mantle part of the lithosphere dark). The arrows indicate convergent motion. The labels on the schematic geotherms in the little T-z diagrams on the right are: $b =$ before thickening, $a =$ after thickening. T_l is the temperature at the base of the lithosphere and z_l is the thickness of the lithosphere prior to deformation. **b** Delamination of the thermal boundary layer from the mantle lithosphere. Note the uplifted surface and the dramatically changed thermal profile. **c** Subsequent thermal equilibration will cause the mantle lithosphere to thicken again. This process counteracts the simultaneously occurring rapid extension that occurs in response to the increased potential energy of the orogen. The two geotherms in the T-z diagram are: $t =$ during thermal equilibration, $f = a =$ final stable stage

the mantle part of the lithosphere may even be a cyclic process that occurs more than once during the evolution of a collisional orogen. Recent studies of Houseman and Molnar (1997), Molnar et al. (1998), Conrad and Molnar (1999) and others have shown that the non-linear viscous behavior of the lithosphere may cause asymmetries between the location of maximum thickening and the location of convective removal.

Mechanical Consequences of Convective Removal. The removal of the dense root of the mantle lithosphere of an orogen causes rapid isostatic uplift of the overlying orogen. The amount of surface uplift depends on the thickness of the removed thermal boundary layer. England and Houseman (1989) estimated about 1–3 km surface uplift, which in turn causes an increase of the potential energy of the orogen by about 2–$10 \cdot 10^{12}$ N m^{-1}. This increase is comparable to the magnitude of several plate tectonic driving forces and will therefore significantly influence the evolution of deformation events in the orogen. In particular, it is very likely that the removal of the lithospheric root may trigger the onset of extension in convergent orogens (sect. 6.1.4).

Thermal Consequences. The rapid removal of a lithospheric root has the consequence that hot asthenospheric material is brought much closer to the Moho than before (Fig. 6.31b). This causes increased heat flow through the Moho and ultimately partial melting in the lower crust. The amount and chemistry of partial melts that may form due to decompression in the mantle and due to partial melting of the lower crust is discussed by McKenzie and Bickle (1988) as well as White and McKenzie (1989).

However, it is not trivial that the increased heat flow at the Moho can also lead to high temperature metamorphism in the middle crust. Because of the slow rates of heat conduction on crustal length scales it may take up to tens of my until the middle crust "feels" the thermal effects at the base of the crust. In this time span extension may also have caused an increase of the geothermal gradient. Thus, a metamorphic event that occurred synchronously with removal of a lithospheric root may occur due to rapid extensional processes rather than heat conduction. Platt and England (1994) showed that if the extensional processes are short-lived, then metamorphism caused by heat flow changes at the Moho may be characterized by isobaric heating and cooling.

6.3.3 Low Pressure - High Temperature Metamorphism

In many regions of this planet, in particular on the Precambrian shields, we can find metamorphic terrains that experienced peak metamorphism at unusually high temperatures, if compared with the depth of metamorphism. In other words, the ratio of peak pressure to peak temperature in these terrains is much higher than that corresponding to a "normal" geothermal gradient or that predicted by models for regional (Barrovian) metamorphism (e. g. sect. 6.2.1). Such terrains are generally called "low-pressure-high-temperature", or short LPHT- terrains. LPHT terrains occur at all grades, ranging from greenschist facies metamorphism at less than a kilobar peak pressure (e. g. Xu et al. 1994) to granulite facies metamorphism at less than 3 or 4 kilobars (Greenfield et al. 1998). The heat sources of metamorphism in these terrains are intensely debated. In principal there are two fundamentally different heat sources that might be considered. We discuss these in the next paragraphs under the headings "external" and "internal" heating:

External Heat Sources. One school of thought argues that the $T - P$ ratio of peak metamorphic conditions in LPHT terrains is much too high to be possibly attainable by a conductive geotherm. Thus, so it is argued, the heat sources must originate from "outside" the terrain under consideration (the heat sources are: "external heat") (e. g. Bohlen 1987; Lux et al. 1986; Sandiford et al. 1991). Examples of external heat sources would be heat sources that are advected from larger depths into the terrain, for example magma or fluids. This process can be considered as "contact metamorphism" in the widest sense. The most important arguments in support of this external heating model are:

- If the terrain was heated by conductive response of the lithosphere to a change thickness geometry of crust and mantle lithosphere, then this implies that the measured PT ratio corresponds more or less to a geothermal gradient (as curve a in Fig. 6.32). Typical PT ratios of LPHT terrains (dark shaded region in Fig. 6.32) imply that a geotherm would reach $1\,200\,°C$ at a depth of about 30 km. Considering the thermal definition of the lithosphere, this implies that the lithosphere is only 30 km thick. Today, we observe such small lithospheric thicknesses only in regions of active extension and intra continental rift zones. In contrast, LPHT terrains are usually characterized by convergent structures and rift volcanics or other lithologies indicating a plate margin setting are usually absent. Thus, monotonously rising temperatures with depth are unlikely.

    Because of the absence of plate margin features it is expected that the lithospheric basis (defined by the $1\,200\,°C$ isotherm) is at "normal depths" between 50 and 200 km and that the geotherm of LPHT terrains rather has the shape of curve b or c in Fig. 6.32. Such geotherms can only form if the heat in the lithosphere is *actively* redistributed.
- In many LPHT terrains metamorphism occurred synchronously with deformation. This observation is easily explained if external heat sources are responsible for metamorphism (s. p. 341). However, it is in contrast with the models that explain regional metamorphism as a function of conductive processes affecting the whole lithosphere (s. sect. 6.2.1). A comparison between the typical duration of continental deformation events and the thermal time constant of the crust shows clearly that metamorphism would be expected to occur much later than deformation if the heating process were heat conduction over the scale of the lithosphere (i.e. internal heating) (s. Problem 3.5).
- Many LPHT terrains are characterized by isobaric cooling curves. This observation indicates that the rate of cooling was substantially larger than the rate of burial or exhumation (s. sect. 7.3.2). As the duration of conductive cooling of a terrain is proportional to the square of the size of the cooling region, the rates of exhumation or burial may be used to constrain the length scale of the heated terrain. Assuming *normal* rates of vertical motion of rocks, such estimates indicate that only a region substantially smaller than the entire lithosphere could have been affected by the LPHT event.

Internal Heat Sources. In contrast to the arguments presented above, another school of thought argues that neither enough magmatic bodies nor sufficient evidence for fluid infiltration is found in LPHT terrains to justify external heat transfer into the terrains (s. sect. 3.6.4, Problem 3.19). Thus,

Figure 6.32. Different models for the interpretation of geotherms in LPHT terrains. The dark shaded rectangle indicates the peak metamorphic conditions of many LPHT terrains. The light shaded rectangle indicates the peak metamorphic conditions of "normal"- conductively heated metamorphic terrains. The medium-shaded rectangle is the region in which high pressure low temperature metamorphism occurs. *a* is a monotonously rising geotherm. Such a geotherm implies that the bottom of the lithosphere (dashed line) is located at a depth of only 30 km (arrow). *b* and *c* show two other possibilities for geotherms that are characterized by LPHT metamorphism, but allow normal lithospheric thicknesses. However, geotherms can usually not get such a shape by conductive processes only

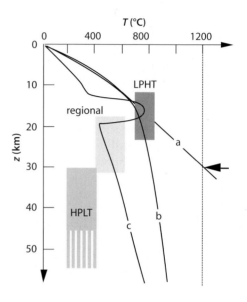

so it is argued, LPHT metamorphism must have similar causes as regional Barrovian type metamorphism (s. Harley 1989). In order to explain the exceptional peak metamorphic PT ratios a series of models have been invoked that all are based on extremely unusual thickness geometries of crust and mantle lithosphere. For example, extreme thinning of the crust *and* the mantle lithosphere may cause conditions appropriate to LPHT metamorphism. Another possibility that has recently received some attention is unusually high radioactive heat production within the crust (Chamberlain and Sonder 1990; Sandiford and Hand 1998). This might lead to a geotherm of the shape of curve *b* on Fig. 6.32.

Spear and Peacock (1989) discuss models of internal and external heating of metamorphic terrains in some detail. Their contribution also includes a series of computer programs that can be used to model these heating mechanisms.

6.3.4 High Pressure Metamorphism

Metamorphic rocks that were buried to depths above 60 km and thus contain high pressure metamorphic minerals are a common minor constituent of the lithological assembly of many orogens. The exhumation of these rocks occurred often simultaneously with convergent deformation phases. The exhumation mechanisms of the rocks are a much discussed topic and remain unexplained with a single unifying theory.

In sect. 4.1.2 we have shown using a simple one-dimensional model, that it is difficult to exhume rocks in convergent orogens from more than 30 km depth. Because of this problem, a series of models have been suggested that can be used to explain exhumation from much larger depths. These models can be grouped according to the nature of the exhuming forces into (s. Platt 1993b):

1. Models that rely on forces that are applied externally to the metamorphic terrain.
2. Models that rely on buoyancy forces caused by density differences between exhumed high pressure rocks and their surroundings.
3. Models that rely on extensional processes caused by gravitational stresses.

The currently most popular model quoted for the exhumation of high pressure rocks is the model of Chemenda (Chemenda et al. 1996; Shemenda 1994), which relies on a complicated interaction of processes from all three model groups in subduction zones. Before we discuss these model groups in some more detail, we note that the observation of high pressure metamorphic parageneses must not necessarily be interpreted as an indicator of large burial depths (e. g. Ernst 1971; Mancktelow 1993, 1995; s. sect. 6.3.5).

1. Exhumation by External Forces. Exhumation mechanisms of the first group may be characterized by the key words *Extrusion, strike slip faulting* and *corner flow* (s. p. 294). The process of vertical extrusion means that material is squeezed out between two hard blocks of rock, for example in a flower structure. This process itself does not help to exhume the rocks, but it brings rocks rapidly into a position where they may be exhumed by erosion. The amount of vertical extrusion as a function of the forces between the two hard blocks may be calculated from the relationships we discussed in sect. 5.1.1, 5.3.1 and 6.2.2 (e. g. eq. 6.22, 5.52).

The *corner flow* model is different. During the continuous deformation of accretionary wedges it is possible that rocks will be exhumed without the removal of a corresponding amount of material from the surface (s. sect. 6.2.3). However, this process can only be invoked as an explanation of exhumation if:

1. The viscosity of rocks is low (Cloos 1982).
2. The exhumed metamorphic rocks occur as isolated blocks in a melange.
3. If there is a spatially fixed indenter.
4. If the high pressure metamorphic rocks occur mainly in close proximity to the indenter.

2. Exhumation by Buoyancy Forces. If high pressure metamorphic rocks have a lower density than their surroundings, then it is conceivable that they rise through the crust only due to their positive buoyancy - a bit like plutons in the solid state. One example where this may occur is if crustal material is brought down into the mantle (Wheeler 1991; Chemenda et al. 1996). However, one of the most common high pressure metamorphic rocks is *eclogite* which is generally embedded into rocks of much lower density and grade. This observation cannot be explained by buoyancy forces. England and Holland (1979) observed that eclogites from the eastern Alps are often embedded in carbonates. They infer that the buoyancy of eclogites *plus* carbonates may be low compared to the density on a regional scale, so that exhumation by buoyancy forces is possible.

3. Exhumation by Extension. Continental extension is by far the most efficient exhumation mechanism for rocks from large depths. Normal faults and crustal scale detachments can bring large areas of high pressure metamorphic rocks practically undeformed to the surface. Despite its efficiency, extension is often not a model that can be used to explain exhumation of high pressure metamorphic rocks as these occur often in convergent orogens where little or no large scale extension is observed. One explanation for this may be that extension at shallow crustal levels occurs simultaneously with convergence at deep crustal levels (Avigard 1992; Platt 1993b; Stüwe and Barr 1998). Platt (1993b) subdivided exhumation processes that occur due to extension in convergent orogens into two groups:

- Extension in connection with underplating in orogenic wedges.
- Extension in collisional orogens (sect. 6.2.2.

Despite the large number of models for the exhumation of high pressure metamorphic rocks to the surface, there seem to be many terrains containing such rocks, where none of these models appear to apply. Several models may have to be combined to explain the exhumation in such areas. However, this research field is still wide open and new models will be invoked in the future.

6.3.5 Tectonic Overpressure

The term "tectonic overpressure" is very loosely used for all non-lithostatic components of pressure (s. sect. 5.1.1; eq. 5.19). We say "loosely" as it is not always easy or meaningful to separate pressure into lithostatic and non lithostatic components. However, for the purpose of the following discussion we assume that the stress field is oriented parallel to the vertical and horizontal directions and that this division may be made. Ernst (1971) and Rutland (1965) suggested that the non lithostatic stresses may form a significant contribution to the pressure measured with geobarometers in metamorphic parageneses. However, today most petrologists assume that this component of pressure is so small that it may be neglected completely when interpreting geobarometric results. Whether this assumption is always justified is not always trivial and needs testing. In the following we begin with a summary of some field observations that may be useful to perform such tests.

Some Field Observations. The fact that non-lithostatic stresses have an influence on the formation of rocks is quite obvious on a large range of scales from the hand specimen scale to the intra plate stress field on the largest scale (Fig. 2.2). On the hand specimen scale we know of the existence of pressure shadows behind minerals, pressure solution, boudinage and folding. In fact, any fabric formation in rocks can all only occur if there is deviations from the isotropic stress state of rocks. For the mechanical analysis of such deviations, numerical methods, in particular the finite element method are now of common use (Ramsay and Lisle 2000). The interested reader is referred to the classics of Strömgard (1973) and Stephansson (1974), more modern studies of Barr and Houseman (1996), Bons et al. (1997) or Tenczer and Stüwe (2001) and – of course – to the wonderful text book of Ramsay and Lisle (2000).

Estimating its Magnitude and Interpretation. The mean elevation of a mountain range may be used to estimate that the mean differential stress in the lithosphere must be of the order of 50 MPa (s. p. 284). This implies that the mean contribution of non-lithostatic components to pressure is

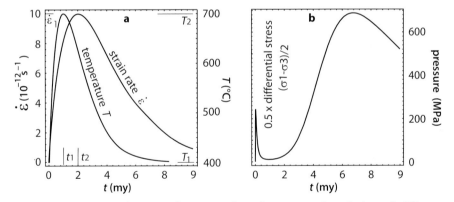

Figure 6.33. This diagram illustrates that the temporal evolution of differential stress need not correlate in any intuitive way with either the evolution of temperature or the evolution of strain rate. This is because differential stress in the ductile regime has such a complicated exponential and power law dependence on temperature and strain rate (eq. 6.30). **a** Schematic illustration of temperature and strain rate in a metamorphic terrain. These curves were calculated with the purely artificial assumptions that temperature may be described by: $T = T_1 + (T_2 - T_1)(t/t_1)e^{(1-t/t_1)}$ and strain rate by: $\dot{\epsilon} = \dot{\epsilon}_1(t/t_2)e^{(1-t/t_2)}$. **b** The differential stress contribution to pressure with the assumed temperature and strain rate evolutions from **a**. Calculated with eq. 6.30 using $A = 2 \cdot 10^{-4}$ MPa^{-3} s^{-1}; $Q = 2 \cdot 10^5$ J mol^{-1}; $n = 3$

about 0.25 kbar. However, considering the rheological anisotropy of the lithosphere, it is very likely that some parts of the lithosphere are significantly weaker than this and others therefore stronger. For example, the rheological model of the Brace-Goetze-lithosphere implies that differential stresses near the brittle ductile transition are of the order of several hundreds of MPa (s. Fig. 5.17 and Fig. 5.20). Thus, the Brace-Goetze-model is in contrast to the wide spread opinion that differential stresses are of negligible influence to the interpretation of geobarometric data.

Let us therefore consider the different parameters contributing to the normal and differential stresses in the ductile regime. For this we consider a simple convergent deformation geometry in which the largest principal stress is oriented horizontally and the smallest vertically and where we can write: $\sigma_2 = (\sigma_1 + \sigma_3)/2$. In this case pressure is the mean of the largest and smallest principal stress and may be described as follows: (s. sect. 5.1.1, eq. 5.19 and eq. 5.41):

$$P = \frac{\sigma_1 + \sigma_3}{2} = \sigma_3 + \frac{\sigma_1 - \sigma_3}{2} = \sigma_3 + \frac{\sigma_d}{2} = \rho g z + 0.5 \left(\frac{\dot{\epsilon}}{A}\right)^{(1/n)} e^{\frac{Q}{nRT}} \quad .(6.30)$$

The second term in this equation may be called the "non lithostatic component of pressure". It depends on the material constants Q, A and n as well as temperature T and strain rate $\dot{\epsilon}$. As it is very difficult to determine the material constants experimentally, geologist have tried to constrain their magnitude using field observations (e.g. Molnar and England 1990a; England and Molnar 1991; Mancktelow 1993, 1995; Stüwe 1998a).

From eq. 6.30 we can see that temperature and strain rate are related to differential stress by complicated exponential and power law functions. Thus, it is not trivial to see if differential stresses (and therefore the tectonic overpressure) will rise or fall when the temperature or the strain rate in a metamorphic terrain change. This is illustrated in Fig. 6.33a where schematic evolutions of temperature and strain rate are drawn for a terrain where deformation and metamorphism occurred at the same time and where both temperature and deformation slowly wane towards the end of the tectonothermal evolution of the orogen. Inserting these two evolutions into the second term of eq. 6.30 we have a simple analytical description of the temporal evolution of differential stress (Fig. 6.33b). Despite the simple and intuitive shape of the functions in Fig. 6.33a, the resulting evolution of stress as a function of time is completely counter intuitive (Fig. 6.33b). The curve has two maxima, both of which do not correspond to either maxima or minima of the strain rate of temperature curves.

The absolute magnitude of the differential stresses in Fig. 6.33b depend on the material constants. Depending on those, there is two different possible interpretations of this figure. If the constants A and Q are *small*, then the figure is irrelevant and the pressure of eq. 6.30 is largely equal to the vertical normal stress. However, if A and Q are large, then differential stress responsible for a significant proportion of total pressure. We may say that the non-intuitive shape of the curve in Fig. 6.33b should serve us as a warning to attribute too much significance to some complications observed in P-T paths of metamorphic rocks, in particular in the low-pressure high-temperature metamorphic environment (s. Stüwe and Sandiford 1994).

6.3.6 Feedback and Episodicity

Feedback between different geological processes is a common phenomenon in the earth science. In general, we discriminate between two types of feedback processes:

– positive feedback,
– negative feedback.

   In both cases one process has an effect on another, the changes of which influence in turn the first process. Feedback processes are called *positive* if two or several processes "accelerate" each other. Feedback processes are called *negative*, if two or several processes hinder each other. We have encountered positive feedback processes between lithospheric extension and thermal weakening in Fig. 6.12: There, the onset of lithospheric extension causes steeping of the geothermal gradient (thermal weakening), which accelerates the extensional strain rates, which in turn results in accelerated thermal weakening. The process begins to "run away" until rifting occurs. Another currently very topical positive feedback processes occur between global glaciation and CO_2 content of the atmosphere possibly resulting in very rapid formation of a snowball earth (North et al. 1981; Hoffman et al. 1998). Negative feedback is more intuitive, for example the feedback between potential energy and effective driving force during collisional orogensis (Fig. 6.21). There, collision causes an increase of the potential energy of a new orogen which opposes the driving force and therefore causes the collisional strain rate to decrease, which in turn leads to a slowed increase of the potential energy until orogenesis comes to a rapid halt. Because of feedback processes the question on the *cause* or *consequence* of one or the other geological process is often difficult to answer. Most feedback processes in the earth sciences are non-linear, that is, there is not a direct linear relationship between cause and consequence. Before discussing some more geological examples, we therefore begin with explaining what exactly non-linear feedback is.

Non-linear Feedback. Consider a very simple theoretical example. If we iterate the non-linear function $x = x^2$ for many iterations, then the result will depend on our assumption for the starting value of x. For all positive starting values of x that are smaller than 1, this function will converge toward zero. For all starting values of x that are larger than 1 this function will diverge towards $x \to \infty$. Because of the non-linearity, it will do so at an increasingly large rate, the more iterations we perform. Only a single starting value, namely

$x = 1$, separates the two different trends. (If the starting value is not a real number but a complex number, then this critical starting guess separating different evolutions will become a series of starting values forming a line in complex plane. This line often has a complicated fractal shape and is known by the name *Julia set*).

Many geological processes behave that way in that they either "run away" or converge to a steady state due to non-linear feedback. For instance, in the example we discussed in Fig. 6.12, we noted that there is single set of starting conditions separating two completely different orogenic evolutions. These are evolutions terminating in self-limiting extension and those terminating in "run away" rifting processes. Even a very small difference in starting conditions is sufficient to result in these very different evolutions, if the two starting conditions lie on either side of this critical set of starting conditions. The example is therefore equivalent to the simple illustrative iteration of $x = x^2$. We shall discuss several more examples on the next pages.

For some functions convergent and diverging evolutions are *not* separated by a single starting value, but by a whole region of starting values. Within this critical region the evolution may behave *oscillatoric* or *chaotic*. In this context *chaotic* means that the evolution of the function may not be predicted directly from the choice of starting parameters. We will discuss some potential geological candidates for episodic but chaotic behavior due to non-linear feedback on the next pages. However, let us illustrate chaotic behavior first with a simple theoretical example: the Lorentz attractor.

• *Lorentz Attractor.* The so called Lorentz attractor describes the feedback between three independent differential equations:

$$\frac{\mathrm{d}x}{\mathrm{d}t} = -10x + 10y \quad,$$

$$\frac{\mathrm{d}y}{\mathrm{d}t} = -xz + 28x - y \quad, \tag{6.31}$$

$$\frac{\mathrm{d}z}{\mathrm{d}t} = xy - \frac{8}{5}z \quad.$$

All three equations are functions of the same parameters: x, y and z. Also, all three equations are non-linear (the fact that variables occur as themselves *and* as their own derivative within the same equation indicates that they are exponential functions; s. p. 372). Numerical simultaneous solution of these three equations using the method of finite differences is straight forward and results in the irregular temporal evolution of the three variables x, y and z that is illustrated in Fig. 6.34b. Even though this purely mathematical example is of no direct relevance to geological processes, it is instructive in showing us how non-linear feedback between three functions can lead to chaotic evolutions of a parameter. In fact, it is very tempting to interpret the episodic evolution of x and z shown in Fig. 6.34b in terms of having some resemblence to the repeated occurrence of deformation or metamorphic events in orogens.

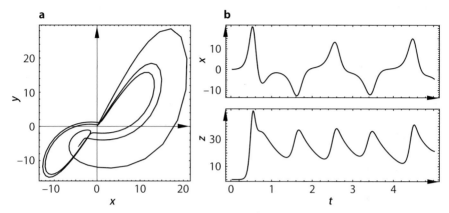

Figure 6.34. a The Lorentz-attractor in the x-y-plane, calculated with the eqs. 6.31. **b** chaotic evolution of the variables x and z plotted as a function of time t

Geological Examples of Non-linear Feedback.

• *1. Surface erosion as part of the orogenic force balance.* One of the more interesting recent developments in geodynamics is the integration of geomorphological and even climatological considerations into tectonic models. There, feedback between climate changes and mantle processes have been inferred (e.g. Molnar et al. 1993; Koons 1990, Harrison et al. 1992; Pinter and Brandon 1997). While the connection between climate and mantle processes is highly speculative, it is much better documented that surface erosion processes play a role in the tectonic and thermal evolution of the *crust*. For example by comparing erosion rates and thermal equilibration rates on the scale of the crust, it may be shown that high erosion rates must be associated not only with the advection of rocks towards the surface, but also with the advection of heat towards the surface. This is well documented in New Zealand, where the difference in heat flow between the west and the east coast of the south island may be correlated with the difference in rain fall on the two sides of the island (Koons 1990). Such observations indicate that erosion processes at the surface may trigger deformation events in the lower crust (Hoffman and Grotzinger 1993).

In a closed orogenic system subjected to a constant tectonic driving force, erosion and mountain building are thus connected by an apparently unsolvable circular argument:

– The deformation rate with which an orogen deforms depends on its thermal structure. High erosion rates may only occur once erosion causes the advection of heat towards the surface and therefore weakening of the crust (s. eq. 3.48, Fig. 3.13).

– Erosion can only set in once a mountain range has reached a certain elevation. For example, the high erosion rate that are caused in the Himalaya by the Monsoon, are only possible once a substantial mountain range has formed.

This paradoxon is discussed in some detail by Molnar and England (1990b). Zhou and Stüwe (1994) showed that this connection of events is only possible for exceptional rates during orogensis. Regardless, while many of these arguments are still being debated it has become clear that the coupled interpretation of tectonic and geomorphological data is an important approach to answer many large scale tectonic problems.

• *2. Deformation and metamorphism – cause or consequence.* Observations from many metamorphic terrains show that there are characteristic temporal relationships between metamorphism and deformation indicating that feedback processes between deformation and metamorphism are likely. In principle, there may be three different timing relationships: the deformation of a terrain may occur pre-, syn- or postdate to metamorphism (s. sect. 7.4.1). Clearly, it is useful to know if there is a genetic relationship between the deformation of a terrain and its metamorphism, in particular if one caused the other.

Models that are used to explain regional Barrovian type conductive metamorphism consider metamorphism to be the *consequence* of deformation (e. g. the model of England and Richardson, s. sect. 6.2.1): the deformation causes thickening which in turn causes a thermal disequilibrium. The subsequent thermal equilibration that causes metamorphism is therefore the result of deformation. The model also implies that the deformation is *pre*-metamorphic (s. sect. 7.4.1; Fig. 7.10).

Conversely, metamorphism can cause a deformation event. For example, if a strong lithospheric plate is thermally-weakened, then it might deform in response to the thermal event, while the far field stress field remains constant (s. eq. 6.24, sect. 5.3.1). In this model, deformation is the result of metamorphism and deformation is likely to be *syn*- metamorphic.

• *3. Orogeny as a non-linear feedback between many processes.* Mountain building processes are similar in that they too are characterized by a series of non-linear relationships between physical parameters. Amongst geophysicists this is well known to be one of the causes of chaotic motions during mantle convection processes. However, it is much more recent that it has been recognized that similar non-linear feedback may also be responsible for the discrete events that structural and metamorphic geologists observe in the field (e. g. Malanson et al. 1992; Hodges 1996; Stüwe et al. 1993). Fig. 6.35 gives an overview over some non-linear processes and their interaction during orogenesis. Many of the processes illustrated there may be described by mathematical relationships not unlike those of Eq. 6.31. A correlation between the evolutions of x, y and z on Fig. 6.34b and strain rate of thermal events may therefore be not completely wrong.

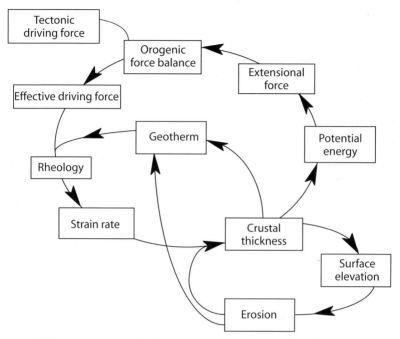

Figure 6.35. Interaction and feedback between some selected processes during orogenesis. Note that the functional relationship between several of the processes is non-linear. For example, the strain rate is an *exponential* function of temperature, a *cubic* function of stress (eq. 5.41) and potential energy is a quadratic function of thickness (eq. 5.44). It is therefore possible that interaction between the different processes may lead to a chaotic evolution of processes during orogenesis

Episodicity in Geological Events. While the possibilitiy that non-linear feedback processes are responsible for episodic occurrence of structural and metamorphic events is a highly exiting and rarely considered possibility, there are two much more common mechanisms that cause the episodic occurrence of geological events. The fact that geological events do in fact occur episodically is widely documented: Field geologists map out *discrete* events of repeated deformation and metamorphism (s. Fig. 7.8); volcanoes erupt cyclically; earthquakes, mass extinctions, glaciations, magmatic activity and orogenesis all occur over time spans that are short compared to the periods of quiescence before they occur again (e. g. Waschbush and Royden 1992; Malniverno and Pockalny 1990). In a very general way, the episodicity of a process may be explained in three different ways:

– Episodicity at the boundaries,

– threshold mechanisms,
– non-linear feedback.

The last of these three mechanisms was dealt with in the last section. The first mechanisms ("episodicity at the boundaries") simply refers to an explanation of the cyclicity of events by cyclicity of events elsewhere. For example, the cyclic nature of tectonic events in the western US has been interpreted as the consequence of the episodic subduction of the Pacific plate under the North American continent (Elison 1991). This interpretation explains the cyclic occurrence of events in North America, but it does not explain episodicity as such. Rather, it defers the causes of episodicity to an area outside the region of consideration - in this case outside North America.

• *Threshold mechanisms.* The most common way how continuous processes are broken up into discrete episodically occurring events is by *threshold mechanisms*. In principle this means that a certain value of a parameter – the threshold – needs to be exceeded by a continuous evolution before a process can occur (Tong 1983). Two examples:

– Earthquakes occur if a certain *threshold stress* is exceeded. During deformation elastic stresses are cumulatively built up until the failure stress of a rock is exceeded. An earthquake dissipates these stresses then by deformation. This is followed by a period of quiescence during which stresses build up again, until the threshold stress is exceeded again.
– Partially molten rocks often intrude into higher crustal levels. For a melted rock to leave its source rocks and form an intrusion a *threshold melt volume* must be exceeded in the source rock (Wickham 1987).

Both the above examples are characterized by a regular, but episodic evolution, even though the applied boundary conditions remain constant. However, in reality, earth quakes and magmatic activity occur much more randomly than the above models and thresholds suggest. This is because the thresholds in question are coupled with a spatial length scale. For example, an earth quake will not occur if the necessary threshold stresses are only exceeded along a single micro fracture. Instead, it will be necessary that the mean stress over a certain area exceeds the threshold stress. Similarly, magma segregation will not occur if a critical melt volume is exceeded on the scale of a thin section. Instead, it will be necessary that melt can only segregate once the critical melt volume is exceeded over a certain length scale.

Thus, in order to be able to predict the occurrence of earth quakes or magmatic intrusion (or any other process relying on threshold mechanisms), it is necessary to know the relationship between the magnitude of the threshold and the length scale over which this threshold needs to be exceeded. Sadly, this relationship is a function of so many variables that is practically impossible to know it. The problem is akin to the problem of meteorologists, who would need to know the motion of every gas molecule in the atmosphere in order to be able to predict weather patterns with any accuracy. The fact that infinitely small variations of a single small parameter are sufficient to bring the whole system out of predictability, this called the *butterfly effect*, reminding us of the possibility that the air current caused by the flight of a butterfly may cause complete weather patterns to change. This effect is closely related to the third explanation for episodicity: non-linear feedback.

6.4 Problems

Problem 6.1. *Extension: uplift or subsidence? (p. 252)*:
Fig. 6.1 illustrates that continental extension can lead to surface uplift or to surface subsidence, depending on the initial thickness ratio of crust and mantle lithosphere and their densities. Use eq. 4.29 and Fig. 4.16 to understand which parameters control whether uplift or subsidence will occur (s. also Problem 6.7).

Problem 6.2. *Porosity estimates (p. 256)*:
Use eq. 6.1 to estimate the depth at which the porosity of sandstone is the same as that of shale. Use the parameter values given in the caption for Fig. 6.3.

Problem 6.3. *Compaction of sediments (p. 257)*:
Use eq. 6.4 to calculate the thickness of a 100 m thick sandstone unit at 2 000 m depth at the time of its deposition. For the porosity data of sandstone use those in Fig. 6.3.

Problem 6.4. *Isostasy of sedimentary basins (p. 260)*:
Derive eq. 6.5 using Fig. 6.5 and eq. 4.17.

Problem 6.5. *Isostasy of sedimentary basins (p. 260)*: Consider three different sedimentary basins that all subsided because of the same thinning process in the lithosphere. Underneath all three basins the crust is now 20 km and the mantle part of the lithosphere is 50 km thick. The underlying asthenosphere has a density of $\rho_m = 3\,200$ kg m^{-3}. The basins are large enough to be isostatically compensated. The first basin is a 2 km deep valley (i.e. it is filled by air for which we assume: $\rho_{air} = 0$). The second basin is a lake (it is completely filled by water with $\rho_w = 1\,000$ kg m^{-3}). The third basin is completely filled by sediments with the density $\rho_s = 2\,300$ kg m^{-3}. How deep is the lake and how thick is the sedimentary basin fill in the second and third basins, respectively? The problem can be solved using the principle of Fig. 6.5: you must compare the mass of the vertical columns between the three basins.

Problem 6.6. *Subsidence analysis (p. 262)*:
Describe the subsidence evolution of the sedimentary basin characterized below. What tectonic process might have formed the basin and discuss field

observations you would look for to test your model? What subsidence mechanism might have been responsible for the evolution of subsidence you have derived? Note that your subsidence analysis must include back stripping to answer these questions.

The basin we consider is filled by a 6 km thick pile of sediments. A drill hole has shown that the pile is made up of five strata which we number from the base to the top with $i = 1$ to $i = 5$. Fossils were found in each layer and allowed to determine a detailed record of the sedimentary succession: The 1. layer is a 3 km thick sandstone layer that was deposited during an increasing water depth from 500 m to 800 m between 55 my and 60 my. The second layer is a 500 m thick shale layer that was deposited in a water depth of 800 m between 40 my and 55 my. The third layer is a 1 km thick sandstone unit, deposited during decreasing water depth from 800 m to 300 m in the time between 40 my and 30 my. The 4. layer is a 500 m thick shale unit that was deposited during a decrease of water depth from 300 m to 0 m in the time between 30 my and 15 my. The 5. layer was deposited between 15 my and present in the tidal environment.

For your analysis use Eq. 6.1, 6.4 and 6.5 (note that you will also need Eq. 6.7 for Eq. 6.5). Use the values for densities and porosities from Fig. 6.3. The density of the asthenosphere is: $\rho_a = 3\,200$ kg m^{-3}.

Problem 6.7. *Extension: uplift or subsidence? (p. 262):*
According to McKenzie (1978) homogeneous stretching of the lithosphere will only lead to subsidence if the initial ratio of crustal thickness to lithospheric thickness exceeds a certain ratio. Derive how large this ratio is using Eq. 6.10 and making the following assumption for the physical parameters. $\rho_0 = 3\,300$ kg m^{-3}; $\rho_c = 2\,750$ kg m^{-3}; $\rho_w = 1\,000$ kg m^{-3}; $\alpha = 3 \cdot 10^{-5}$ °C^{-1}; $T_l = 1\,280$ °C and $z_l = 130$ km. (See also Problem 6.1.)

Problem 6.8. *Estimating thermal sag (p. 264):*
Estimate roughly the total duration of the thermal sag phase that may be expected in a hypothetical basin forming in Europe, where the lithosphere is roughly 100 km thick and another one in southern Africa, where the lithosphere is almost 200 km thick. Use eq. 6.11.

Problem 6.9. *Radioactive contribution to heat flow (p. 274):*
Table 6.1 lists a range of geologically realistic radiogenically caused heat flows q_{rad} and mantle heat flows q_m for the continental lithosphere. Calculate the expected equilibrium surface heat flow in lithosphere that is (a) doubled in thickness and (b) halved in thickness using the logic outlined in eq. 6.13 to eq. 6.15

Problem 6.10. *Understanding orogenic force balance (p. 280 – p. 285):*
A mountain belt has formed in response to the collision of two plates. How high can the (isostatically supported) mountain range grow if the driving force

for collision is $F_d=5 \cdot 10^{12} \, \mathrm{N\,m}^{-1}$? Use eq. 6.22. Tackle the problem in 3 different ways and compare the answers: (a) Solve the problem graphically by using the logic of eq. 6.22 and comparing Fig. 4.16 and Fig. 5.32 (assuming that only the crust thickens but that the complete lithospheric thickness remains unchanged, i.e. $f_l = 1$ at all times). (b) Calculate the answer using eq. 5.52 and $z_c = 30$ km, $\rho_c = 2\,700 \, \mathrm{kg\,m}^{-3}$, $\rho_m = 3\,300 \, \mathrm{kg\,m}^{-3}$ and $g = 10 \mathrm{ms}^{-2}$. (c) Refine that answer of (b) by using eq. 5.54 (using $\alpha = 3 \times 10^{-5} {}^\circ \mathrm{C}^{-1}$, $T_l = 1\,200$ $^\circ$C and $z_l = 100$ km as additional variables) to calculate the buoyancy force and eq. 4.25 to calculate the isostatically supported surface elevation (assuming again that only the crust thickens but that the complete lithospheric thickness remains unchanged), (s. also Problem 5.18.)

Problem 6.11. *Estimating the Argand number (p. 291):*
Use eq. 6.28 to make a rough estimate how an orogen will look like in which the thickness increase is $L = 50$ km, $\rho_c = 2\,700$ kg m^{-3}, $\rho_m = 3\,300$ kg m^{-3} and the collision rate is 5 cm per year. For the parameter B use the relationship given in the text following eq. 6.26 (see eq. 5.41) and the values for the material constants of olivine given in Table 5.3. Calculate the Argand number for 400 $^\circ$C and 500 $^\circ$C and discuss what difference this makes.

Problem 6.12. *Estimating tectonic overpressure (p. 310):*
How large is the tectonic overpressure in a quartz dominated rock in 10 or 15 km depth if the thermal gradient from the surface to that depth is 30 $^\circ$C km^{-1} and the strain rate is large enough to double the thickness of the crust in 5 my? Use eq. 6.30 and the rheological data for quartz given in Table 5.3. Assumed eq. 5.41 to be the governing constitutive relationship.

7. *P-T-t-D*-Paths

One of the basic data sets used by geologists for the geodynamic interpretation of a metamorphic terrain is the spatial and temporal evolution of pressure P, temperature T and deformation D that the rocks experienced: the metamorphic evolution of the rocks. Data on the metamorphic evolution are particularly important when interpreting ancient orogens where it is impossible to measure many other parameters directly (e.g. surface elevation, surface heat flow, gravity etc.). The relative evolution of pressure, temperature and deformation may well be illustrated as curves in P-T-space. Such curves are called P-T-paths or P-T-t-D-paths, if the path is labeled for deformation events.

Following a basic introduction, we discuss in sect. 7.2 some basic petrological principles. In the remainder of this chapter we discuss the interpretation and modeling of timing relationships mapped in the field or derived from metamorphic P-T-paths. For detailed treatment of thermodynamics we recommend:

– Anderson and Crerar (1993) Thermodynamics in Geochemistry.
– Atkins (1994) Physical Chemistry.

For more petrologically oriented texts with geodynamic applications we recommend:

– Spear (1993) Metamorphic Phase Equilibria and Pressure Temperature Time Paths.
– Spear and Peacock (1989) Metamorphic Pressure Temperature Time Paths.

7.1 Introduction

Most tectonic processes are characterized by process-specific spatial and/or temporal relationships between the evolutions of pressure, temperature and deformation. Thus, when inferring tectonic processes from such relationships, it is important to subdivide these three parameters in as much detail as possible: The thermal evolution may be subdivided into a heating phase and a cooling phase (or several thereof), the baric evolution may be subdivided into a phase of increasing pressure and into one of decreasing pressure and the evolution of deformation may be subdivided into phases of increasing and phases of decreasing strain or strain rate. When collecting field observations it is important to discriminate between two different sets of data pertaining to two different questions:

- What is the temperature, pressure and deformation evolution of a single rock? (*temporal* evolution)
- How does this temporal evolution change in space across the terrain in question; for example as a function of a regionally decreasing or increasing metamorphic or strain gradient? (*spatial* structure)

The more detailed these questions can be answered in the field or in the laboratory, the easier it is to constrain the tectonic processes that formed the terrain in question. However, answering these questions is not always easy. An observed regional strain gradient could be the consequence of both a regional change in the *duration* of deformation or a regional change in *strain rate*. In other words, it remains unclear if the observation should be interpreted in terms of a *spatial* or *temporal* change. Corresponding problems may arise during the interpretation of pressure or temperature changes: A rise of temperature need not imply heat conduction and may be related to other thermal processes like those discussed in sect. 3.1, 3.2 and 3.3. A change in pressure need not correlate with a change in depth. Finally, the interpretation of these questions is complicated by the fact that the evolution of temperature, pressure and deformation may not be independent. Thermal expansion may have an influence on the pressure field, pressure changes may be followed by adiabatic heating or friction heat released during deformation may influence the thermal evolution.

Fortunately, such couplings do not necessarily hinder the geodynamical interpretation. In contrast, this coupling may be the very reason for a process specific and characteristic temporal or spatial relationship of these different evolutions which may be so well illustrated on *P-T*-paths (sect. 7.4.1, 7.2.2).

7.1.1 What Exactly are *P-T*- and *P-T-t-D*-Paths?

P-T-paths are curves on a diagram in which pressure and temperature form the axes. Thus, a *P-T*-path illustrates the relative change of pressure and temperature in a rock, but it cannot show the temporal evolution of either of them. *P-T*-diagrams are *parametric* diagrams because they show the relationship between two independent functions of the same variable (in this case: time). In rocks it is much easier to document the *relative* evolution of pressure and temperature, than the absolute temporal changes of these two parameters. Thus, *P-T*-diagrams have become one of the standard tools for the interpretation of metamorphic rocks. If absolute timing information is available, or the timing of deformation relative to pressure and temperature is known, then we speak of *P-T-t*- or even *P-T-t-D*-paths.

Relationship Between Pressure and Depth. The interpretation of P-T-paths is often made with the basic assumption that pressure is only caused by the weight of the overlying rock column (i.e. it is *lithostatic*) and that - therefore - pressure correlates directly with depth. This assumption implies that deviatoric stress does not play a role (see however sect. 6.3.5). Then, the pressure at depth z corresponds to the vertical normal stress, which may be calculated by integrating the density over depth (s. also sect. 4.2.1):

$$P = \sigma_{zz} = \int_0^z \rho_{(z)} g \mathrm{d}z \quad . \tag{7.1}$$

If the density is independent of depth, then this integral is easily solved. It is: $P = \rho g z$. This may be understood from basic physical principles: stress = force per area, force = mass times acceleration and mass = density times volume. For a 10 km thick column of rock with the density $\rho = 2\,700$ kg m^{-3}, we get a mass per area of $2\,700 \times 10\,000$ kg m^{-2}. Multiplied with the gravitational acceleration ($g \approx 10$ m s^{-2}), this gives a pressure of $2.7 \cdot 10^8$ kg m^{-1} s$^{-2} = 2.7 \cdot 10^8$ Pa $= 270$ MPa $= 2.7$ kbar. In words: the depth in kilometers times ≈ 0.27 gives us the pressure in kbar. Correspondingly, the pressure in kilobar times ≈ 3.7 gives us the depth in kilometers. The errors that arise in this estimate from rounding of the gravitational acceleration or the density changes with depth are significantly smaller than the accuracy of geobarometers and may therefore be neglected. Note that pressure is generally given in kilobars in the geological literature, in energy per volume by thermodynamics community and in Pascal by the geophysicists. The conversions between these units are given by: 1 kbar $= 100$ MPa $= 10^8$ Pa $= 10^8$ J m^{-3}.

7.2 Basic Principles of Petrology

Much of petrology is concerned with the physical and chemical conditions that prevailed at the time of formation of a rock. As such, petrology is inseparably connected with structural geology. Textural interpretation of microstructures and reaction textures in thin section is one of the basic tools to infer both the structural and the metamorphic evolution of a rock. Any reader thinking of him- or herself as either a petrologist or a structural geologist only, is warned when embarking on field or laboratory work with the aim of explaining the tectonic evolution of a region. Because petrology and structural geology are inseparable field tools for a geologist trying to understand the tectonic evolution of a terrain, a section on basic petrology finds its place in a geodynamics text book.

Classic phase petrology is based on the principle of thermodynamic equilibrium. In contrast, most of the geodynamic processes discussed in this book are inherently out of equilibrium, for example the thermal evolution of orogens. The large success that equilibrium thermodynamic considerations have had in determining metamorphic conditions largely relies on the Arrhenius relationship which we will discuss in some detail on p. 325. However, here we want to begin by stating that petrology may be seen to involve:

– equilibrium considerations,
– non-equilibrium considerations.

We will deal with these two fundamentally different approaches in sect. 7.2.1 and sect. 7.2.2, respectively. Both rely on the degree to which chemical reaction is completed in rocks. We begin therefore with some basic concepts of chemical reaction. Throughout this chapter we remind the reader that all topics are dealt with in a very sweeping way and refer to the wide range of excellent teaching texts in petrology (e. g. Winter 2001; Spear 1993; Anderson and Crerar 1993; Yardley 1989; Powell 1978).

Chemical Reaction. Microstructural and chemical changes in rocks depend on two factors:

1. The rate of diffusion with which atoms are brought from one point of the rock to another.
2. The rate of reaction or nucleation which actually binds the atoms at the new location structurally into the crystal lattice.

These two processes depend on very different parameters. The rate of *diffusion* depends very strongly on temperature (in a manner described by eq. 7.4). The rate of *nucleation* and *reaction* is independent of temperature, but depends on how far the reaction is overstepped. The slower of the two processes will be the bottleneck for the overall process and will determine the process rate (Fischer 1973; Joesten 1977, Putnis and McConnell 1980). Petrologists refer to *diffusion controlled* procsses and *reaction controlled* processes. At geologically relevant temperatures the rate determining factor is usually diffusion and we discuss chemical diffusion therefore in the next paragraph.

• *Chemical diffusion.* In the most simple case, one-dimensional diffusion of a single element through a crystal lattice may be described by:

$$\frac{\partial C}{\partial t} = D_{(T)} \frac{\partial^2 C}{\partial x^2} \quad . \tag{7.2}$$

There, C is the concentration of a given element in a given mineral, t is time and x is a spatial coordinate, for example the distance from the center of a garnet crystal to its surface. $D_{(T)}$ is the cation diffusivity and we note already here that it is *not* a constant, but a strong function of temperature. Other than that, element diffusion in minerals is completely analogous to the diffusion of heat (sect. 3.1, eq. 3.6), or the diffusion of mass on a larger scale (e.g. sect. 4.3.2) (Stüwe 1998b). As for those, analytical solutions of eq. 7.2 may be found for a large range of initial and boundary conditions in Carslaw and Jaeger (1959). The specifics of element diffusion is described in detail by the excellent text book of Crank (1975).

The application of eq. 7.2 is not only hindered by the temperature dependence of the diffusivity. It is also complicated by that fact that diffusion rates depend on the diffusion path. For example, diffusion of atoms along grain boundaries is orders of magnitude faster than volume diffusion through the crystal lattice. Thus, volume diffusion is the process rate limiting factor. The rate of volume diffusion of chemical elements through a crystal lattice itself depends on many factors and is – akin to the determination of creep constants for power law creep – difficult to determine experimentally. Also, if several elements diffuse at the same time in the same crystal, then they may influence each others diffusion rates (Onsager 1931).

Nevertheless, eq. 7.2 may be used to make some important estimates on the degree to which chemical equilibration of a rock has gone to completion, for example by using the same diffusive time constant argument we have used on p. 57 for the diffusion of heat (sect. 3.1.3 and 3.1.4, eq. 3.18). In analogy to there we can write here:

$$\tau \approx \frac{l^2}{D} \ . \tag{7.3}$$

There, l is the length scale of diffusion, which a measure of the spatial distance over which elements diffuse in time τ. Eq. 7.3 may be used to estimate the time it takes for a garnet crystal of a given size to equilibrate at a given temperature, or vice versa. For example, we will show on p. 325 that the diffusivities of iron through a garnet lattice at 400 °C is about: $D_{400} \approx 2.7 \cdot 10^{-27}$ m^2 s^{-1} and at 800 °C it is about $D_{800} \approx 2.3 \cdot 10^{-20}$ m^2 s^{-1}. Using eq. 7.3 we can estimate that - if a metamorphic event of 800 °C temperature lastet of the order of 10 my – only garnets smaller than 2.7 mm can be equilibrate by volume diffusion in this time. Correspondingly, we can judge that garnet crystals of 2 mm diameter must be heated for at least 5 my to 800 °C or at least for $4.7 \cdot 10^7$ my to 400 °C, so that they can equilibrate.

Why does equilibrium work?. The fantastic success of equilibrium thermodynamic considerations in petrology may be largely accredited to the Arrhenius relationship. The Arrhenius relationship states that diffusion processes are an exponential function of temperature:

$$D_{(T)} = D_0 e^{\left(-\frac{Q+VP}{RT}\right)} \ . \tag{7.4}$$

In this equation, $D_{(T)}$ is the diffusivity of elements as a function of temperature (in m^2 s^{-1}), D_0 is a pre exponential factor, Q is the activation energy, R is the gas constant and T the absolute temperature. The numerator of the exponent also includes the product of the activation volume V and pressure P. However, the activation volumes of petrological processes are generally so small, so that the pressure dependence of element diffusion is generally neglected. According to eq. 7.4 the diffusivity goes towards zero when temperature goes towards zero. In other words, all equilibria that rely on diffusion freeze at low temperatures. At higher temperatures, the diffusivity rises and goes asymptotically towards D_0. Thus, with higher temperatures, it becomes increasingly likely that different minerals in a rock can communicate with each other and are in chemical or structural equilibrium. The Arrhenius relationship applies not only to the diffusion of ions on a crystal lattice but also to the diffusion of radioactive isotopes or lattice dislocations. The former is highly relevant to geochronologists, the latter is relevant when discussing flow processes and microstructures. In this context we have met the Arrhenius equation already when we discussed deformation mechanisms (sect. 5.1.2).

• *The irreversibility principle.* Because of the temperature dependence of diffusion processes (eq. 7.4), only small parts of a metamorphic *P-T* path are actually preserved in a rock. According to which parts of a *P-T* path are preserved in a rock and which parts are not, the metamorphic evolution of rocks may be divided into 5 different sections (Fig. 7.1). In the 1st section there is no chemical reaction or mineral growth. Temperatures and diffusion rates are too slow. In the second part chemical equilibration does occur. However, because temperature and diffusion rates continue to increase, each equilibrium will be erased and superceded by that of the next higher temperature. The 3rd section of the *P-T*-path is reached at the metamorphic temperature peak where - by definition - the rate of temperature change is small. There, chemical and textural equilibrium is reached to the best degree. The 4th part of the *P-T* path occurs during the initial cooling history. Diffusion rates decrease and the volume of the rock which can be in equilibrium at any given temperature decreases. Thus, larger and larger parts of the rock will cease to equilibrate with their surroundings and preserve larger and larger parts of the rocks. Partial retrogression will occur in this 4th section. In the 5th section of the *P-T* path all equilibria are frozen and reaction has terminated.

Fig. 7.1a summarizes this information. It shows that only the thermal peak and a small section of the path thereafter are likely to be preserved by rocks, while the heating phases is likely to be only preserved in relics. Because temperature is so much more important to the equilibration process than pressure (see discussion of eq. 7.4) the *thermal* maximum (rather than the *baric* maximum) is generally referred to as the metamorphic peak. However, it is emphasized that pressure and temperature peak of a *P-T* path must not coincide (Fig. 7.1b). Indeed, valuable tectonic information may be extracted from the temporal relationship between pressure and temperature maximum on a metamorphic *P-T* path.

7.2.1 Equilibrium Information: Thermobarometry

If we assume that rocks do, in fact, record some sort of chemical equilibrium from some stage of their metamorphic evolution it is possible to use the principles of equilibrium thermodynamics to infer their formation conditions. Very often this concerns the derivation of metamorphic pressures and temperatures using geothermometers and geobarometers. In general, the term *geothermobarometry* summarizes the methods used. It is useful to discriminate between:

— petrological thermobarometry,
— mineralogical - crystallographic thermobarometry and:
— structural thermobarometry.

An example of a *structural barometry* are the palaeopiezometric methods (e. g. Christie and Ord 1980; Dunning et al. 1982) and an example of *structural thermometry* is textural analysis like lattice preferred orientations of quartz (e. g. Jessel and Lister 1990). *Mineralogical thermobarometry* uses the pressure and temperature dependence of parameters hat may be measured in a single mineral, for example the lattice parameters, the concentration of a given element or the composition of fluid inclusions.

  Petrological thermobarometry is based on the fact that the distribution of elements between minerals is a function of pressure and temperature. For energetic reasons, this distribution always aims to be in chemical equilibrium. In particular with high grade metamorphic rocks, petrological thermobarometry has become the standard method to determine the physical conditions of metamorphism. However, for a meaningful application it is crucial that chemical equilibrium was in fact reached and that reaction textures between different minerals are interpreted correctly (Cooke et al. 2000). About the methods and problems of textural interpretation of mineral parageneses there is abundant literature (e. g. Spear and Florence 1992; Robinson 1990; Stüwe 1997).

The Chemical Equilibrium Condition. Most of the *petrological thermobarometry* is based on the assumption of chemical equilibrium in rocks. Then,

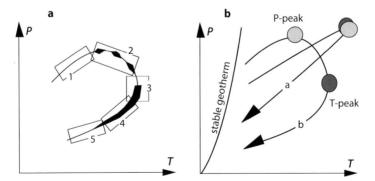

Figure 7.1. a Schematic *P-T*-path that is divided into five different sections that are recorded in different detail in a rock. The different sections are discussed in detail in the text. The sections of a *P-T*-path that are well recorded by a rock are shown thick **b** *P-T*-path in which pressure and temperature peak coincide (*a*); and *P-T*-path in which pressure and temperature peak do *not* coincide (*b*). A schematic stable geotherm is shown for reference

the thermodynamic equations describing equilibrium can be used to predict the distribution of a chemical element between the different minerals in a rock. The *thermodynamic equilibrium condition* may be formulated as:

$$0 = \Delta G^0 + RT\ln K \quad . \tag{7.5}$$

In this equation, R is the gas constant, ΔG^0 is the change in Gibb's free energy between all phases involved in a reaction in their standard states and K is the thermodynamic equilibrium coefficient. This coefficient may be derived from the activities of different mineral end members in a mineral. T is the absolute temperature. For a simple exchange reaction of the element C between the phases A and B, K is given by the relationship:

$$K = \left(\frac{1 - X_C^A}{X_C^A}\right) \left(\frac{X_C^B}{1 - X_C^B}\right) \quad . \tag{7.6}$$

There, X_C^A is the mol fraction of the element C in mineral A and X_C^B is the corresponding mol fraction in mineral phase B. Thus, K may be measured directly from the compositions of different minerals in a rock, for example by electron microprobe. The expression ΔG^0 is the difference between the free enthalpies of the reactants and the products (in this case the pure phases A and B) in their standard state. These free enthalpies are functions of material constants like heat capacity, formation energy, pressure and temperature. Thus, eq. 7.5 may be solved for pressure as a function of temperature or vice versa, if all the other constants are known. If eq. 7.5 describes a curve with a shallow slope in a pressure-temperature diagram, then this reaction is called a *barometer*. Correspondingly, if the slope of this curve is steep, then it is called a *thermometer*. Clearly, the information given in this short paragraph is by no means sufficient to learn about the thermodynamics of thermobarometry. It is just meant to indicate that an entry into this field is not all this hard. For details see an abundance of literature on the thermodynamics of rocks, for example Holland and Powell (1990), Atkins (1994), Powell (1978), Spear (1993) or Will (1998).

The Phase Rule. The concept of the phase rule is (while often considered as only of petrological interest) is a much undervalued tool for the geodynamically oriented field geologist. This section will show how it can be used in the field. We need to start with a few sentences of theory. Again, we refer to the same abundant literature on the subject as above for more detail.

The phase rule expresses the relationship between the number of degrees of thermodynamic freedom, the number of exchange components and the number of mineral phases. It may be formulated as:

$$\text{Phases} = \text{Components} - \text{Degrees of Freedom} + 2 \quad . \tag{7.7}$$

For example, a reaction involving the chemical exchange of *one* chemical component between *two* mineral phases will have *one* degree of thermodynamic freedom. We may freely choose the pressure where this reaction is

supposed to occur, but then the temperature and compositions of all phases
are predetermined by this assumption - they are fixed. This is the example we
discussed in the last section where we discussed the exchange of component
C between the mineral phases A and B and said that it can be used as a thermometer of barometer. If the exchange of the component C occurs between
more than two mineral phases, then the reaction has a correspondingly larger
number of degrees of freedom (s. Zen 1966).

Depending on the use of the phase rule, the term "phase" may be used
either to describe the number of pure mineral end members (e.g. pyrope),
or number of minerals as found in the field (each of which may be composed
of several end members e.g. garnet). Which way the term "phase" is used
depends on whether the phase rule is used (a) to estimate the number of
possible thermobarometers that we could apply to a given paragenesis (in
which case "phase" would be the number of end member phases of an assemblage in question), or (b) to estimate in the field how well we can constrain
the formation conditions qualitatively on a petrogenetic grid (in which case
"phase" would be the number of minerals in a rock) (s. eq. 7.5).

- *The use of the phase rule in the field.* Let us reconsider eq. 7.7. We can see
that the chemical equilibrium between three minerals but only a single exchange component has zero degrees of freedom. For example, the three phases
ice, water and steam (all being made up of the single component H_2O) or
sillimanite, andalusite and kyanite may only occur together (in equilibrium)
at a single pressure and a single temperature. There is no choice (freedom)
in picking either one of these physical conditions freely. Also, according to
eq. 7.7 it is impossible for more than 3 phases to exist in stable equilibrium
if there is only a single exchange component available.

In real rocks we generally are confronted with six or more exchange components. According to eq. 7.7 rocks with six components cannot have more
than 8 equilibrium phases in them. Thus, if we find a rock that is made up
of six components in the field (like most pelites), that contains more than
eight minerals, we can immediately conclude that this rock has experienced
more than one metamorphic event and is *not* in equilibrium. This may be
extremely useful in the field if we are trying to correlate events of mineral
growth with deformation phases.

The phase rule can also be used to estimate how useful this rock may be
back in the laboratory for thermobarometry. For this we need to estimate
how many mineral *end members* we can extract from its parageneses. For
example, a rock containing garnet, staurolite, biotite, muscovite, kyanite,
chlorite, quartz and water is thermodynamically invariant if it consists of only
six components. However, it allows us to extract far more than 10 mineral
end members and their activities (pyrope, almandine......). This means that
we will be able to apply quite a number of thermometers and/or barometers
to this rock to constrain its formation conditions.

Using several thermobarometers simultaneously is the principle underlying the idea of formulating *internally consistent thermodynamic data sets* that can then be used to do average calculations between many thermobarometers (e. g. Holland and Powell 1990, Berman 1988). Because of these reasons it is generally useful to collect in the field rocks that contain as many phases as possible for as few minerals as possible.

7.2.2 Non-equilibrium Information: Kinetics

While equilibrium thermodynamic methods are the most widely used petrological tool, much additional information can be gained by using the non-equilibrium information recorded by rocks. Very loosely, it may be said that this is done using two different approaches:

− using *non-equilibrium* thermodynamics,
− using *equilibrium* thermodynamics and "being careful".

The use of non-equilibrium thermodynamics in petrology was pioneered in the seventies (e.g. Fischer 1973, Joesten 1977), but − while undoubtedly the rigorous treatment of the subject − it has not found wide application by the field geology community. In contrast, the application of equilibrium considerations to the interpretation of metamorphic non-equilibria (e.g. partial reaction textures) by "being careful" is widely applied by geologists to the interpretation of metmorphic *P-T* paths. In particular the consideration of a time scale in diffusion processes has led to a lot of progress which may be summarized under the term "geospeedometry" (Lasaga 1985). the following section presents a few highlights form this field.

Working with Changing Diffusivities. Let us begin by returning to eq. 7.4 to determine the rate of volume diffusion of cations in garnet crystals. The magnitude of the diffusion rate of magnesium in garnet is roughly known. It is described by the material constants $Q = 239\,000$ J mol^{-1} and $D_0 = 9.81 \cdot 10^{-9}$ m^2 s^{-1} (Cygan and Lasaga 1985). Using eq. 7.4 we can derive a diffusivity of $D_{400} \approx 2.7 \cdot 10^{-27}$ m^2 s^{-1} or $D_{800} \approx 2.3 \cdot 10^{-20}$ m^2 s^{-1} at a temperature of 400 °C or 800 °C, respectively. We can see that the diffusivity changes over many orders of magnitude over this geologically relevant temperature interval

• *The concept of closure temperature.* From the above (and eq. 7.3) it should be clear that a crystal of a given size can *only* equilibrate in a given time if its *P-T*-path exceeds a certain temperature over a certain time. If the temperature decreases below this temperature, then full equilibration is not any more possible and the crystal will only equilibrate partially. In other words, if the diffusive length scale (given by l in eq. 7.3), is substantially smaller than the radius of a crystal, then at least part of the crystal center will not be able to keep up with the processes on the crystal surface and it will cease to react with the surroundings. Its composition will freeze. This phenomenon is called

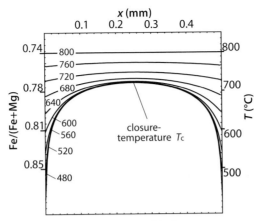

Figure 7.2. Development of a zoning profile in a garnet crystal during cooling and the definition of closure temperature T_c. The horizontal axis is a profile through a garnet crystal that is 0.5 mm in diameter. The grain center is located at $x = 0.25$ mm. The left hand vertical axis is labeled with the Fe concentration in the crystal (given as mol fraction Fe / (Fe + Mg)). In equilibrium this concentrations may be directly correlated with temperature (right hand vertical axis). The surface of the crystal is assumed to be in chemical equilibrium with a much larger biotite crystal characterized by an Fe concentration of Fe / (Fe + Mg) = 0.5 at all times. A cooling rate of $100\,^\circ\mathrm{C}\,\mathrm{my}^{-1}$ was assumed. Calculated by numerical coupling of eqs. 7.2, 7.4 and 7.6 and the garnet - biotite thermometer equation of Holland and Powell (1990)

closure and the temperature at which this occurs is called the *closure temperature*. The successive closure of the chemical composition of a crystal leads to the development of a zoning profile and can be observed in the minerals of many rocks (Fig. 7.2). The chemical non-equilibrium that is frozen into such zoning profiles is an important piece of evidence for the interpretation of the cooling history of a rock.

• *Mean diffusivities.* Even if a mineral grain is open to diffusive equilibration (i.e. it is above its closure temperature) diffusivities are very strong functions of temperature so that it is difficult to use eq. 7.3 without having found an appropriate mean diffusivity that is representative for any temperature changes during. As the Arrhenius relationship is strongly non linear this is not a trivial exercise. However, if the cooling rate is constant, then the mean diffusion rate may be estimated with the following relationship (Itayama and Stüwe 1974):

$$\overline{D} \approx \frac{D_A}{\left(\frac{Q}{RT_E}\right)(1 - T_E/T_A)} \qquad (7.8)$$

There, $\overline{D}$ is the mean diffusion rate averaged between the starting temperature T_A and the final temperature T_E (e. g. the metamorphic peak temperature and the temperature of the earth's surface). D_A is the diffusion rate at the starting temperature and may be calculated from eq. 7.4.

Determining Cooling Rates. From the above we can summarize, that the following parameters have a close linking in diffusive processes:

– Diffusive length scale (grain size),
– cooling rate (temperature, time),
– diffusivity (material constants).

For simple exchange systems and simple geometries of the crystal eq. 7.2 may be solved analytically to describe this relationship:

$$s = D_0 e^{\left(-\frac{Q}{RT_c}+G\right)} \left(\frac{RT_c^2}{(l/2)^2 Q}\right) \quad . \tag{7.9}$$

There, s is the cooling rate, T_c is the closure temperature, (defined in Fig. 7.2) and l is the radius of a crystal. The constant G is a parameter that describes the geometry of the crystal in which diffusion occurs. For spherical crystals (as is well-approximated by garnet) this constant is: $G = 1.96$. The variables Q, D_0 and R were defined in eq. 7.4. Eq. 7.9 was derived by Dodson (1973) for application in geochronological systems but it is equally applicable to cation diffusion in minerals where it has become an important tool to estimate cooling rates of rocks using data that are easily obtained using optical microscopy (to determine l) and electron microprobe (to determine T_c) (Ehlers and Powell 1994; Ehlers et al. 1994a). The relationship between grain size, closure temperature and cooling rate defined by eq. 7.9 is illustrated in Fig. 7.3a.

• *Tectonic use of cooling rate information.* The slope and curvature of cooling curves is characteristic of the underlying cooling process. Thus, determination of cooling curves is an important tool for the interpretation of geodynamic processes. For example, a comparison of Fig. 3.14 with 3.30 (as contrasted in Fig. 7.3b) shows, that cooling curves terminating a regional metamorphic cycle are fundamentally different from those that characterize cooling after contact metamorphism (Harrison and Clark 1979). If cooling occurs because the rocks are exhumed, then the cooling rate *increases* with decreasing temperature. In contrast, the cooling rate *decreases* with decreasing temperature towards the end of contact metamorphism (see Fig. 7.3b).

Fig. 7.3a illustrates the different relationships between closure temperatures and grain size that are expected in two rocks that cooled according to the two cooling curves shown in Fig. 7.3b. Overlaying Fig. 7.3a onto Fig. 7.3b shows that these relationships are different! Ehlers et al. (1994b) have documented closure temperatures between 400 °C and 700 °C in a single thin section and were able to correlate these closure temperatures with grain size. This range of closure temperatures is sufficient to document crucial parts of a cooling curve.

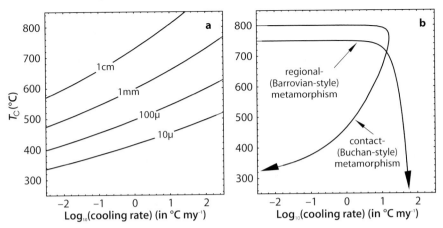

Figure 7.3. Relationships between cooling rate and temperature. **a** Cooling rate plotted against closure temperature T_c of crystals with different diameters. The curves were calculated with eq. 7.9 assuming a spherical grain geometry. The diffusion data of Cygan and Lasaga (1985) for garnet were used. **b** Cooling rate against temperature for two different cooling processes. The curve for contact metamorphism was calculated with eq. 3.89 using $T_i = 800\,°C$, $T_b = 300\,°C$, $l = 60$ km, $\kappa = 10^{-6}\ m^2\,s^{-1}$ and $z = 0$. The curve for regional metamorphism was calculated with eq. 3.48 and eq. 3.49 using $g = 30\,°C\,km^{-1}$, $z_i = 25$ km and an erosion rate of $u = -1\ km\,my^{-1}$. While the assumptions underlying these equations are very much simplified from real settings, they do show some typical characteristics. For example, note that during cooling from regional metamorphism the cooling rate *increases* with decreasing temperature, while the cooling rate *decreases* with decreasing temperature following contact metamorphism. Such differences may be extracted, at least in principle, from zoned crystals

• *A circular argument in geochronology.* A common method to document cooling curves uses isotopic dating of a series of isotopic systems in minerals with different closure temperatures. Closure temperature of the various isotopic systems is then plotted against the radiometrically determined age and a cooling curve is constructed. However, in the previous section we have shown that the closure temperature depends on cooling rate. Thus there is an obvious circular argument: Cooling rates obtained by this approach can – strictly taken – only be estimated once a cooling rates is assumed. Only then the closure temperature of the system is known.

The successful application of geochronological methods to the determination of cooling curves indicates that the variability of closure temperatures of isotopic systems is small – even for a large range of cooling rates – in comparison to many other factors influencing closure of isotopic systems during cooling. However, it is important to be aware of the implicit circularity in the uncritical acceptance of a certain value for closure temperature of an isotopic system.

7.3 Documentation of *P-T*-Paths

7.3.1 Qualitative Shape of *P-T*-Paths

P-T-paths are generally divided into two groups, according to their slope following the metamorphic temperature peak:

- 1. *P-T*-paths which are characterized by a decrease in pressure before cooling commences.
- 2. *P-T*-paths which are characterized by a decrease in temperature before decompression commences and *P-T*-paths where pressure increase accompanies cooling (Fig. 7.4, Table 7.1).

   In a diagram in which the positive temperature axis is drawn to the right and the positive pressure axis is drawn upwards, these two paths follow a clockwise and an anticlockwise curvature, respectively (Fig. 7.4a). Thus, the two path groups 1. and 2. are referred to as clockwise *P-T*-paths and anticlockwise *P-T*-paths, respectively.

   However, in the geological literature it is common to plot the pressure axis downwards in order to assist the intuitive understanding that pressure increases with depth in the crust (Fig. 7.4b). Obviously, the direction of a

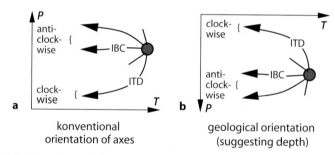

konventional   geological orientation
orientation of axes   (suggesting depth)

Figure 7.4. Directions of *P-T*-paths in different presentations. **a** In the conventional representation, isothermal decompression paths follow a clockwise path and isobaric cooling paths have an anticlockwise shape. **b** In the geologically most commonly used representation pressure is usually plotted to increase downwards to indicate that it increases with depth in the crust. Then, "anticlockwise paths" actually follow a clockwise path and vice versa. The dashed extensions of the paths indicate the overall shape of the paths, but these sections are rarely preserved

Table 7.1. Classification of *P-T*-paths

1. clockwise	ITD (IsoThermal Decomp.)	*P* decreases after peak
2. anticlockwise	IBC (IsoBaric Cooling)	a) cooling with no *P* decrease b) *P* increases during cooling

P-T-path in such a diagram is reversed from that in the conventional representation. Thus, it is preferred to use the less ambiguous description of ITD- (**IsoThermal Decompression**) or IBC-paths (**IsoBaric Cooling**) instead of clockwise and anticlockwise (Table 7.1).

This qualitative division of P-T-paths may be formulated a bit more rigorously. At the thermal peak, the rate of temperature change is, by definition, zero: $dT/dt=0$. Thus, differently shaped P-T paths may be defined on the basis of the qualitative nature of the pressure change at the temperature peak. We can write:

$$ - \left. \frac{dP}{dt} \right|_{(dT/dt=0)} = \text{negative:} \qquad \text{clockwise (ITD)} $$

$$ - \left. \frac{dP}{dt} \right|_{(dT/dt=0)} = 0: \qquad \text{anticlockwise (IBC)} $$

$$ - \left. \frac{dP}{dt} \right|_{(dT/dt=0)} = \text{positive:} \qquad \text{anticlockwise} $$

While this is quite a rigorous classification, this definition does not state how long this condition is held upright. For example, it is possible that the P-T-path is characterized by pressure increase only at the very temperature peak, but that most of the time the path occurred during decompression (Fig. 7.5a). In short, it remains difficult to define a rigorous quantitative classification of P-T-paths.

• *Continuous or discontinuous P-T-paths?.* The most common cause for misinterpretations (and also much discussed issue) of the shape of P-T-paths is illustrated in Fig. 7.5b. There, two black dots indicate the P-T-conditions for the formation of two overprinting metamorphic parageneses in a single rock. Usually it is very difficult to see in a rock if the two parageneses formed in a continuous P-T-evolution (ii), or as the consequence of two completely independent metamorphic events (i) (Fig. 7.5b).

7.3.2 Slope and Curvature of P-T-Paths

The slope and curvature of metamorphic P-T-paths contains important information on the relative and absolute rate of different geological processes. The two diagrams on the left in Fig. 7.6 show the temporal evolution of pressure a rock experienced (bottom diagram) and three different possibilities of the temporal evolution of temperature in the same rock. The diagram on the right shows three P-T-paths that correspond to the temporal evolutions shown at left. The temporal evolution of neither pressure nor temperature may be inferred from the P-T-path, but their *relative* temporal evolution is. This may be inferred from the slope of the three different P-T-paths. From the IBC-path c we can infer that the rate of cooling was rapid compared to the rate of pressure change. Correspondingly, we can conclude from the ITD-path a that the rate of decompression was rapid compared to the rate of cooling. This information is already an important piece of information for the

geodynamic interpretation of a metamorphic terrain. Note that we extracted
this information solely on the basis of its *P-T*-path.

If pressure and temperature are not linear functions of time, then the
relationships discussed above need not apply. For example, if the temporal

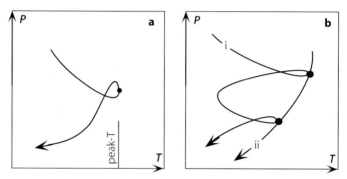

Figure 7.5. Examples of *P-T*-paths that are easily misinterpreted. **a** *P-T*-path
which – according to conventional classification – would be interpreted to be anticlockwise, but which has the overall form of a ITD-path. **b** Example of a fundamental problem of the interpretation of *P-T*-paths. If two metamorphic parageneses are
observed in a rock that indicate the two *P-T* conditions shown by the two black
dots, then it is usually very difficult to discern if the path between the two was
characterized by two different events with IBC-paths (shown by path *i*) or a single
ITD path (shown by path *ii*)

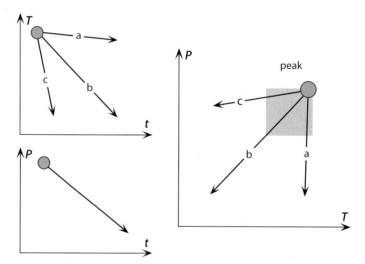

Figure 7.6. Different shapes of *P-T*-paths (diagram on right), as they occur as a
function of different cooling rates of rocks (top left) that all experienced the same
decompression rate (bottom left). The *P-T*-paths *a*, *b* and *c* in the diagram on
the right correspond to the three different cooling curves as labeled on the left.
The shaded rectangle indicates schematically the *P-T*-region which is likely to be
preserved in a rock

Figure 7.7. Possibilities for *P-T*-paths as they may arise as a consequence of various non-linear evolutions of pressure and temperature. The figure is analogous to Fig. 7.6

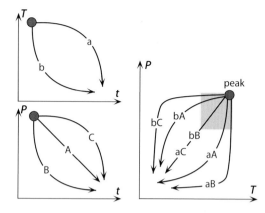

evolution of pressure and that of temperature are the same, then the *P-T*-path will be linear, irregardless of *what* temporal evolution pressure and temperature undertake. This is shown on Fig. 7.7 where both the combination of temperature evolution *a* with pressure evolution *C* and the temperature evolution *b* with pressure evolution *B* produce the same linear *P-T*-path.

An Example of Slope Interpretation of a *P-T*-Path. Let us illustrate the enormous geodynamical relevance of the considerations of the last sections with an example. In many high grade metamorphic terrains in the Precambrian shields IBC-curves have been documented (e. g. Harley 1989). We can conclude that the cooling rate of all these terrains must have been much more rapid than their rate of exhumation or burial during the thermal evolution (assuming that all pressure change is related to depth change). As most of these terrains are characterized by intense synmetamorphic deformation (both extensional and compressional) it is likely that the terrains did indeed exhume or get buried during their thermal history. Thus, if the rates of exhumation and burial during erosion or continental deformation were known, the we can use the fact that a terrain is characterized by an IBC paths to constrain minimum values for the rate of cooling. During normal continental deformation rates of $\dot{\epsilon} = 10^{-13}$–$10^{-14}$ s^{-1} the rate of exhumation or burial is of the order of some millimeters per year. At this rate of depth change an IBC-path can only occur if the cooling phase of the terrain lasted a mere few millions of years. For example, if we just take an arbitrary number of 3 millimeters per year of depth change and assume that a *P-T*-path will look like an IBC path if it cools more rapidly than 300 °C per kbar of pressure change, then this implies a cooling rate of at least 243 °C my^{-1}.

If we now recall sect. 3.1.4 where we have shown that the duration of conductive processes is proportional to the square of the length scale of the conducting body, then we can use the estimate for the total duration of cooling to infer the spatial size of the terrain in question, at least to the order of magnitude. It turns out that for the total duration of the cooling phase as estimated above, the size of the heated region is a mere few kilometers. We can conclude from our derived IBC-path that it is unlikely that the metamorphic event was caused by a process that affected the entire lithosphere. It is more likely that the heating mechanism was localized. Indeed, it is this very argument that has often been used to interpret that IBC-paths are atypical for regional metamorphic events and are more likely to be associated with contact metamorphism around magma bodies (e. g. Lux et al. 1986; DeYoreo 1991; see. sect. 6.3.3). You may argue that all the numbers used here are wrong or not well constrained, but the thought process carried out here is independent of the numbers and in an example where they *are* known, it may extremely useful.

7.4 Interpretation of *P-T-t-D*-Relationships in Orogens

In the past sections we have been concerned with the classification and documentation of metamorphic *P-T*-paths. We will now go into some detail how to interpret spatial and temporal field relationships and *P-T-t-D*-paths. Conceptual interpretations like those presented below are particularly important in ancient or badly exposed metamorphic terrains, where little is known about the geometry, style or tectonic setting of the underlying orogenic event. However, even in well known orogens like the European Alps, such conceptual interpretations may help geologists to free the eye from a huge abundance of detailed regional knowledge and help to understand some fundamental background of orogenic processes. In general, it is important to discriminate between two different types of relationships:

– 1. The *temporal* relationship between deformation and metamorphism as well as the temporal relationship between pressure peak and temperature peak. (The former can be mapped in the field, the latter can usually only be derived with detailed petrological work).
– 2. The *spatial* change of the relationships documented in 1. across the terrain in question.

Such relationships, both in space and time, may be well illustrated on an event diagram (Fig. 7.8). When mapping such changes in the field it is important to be aware of the fact that different mapped parameters may have completely different strike directions in the field. For example, the lithological boundaries may strike completely different from the direction of the metamorphic isograds or from lines of constant strain or those of constant age.

Terminology Definition. When discussing spatial and temporal relationships between metamorphic pressure, temperature and time across a terrain, or within a single rock, it is easy to confuse a number of quite different issues each of which contains important interpretative information. We therefore define the following terms:

• *Metamorphic field gradient.* The term "metamorphic field gradient" is used strictly to describe the change of metamorphic grade with distance as observed in the field (measured normal to the metamorphic isograds). As such, metamorphic field gradients have the units of dT/dx or dP/dx depending whether the gradient describes the change of metamorphic temperature or pressure with lateral distance x. It is important to realize that the ratio of the metamorphic temperature gradient to the metamorphic pressure gradient does *not* need to document a metamorphic geotherm (although it has the units of dT/dP and may be easily converted into dT/dz assuming the conversion discussed in eq. 7.1). In contrast, the T/P ratio of a single rock will *always* record one point on a metamorphic geotherm (by definition - see below).

• *Metamorphic geotherm.* The term "metamorphic geotherm" is used to describe the relationship between temperature and depth at a chosen point in time during metamorphism. As such, a metamorphic geotherm is a transient feature and contrasts the term *stable geotherm.* Very generally, it may be said that metamorphic geotherms change at rates of $\gg 10\,°C\ my^{-1}$, whereas stable geotherms change at rates of $\ll 10\,°C\ my^{-1}$. During regional (Bar-

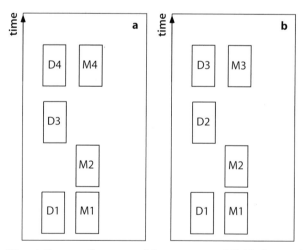

Figure 7.8. "Event diagrams" are a useful way to record field observations. **a** and **b** show two different nomenclatures used for event diagrams for the same set of field observations. Both diagrams are drawn for a terrain in which three deformation phases (D) and three metamorphic events (M) have been mapped in the field. In **a** the different events were numbered according to their temporal sequence, but without consideration to whether the event is a deformation event or a metamorphic event. In **b** deformation events and metamorphic events are numbered separately. Both according to their respective temporal sequence, but with no consideration of the temporal relationship *between* deformation and metamorphic events. Which representation is more useful depends on the question that is to be answered with the illustration

rovian) metamorphism, metamorphic geotherms are generally characterized
by a monotonous increase of temperature with depth (England and Thompson 1984). However, during contact metamorphism or in staked nappe piles,
metamorphic geotherms may transiently be characterized by some crustal
sections in which temperature decreases with increasing depth.

• *Piezothermal array.* The term "piezothermal array" was clearly defined by
Richardson and England (1979) as the line that connects the metamorphic
temperature peaks of all rocks in a vertical crustal section through depth and
time. Piezothermal arrays are diachronous in time and they cross metamorphic geotherms. In general, it is a piezothermal array that will be preserved
by an obliquely exposed crustal section, but piezotherms may coincide with
metamorphic geotherms for some tectonic settings, for example if exhumation
is practically instantaneous.

7.4.1 Interpreting Temporal Relationships

Here, we focus on how the temporal relationship between *metamorphism* and
deformation may be interpreted. Interpretations of the temporal relationships between pressure and temperature peak were discussed in the context
of Fig. 7.1. The relative temporal relationship between deformation and metamorphism is usually well-preserved in metamorphic rocks, but the value of
this easily accessed information is often under appreciated (Fig. 7.8). In principle, there is three different possible relationships between deformation and
metamorphism:

1. Deformation of rocks precedes metamorphism,
2. deformation and metamorphism occurred contemporaneously,
3. deformation occurred *after* metamorphism.

Clearly, it is always possible that the causes for deformation are unrelated
to the causes of metamorphism. However, several of these timing relationships occur because of the intimate coupling of structural and metamorphic
processes. Thus, in the following, we discuss possible interpretations of these
timing relationships in terms of a single underlying cause.

• *Deformation precedes metamorphism.* This timing relationship is characteristic for regional metamorphism, where metamorphism was caused by heat
conduction and radiogenic heat production in overthickened continental crust
(sect. 6.2.1). This is because continental deformation rates are typically about
one order of magnitude more rapid than thermal equilibration over the length
scale of the crust (s. sect. 3.1.4, 6.3.6). Thus, thickening of the crust at "normal" collision rates is usually completed before conductive equilibration may
catch up. As a consequence, it may expected that metamorphism is separated
from peak metamorphism by up to several tens of millions of years.

• *Deformation occurred contemporaneously with metamorphism.* Deformation events that occur contemporaneously with metamorphism are typically observed in low-pressure high-temperature metamorphic terrains. Clearly, the temporal synchroneiety invites an interpretation in terms of a common cause. Three possible relationships may be thought of (s. also p. 314):

1. Deformation is caused directly and instantaneously by the heating. This is possible if the terrain in question is subjected to a constant plate boundary driving force. Then, heating of the terrain may weaken the rocks sufficiently so that strain rate rises rapidly in response to the far field stress (Sandiford et al. 1991). In the field this may be seen as a sudden deformation event.
2. Heating is caused directly and instantaneously by deformation. This is only possible if the rocks are strong enough so that shear heating causes the metamorphism. This is widely-observed on a local scale, for example around pseudotachylites, but the possibility for regional shear heating is subject to debate (sect. 3.2.2).
3. Both heating and deformation may have been caused by rising magma bodies.

• *Deformation occurs after metamorphism.* Field examples where deformation of a terrain occurred after its metamorphic peak, are typically interpreted in terms of two independent events. That is, deformation and metamorphism are unrelated. Unlike the other two timing relationships discussed above, there is currently no elegant models which allow us to interpret this timing relationship in terms of a single tectonic process.

7.4.2 Interpreting Spatial Relationships

In many metamorphic terrains the metamorphic grade changes across the terrain. This change in timing is often accompanied by a change of the timing of peak metamorphism. For example, high grade parts may experience their metamorphic peak earlier or later than lower grade parts, either relative to an absolute marker (e.g. a dike swarm), or as determined by absolute geochronology. Such spatial changes in metamorphic grade and timing may be interpreted in three fundamentally different ways:

1. Independent heating mechanisms heated different parts of the terrain in question at different times. In this case, it is expected that there is a discontinuity in metamorphic grade or metamorphic age somewhere in the terrain; for example, because the rocks of different metamorphic age where juxtaposed much later.
2. Lateral variations in the physical parameters, for example, thermal conductivity or heat production rate may cause different parts of a terrain to heat at different rates and to different degrees (Sonder and Chamberlain 1992).

3. The changes in the timing of metamorphism across the terrain are inherent to the nature of a single heating process. A typical example for this is heat conduction: because of the time it takes for a terrain to equilibrate conductively, it is possible that some parts of a terrain experienced metamorphism earlier than others (e. g. s. sect. 3.1.4). In this case, changes in metamorphic timing and grade are expected to be continuous across a terrain.

The curve that must be interpreted to understand some of these relationships is the *piezothermal array*.

Piezothermal Arrays. Piezothermal arrays are curves that connect the metamorphic temperature peaks of rocks from all crustal depths through grade and time (England and Thompson 1984; Fig. 7.9). Accordingly, piezothermal arrays may be plotted in pressure-temperature, in temperature-time, or in pressure-time diagrams. As the slopes of the piezotherms in two of these different diagrams are independent, it is useful to discriminate between two different features of a piezotherm. These are: 1. their temporal characteristics and 2. their thermal perturbation characteristics.

● *Temporal characteristics of piezotherms.* The temporal characteristics of a piezotherm describe if the metamorphic peak of high grade rocks occurs earlier or later than that of their lower grade equivalents (Fig. 7.9a). These timing relationships may be characterized by the slope of the piezotherm in a temperature-time diagram (e. g. Stüwe et al. 1993, 1997). Three qualitatively different relationships may be discerned:

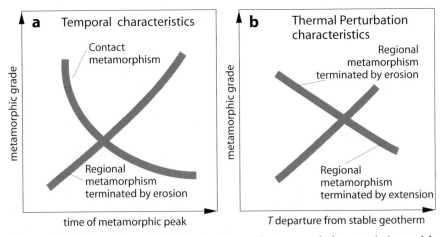

Figure 7.9. Schematic cartoons illustrating **a** the temporal characteristics and **b** the thermal perturbation characteristics of piezothermal arrays for some different tectonic settings and metamorphic heat sources. As the slope of the piezotherms in these diagrams are mapable in the field, these functions are extremely useful interpretative information

1. If the metamorphic peak in high grade rocks occurred later than in low grade rocks, then the slope of the piezothermal array in a temperature-time diagram is positive. Piezothermal arrays with a positive slope are typical of regional metamorphism (sect. 6.2.1, Fig. 7.9a). Regional metamorphism is caused by heat conduction in the crust. Rocks from shallow crustal levels will not heat significantly because of their proximity to the surface and will be the first to cool when exhumation commences (Fig. 6.14). Rocks from deeper and deeper crustal levels have more and more time to equilibrate, even while exhumation commences and will reach their metamorphic peak therefore later (Fig. 7.10a).
2. If the metamorphic peak in high grade rocks occurred earlier than in low grade rocks, then the slope of the piezothermal array in a temperature-time diagram is negative. Piezothermal arrays with a negative slope are typical of contact metamorphism (Fig. 3.30, 7.9). Near the contact of a heat source, rocks are heated very rapidly to high temperatures. With increasing distance from the heat source, the thermal effects of contact metamorphism are felt later and the metamorphic grade is lower. Details of these relationships were already discussed in sect. 3.6.2.

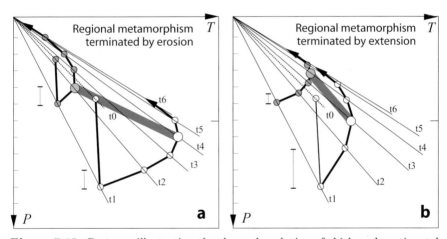

Figure 7.10. Cartoons illustrating the thermal evolution of thickened continental crust during: **a** exhumation by erosion (all rocks in a vertical column move upwards by the same amount during each time step as indicated by the vertical bas) and **b** exhumation by pure shear extension (the exhumation rate is depth dependent). Shown are: stable geotherms prior to thickening (dotted lines), *metamorphic geotherms* (thin lines labeled from time t_1 immediately after thickening to t_6 near the final stages of exhumation), *piezothermal arrays* (shaded bars) and *PTt* paths (thick lines). Note that in **a** the slope of the piezothermal array has a higher T/P gradient than any one geotherm but in **b** it has a lower T/P gradient. Moreover, the temporal relationships between high and low grade rocks are reversed between the two exhumation mechanisms

3. If the time of metamorphism is independent of grade, then the piezothermal array forms a vertical line in a temperature-time diagram. It is unusual that all rocks from a terrain with variable metamorphic grade experienced their metamorphic peak at the same time, even if the temporal variation is too small to be discernable in the field. One mechanism that *would* allow the metamorphic peak to occur simultaneously in all areas independently of metamorphic grade, is if metamorphism is caused by mechanical heat production in a region with a strong strain gradient. Then, heating is expected to be everywhere synchronous with deformation and the amount of heating will depend on the amount of deformation.

While the slope of the piezothermal array contains a lot of information on the underlying thermal event, the age difference between metamorphism of the highest grade rocks and that of the lowest grade rocks is often too small to be resolved with geochronological methods. Thus, in order to document the slope of a piezothermal array, we are often limited to interpret it from detailed observations on the relative timing of metamorphic parageneses in different parts of a terrain, for example relative to a pervasive fabric or some other structural time marker. It should also be said that the interpretations discussed here choose an extremely simplistic point of view, for example by implying that only a single event is responsible for the observed timing relationships. Some of the complications that may arise due to superposition of different heating and cooling mechanisms were discussed by Stüwe et al. (1993).

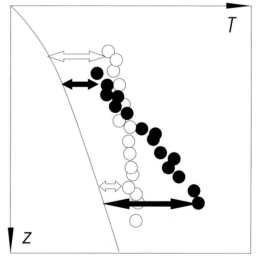

Figure 7.11. *T-z* diagram showing two hypothetical different datasets collected from metamorphic terrains with a field gradient (large dots). The dataset shown by the black dots has positive thermal perturbation characteristics: low grade rocks are less thermally pertubed from stable metamorphic conditions than high grade rocks. For the data set shown by the white dots this relationship is reversed: high grade rocks are less thermally perturbed than low grade rocks. The information of this figure is equivalent to that shown in Fig. 7.9b

• *Thermal perturbation characteristics of piezotherms.* The thermal perturbation characteristics of a piezotherm describe if high grade rocks are more or less thermally perturbed from stable geothermal conditions than their lower grade equivalents. This may be characterized by the slope of the piezotherm in a PT diagram. For example, during regional metamorphism according to the classic description of England and Thompson (1984), high grade rocks may experience their peak at large depths and substantially derived from stable geothermal conditions (Figs. 7.9, 7.10a, black dots on Fig. 7.11). In contrast, many metamorphic terrains show reversed relationships between grade and magnitude of thermal perturbation (white dots on Fig. 7.11). For example, in the eclogite type locality in the eastern Alps, eclogites equilibrate at 650°C and almost 20 kbar, (which is not very perturbed as it lies near a "normal" geothermal gradient around 9°C km^{-1}), but amphibolite facies rocks further north equilibrated at 650°C and 6 kbar, which places them of the order of several hundreds of degrees from normal geothermal conditions at depths corresponding to these peak pressures.

The problems that such relationships present for the interpretation of metamorphic heat sources were discussed by, for example, Stüwe (1998a). Fig. 7.10b shows that regional metamorphism terminated by extension is one mechanism that may result in a PT slope of the piezotherm that is lower than that of any one metamorphic geotherm.

7.5 Problems

Problem 7.1. *Converting pressure and volume (p. 323):*
a) What is the mean density of the atmosphere assuming that it is 10 km thick and that the pressure on the surface is 1 atmosphere? b) What is the barometric pressure in kbar on the surface of the earth?

Problem 7.2. *Converting pressure and volume (p. 323):*
The molar volume of almandine-garnet is about 11.5 J bar^{-1}. Convert these units of volume (volume = energy/pressure) into cubic centimeters. Check your result by comparing the weight of one formula unit of almandine with its density. (The garnet almandine has the formula $Fe_3Al_2Si_3O_{12}$ and a density of about 4 g cm^{-3}.

Problem 7.3. *Estimating cation diffusivities (p. 324):*
Use the Arrhenius relationship (eq. 7.4) to estimate the diffusivity of cations through a garnet lattice at 300°C, 500°C and 1 000°C. For the material constants use: $Q = 239\,000$ J mol^{-1}, $D_0 = 9.81 \cdot 10^{-9}$ m^2 s^{-1} and assume the activation volume is negligible (Cygan and Lasaga 1985). The value of the gas constant may be found in Table C.4. Discuss the significance of the enormous difference between these cation diffusivities and the thermal diffusivity $\kappa \approx 10^{-6}$m^{2}s^{-1}.

Problem 7.4. *Time scale of chemical equilibration (p. 325):*
How long does it roughly take for a 1 mm large garnet crystal to equilibrate chemically at 600°C ? Use eq. 7.3 and the data from Problem 7.3.

Problem 7.5. *Estimating mean diffusivity (p. 331):*
The Arrhenius relationship and the results of Problem 7.3 show a strong exponential dependence of cation diffusivity on temperature. In metamorphic processes cations in minerals diffuse typically during temperature change. It is therefore useful to know some sort of mean diffusivity that characterizes the diffusion rate during the entire thermal evolution. Use eq. 7.8 to estimate this mean diffusivity of cations in garnet between the starting temperature $T_A = 700\,°C$ and $T_E = 400\,°C$. Use the data from Problem 7.3 and compare your result.

Problem 7.6. *The meaning of "metamorphic peak" (p. 327):*
Figure 7.1 shows in *a* a *P-T*-path in which the thermal peak coincides with the pressure peak and in *b* a *P-T*-path in which pressure and temperature peak occur at different times. The case of *b* is more common. In the literature we often read simply about the "metamorphic peak" of a terrain. a) Does this mean the temperature peak or the pressure peak? b) Why is this imprecise usage of the term "metamorphic peak" so common? c) Draw a *P-T*-path in which the temperature peak is reached before the pressure peak. d) What tectonic process can you think of that causes such a path?

Problem 7.7. *Estimating cooling rates (p. 332):*
Assume you have found a garnet that is 2 mm in diameter (1 mm radius) that has a retrograde zoning profile similar to that shown on Fig. 7.2. You have analyzed it and found it to have a closure temperature of $600\,°C$. Estimate the cooling rate of the metamorphic event that caused the zoning profile using eq. 7.9.

Problem 7.8. *Using diffusivity estimates in the field (p. 324-325):*
(This is an integrated problem using knowledge on closure temperatures as well as the time scale of diffusion of both mass and heat. To tackle this problem you should solve problems 7.3, 7.4 and 7.5 first).
   A 2 km thick mafic dike has intruded rocks that were cooling from a previous regional metamorphic event at the time of intrusion. The intrusion temperature of the mafic magma was $T_i = 1\,200\,°C$. The temperature of the cooling host rocks at the time of intrusion was $T_b = 500\,°C$. The dike has a contact metamorphic aureole that overprints the metamorphic parageneses from the previous event. In this contact metamorphic aureole, 5 mm large mica crystals were found at a distance of about 50 m from the dike contact. Microstructural observations show that these crystals grew prior to intrusion during the previous regional metamorphic event.
   Question: Will the isotopic systems of the micas record the older regional metamorphic event or will they have been reset by the contact metamorphic event?
   a) Find an answer by estimating the time scale of the contact metamorphic event (eq. 3.18) ($\kappa = 10^{-6}\,m^2\,s^{-1}$). Compare this with the time scale of diffusive equilibration of the micas at $1\,200\,°C$ and $500\,°C$ (from eq. 7.3 and using

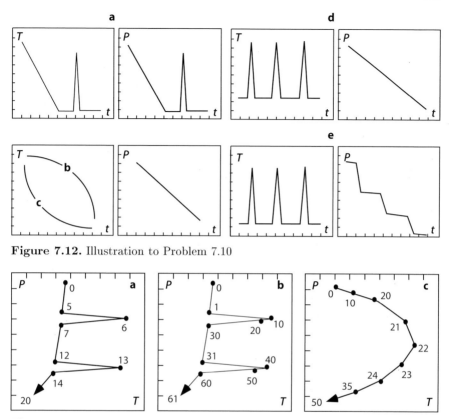

Figure 7.12. Illustration to Problem 7.10

Figure 7.13. Illustration to Problem 7.9

eq. 7.4).Use the following diffusion constants for mica: $Q = 163\,000$ J mol^{-1} and $D_0 = 7.7 \cdot 10^{-9}$ m^2 s^{-1} (Fortier and Giletti 1991).

b) The result from a) should have shown that, in principle, the micas *can* be used to date the intrusion event, but that the estimate we have made is very crude and needs refining. The dating may be improved by calculating the maximum temperature ($T_{\max}$) the mica experienced during the contact metamorphic event (by substituting eq. 3.91 into eq. 3.89). Also, a better value for the diffusive time constant of the micas can be derived by calculating the mean diffusion rate between $T_{\max}$ and $T_{\mathrm{b}} = 500\,°$C using eq. 7.8.

c) Sketch out a flow chart for a computer program that could be used to find a detailed solution for the problem.

Problem 7.9. *Constructing P-T-paths (p. 335):*
Draw pressure-time and temperature-time paths that correspond to the *P-T-t*-paths shown in Fig. 7.13. Discuss some possible interpretations. The numbers shown along the *P-T*-paths are geochronologically determined ages in my.

Problem 7.10. *Constructing P-T-paths (p. 335):*
Construct *P-T*-paths from the pairs of pressure-time and temperature-time curves shown in Fig. 7.12. The *P-T* paths you have constructed contain no information on time. It is therefore possible to interpret them in terms of different temporal evolutions for *P* and *T* from those shown in Fig. 7.12 (if only the *P-T* path is known). Discuss some.

Problem 7.11. *Construction of P-T-paths (p. 335):*
Draw the retrograde *P-T*-path of a rock that experienced the following evolution following its metamorphic peak: Peak metamorphism occurred in 20 km depth at 700 °C. Then, the rock was exhumed at a rate of 1 200 m my^{-1} for 5 my. After that, the exhumation rate decreased to 400 m my^{-1}. The metamorphic temperature remained at 700 °C for another 3 my after the metamorphic peak before cooling commenced. Then cooling commenced a rate of 100 °C my^{-1} and lasted for 3 my. After that, the cooling rate decreased to 20 °C my^{-1} until a stable geotherm is reached. Final cooling occurred along the stable geotherm. Assume that the stable geotherm is characterized by a constant gradient of 20 °C km^{-1}.

Hint: It is easiest to first draw depth-time and temperature-time curves and construct a depth-temperature diagram from those. The conversion from depth to pressure can be done as a final step in the calculation using $\rho = 2\,700$ kg m^{-3} for density and $g = 10$ m s^{-2} for the gravitational acceleration. Non lithostatic components of pressure are to be neglected.

A. Mathematical Tools

Most geodynamic processes are processes that change in space and time. One of the most important tools to describe such changing processes are *differential equations*. This chapter is therefore mainly concerned with the use and interpretation of differential equations. A few selected other important numerical tricks and basic rules are summarized towards the end of this chapter.

A.1 What is a Differential Equation?

The derivative (or: differential) dy/dx is a way to describe the *change* of y with respect to another variable x. It can be interpreted as the slope (or *gradient*) of the function $y = f(x)$. If the slope of this function is constant between two points along the x axis, for example between x and $x + \Delta x$, then we need no derivative and we can write:

$$\text{gradient} = \frac{y(x + \Delta x) - y(x)}{\Delta x} \quad . \tag{A.1}$$

The numerator of the fraction on the right hand side of this equation is given by the difference between the y values of the function at the two points x and $x + \Delta x$. The denominator is given by the distance between the two points on the x axis between which the gradient is measured (Fig. A.1). Their ratio is the slope between the points x and $x + \Delta x$. If we consider a function where the slope is *not* constant between x and $x + \Delta x$, then eq. A.1 would give us only some mean of all the slopes of this function between the two points. However, the smaller we choose our Δx, the better will eq. A.1 describe the exact slope at point x. We can write:

$$y'(x) = \frac{dy}{dx} = \lim \Delta x \to 0 \left(\frac{y(x + \Delta x) - y(x)}{\Delta x} \right) \quad . \tag{A.2}$$

Eq. A.2 is the mathematical definition of a derivative. Note that we used a dash to indicate that y' is a derivative. This is a commonly used notation. The slope of a mountain road is a clear example to illustrate the meaning of *slope*. Assume that H describes the elevation of the road surface as a function

Figure A.1. Diagram illustrating the definition of the first derivative of a function (thick curve) (eq. A.1 and A.2). The slope of the thick drawn straight line may be accurately described by the ratio $(y(x + \Delta x) - y(x))/\Delta x$. The slope of the curved function at x is only moderately well approximated by this ratio. However, the smaller Δx becomes, the better this approximation will be to describe the slope at a single point

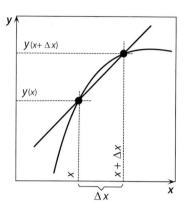

of distance from the valley x: $H = f(x)$. Then, the slope of the road is given by the first derivative of this function f_x':

$$f_x' = \frac{\mathrm{d}H}{\mathrm{d}x} \quad . \tag{A.3}$$

The units of this differential are $\mathrm{m\,m^{-1}}$. In short, it is dimensionless for this example. Familiar geological examples described by first derivatives (first differentials) of functions are the geothermal gradient, describing the change of temperature with depth (in $\mathrm{°C\,m^{-1}}$), or cooling histories of metamorphic terrains that describe the change of temperature over time (in $\mathrm{°C\,s^{-1}}$).

The *second derivative* (or second differential) of a function describes how the slope changes. The familiar name of the second derivative is *curvature*. It is often abbreviated with f_x''. In our example of a mountain road, the vertical curvature of the road is

$$f_x'' = \frac{\mathrm{d}\left(\frac{\mathrm{d}H}{\mathrm{d}x}\right)}{\mathrm{d}x} = \frac{\mathrm{d}^2 H}{\mathrm{d}x^2} \tag{A.4}$$

and has the units of $\mathrm{m/(m\,m^{-2})}$, which is: $\mathrm{m^{-1}}$. It may be read as: "d *two H over* dx *square*". The scheme we have followed to go from first to second derivative may be followed to describe the third, fourth or higher derivatives of functions. Corresponding to the first two derivatives, the third derivative of a function describes the *change* of the curvature of the function and the fourth the curvature of the curvature and so on (Fig. A.2):

$$f_x''' = \frac{\mathrm{d}^3 H}{\mathrm{d}x^3} \qquad \text{and :} \qquad f_x'''' = \frac{\mathrm{d}^4 H}{\mathrm{d}x^4} \quad . \tag{A.5}$$

In our example of a mountain road, the units of the third and fourth derivatives are $\mathrm{m^{-2}}$ and $\mathrm{m^{-3}}$, respectively. Be careful not to confuse these *linear* derivatives of the third and fourth order (in eq. A.5) with the *non-linear* first order derivatives $(\mathrm{d}H/\mathrm{d}x)^3$ and $(\mathrm{d}H/\mathrm{d}x)^4$ (see below)!

Figure A.2. A sine function as an example for the function f_x and its first, second and third derivatives. At the maximum of the function (point A), the slope of the function is $f'_x = 0$ and the curvature has a negative maximum: $f''_x = -1$. Conversely, at the inflection point of the curve (point B), the slope has a minimum the curvature is zero $f''_x = 0$. At the minimum of the function (point C), the slope is also zero and the curvature has a maximum

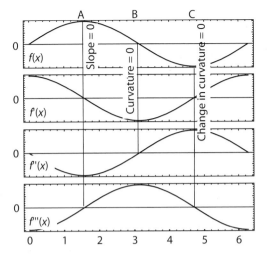

A.1.1 Terminology Used in Differential Calculus

Order. The order of the highest derivative in a differential equation is called the *order of the equation*. For example, eq. 3.58 is a first order differential equation, eq. 3.6 or eq. 3.57 are second order differential equation eq. 4.42 is a fourth order differential equation. Fig. A.2 illustrates the meaning of derivatives of higher order using the example of a simple sine function.

Partial and Total Derivatives. A function may have several variables. For example, the elevation of a point H on the surface of the earth can be described as a function of two spatial coordinates in the horizontal directions x and y, but it may also be a function of time or any other variables that we deem of importance, say vegetation or lithology. If we consider the spatial dependence only, we may be in the situation to describe the elevation of our point as:

$$H = 3x^2 + 4y^3 \quad . \tag{A.6}$$

If we differentiate this function with respect to one of the variables only, then this is called a *partial derivative*. It describes the slope of the function in *one* spatial direction only. When forming a partial derivative with respect to one variable, then all other variables are kept constant during the process and are treated like any other constant of the equation. The symbol for the partial differential is ∂ (say: "del"). However, "del" is no real Greek letter and should not be confused with δ. The partial derivative of eq. A.6 after x is:

$$\frac{\partial H}{\partial x}_{y=const.} = 6x \quad . \tag{A.7}$$

This partial derivative describes the slope of the function in direction x. The *total differential* is the sum of all partial derivatives. In our example it is:

$$\mathrm{d}H = \left(\frac{\partial H}{\partial x}\right)_{y=const.} \times \mathrm{d}x + \left(\frac{\partial H}{\partial y}\right)_{x=const.} \times \mathrm{d}y \ . \tag{A.8}$$

This gives the total derivative of the function from eq. A.6 to be:

$$\mathrm{d}H = 6x + 12y \ . \tag{A.9}$$

This total derivative describes the tangential plane at the point x, y. In our example of surface elevation, eq. A.9 may also be interpreted as the slope of the surface in dip direction. Partial and total derivative are identical if the function contains only one variable. Partial derivatives after time are often called *rates* and symbolized by a little dot above the variable concerned (e. g. strain ϵ and strain rate $\partial \epsilon / \partial t = \dot{\epsilon}$). Differential equations that contain only total derivatives are called *ordinary* differential equations, in contrast to the *partial* differential equations, which contain partial derivatives. A good summary of these simple definitions may be found in Anderson and Crerar (1993) or Zill (1986), as well as many other mathematical hand books.

Linear and Non-linear Differential Equations. A differential equation is said to be linear if it is characterized by two properties: 1.) the dependent variable and all its derivatives are of the first degree, that is, the power of each term involving it is 1; and 2.) each coefficient depends only on the independent variable. In this context, the *dependent* variable is generally the one in the *numerator* of the derivative, while the variable in the *denominator* is called the *independent* variable. For example, in the differential equations in chapter 3, temperature, T or heat H are generally used as the dependent variables, while the spatial coordinates or time are generally the independent variables. Just about all differential equations in this book are *linear* differential equations (e.g. eqs. 3.6, 3.57 or 4.42). On the other hand, hypothetical examples of non-linear differential equations would be:

$$T\frac{\partial^2 T}{\partial x^2} - 5\frac{\partial T}{\partial y} = x \quad \text{or}: \quad \frac{\mathrm{d}^3 T}{\mathrm{d}z^3} + T^2 = 0 \ . \tag{A.10}$$

These are examples of non-linear second and third order partial differential equations, respectively. The first example is non-linear because the coefficient of the first term depends on T, the second because T occurs in the power of 2. The only geologically relevant example of a non-linear differential equation discussed in this book occurs on p. 173 (s. also p. 313).

Analytical and Numerical Solutions. In order to make use of a differential equation we must solve it. Only then, they can be used as a tool to extract numbers that describe some process. There are two fundamentally different ways to solve them.

- *1. Analytical solutions. Analytical* or *closed* solutions of differential equations may be found by integrating them. Let us consider as an example the description of a geotherm by

$$\frac{dT}{dz} = \frac{1.5}{\sqrt{z}} \quad . \tag{A.11}$$

There, T is temperature in °C and z is depth. This differential equation can be integrated without difficulty:

$$T = 3\sqrt{z} + C \quad . \tag{A.12}$$

The integration constant C must be determined using boundary conditions. Eq. A.12 is said to be an "analytical solution of the differential equation eq. A.11". If we assume (as our boundary condition) that the temperature at the earths surface is always zero and we assume a coordinate system where the surface is at $z = 0$, then this constant must be also zero: $C = 0$. Now eq. A.12 can be used to calculate temperatures at any depth of our choice by inserting numbers for z. For example, for $z = 100\,000$ m eq. A.12 gives $T = 949\,°C$.

- *2. Numerical solutions. Numerical solutions* of differential equations are used to extract numbers from differential equations *without* having to solve (integrate) them. With their aid we can arrive at the result that eq. A.11 describes a temperature of $T = 949\,°C$, at 100 km depth if the surface temperature is zero without having to solve the differential equation, i.e. without having to go from eq. A.11 to eq. A.12. However, numerical solutions are not exact. Numerical approximations are always only approximations and they are plagued by stability and accuracy problems (s. p. 358). The numerical solution of partial differential equations is a science on its own (sect. A.2, A.4). The two most important methods that are in use are the *finite difference methods* and the *finite element methods*.

The *finite element method* has the advantage that it is much more elegant to use it for the description of deformation on Lagrangian coordinates. The principal disadvantage of the finite element method is that it is quite a complicated method not amenable to the understanding of a field geologist.

The *finite difference method* has the enormous advantage that it is quite intuitive, easy to implement on a computer (even by inexperienced mathematicians) and easily adaptable to many different problems. Its principal problems are those of instability, and that they are quite cumbersome when it comes to the treatment of discontinuous boundary conditions and deformed grids (sect. A.2).

- *Advantages and disadvantages.* Numerical and analytical solutions have both their advantages and disadvantages. The enormous advantage of numerical solutions is that they allow us to arrive at results without having to know enough differential calculus to be able to integrate the equation in question. In fact many geological problems can be simplified enough to be able to formulate them into an equation, but are too complicated so that an analytical solution even exists. In such cases, numerical solutions are the only way to obtain results.

Analytical solutions have the advantage that they are much more useful to understand the nature of a geological process. For example, eq. A.12 may be used directly to infer that the temperature in the crust rises with the square root of depth. If this model corresponds well with our observations in nature, then we can continue to think about the significance of this quadratic relationship. Such considerations are difficult with numerical solutions as they only deliver numbers.

Initial- and Boundary Conditions.

• *Boundary conditions.* When solving differential equations, *boundary conditions* are necessary in order to determine the integration constants. This is true for both numerical and analytical solutions. For differential equations of the first order we need one boundary condition, for those of the second order two and so on. The term *boundary condition* is exactly what it implies: it is a condition at the boundary of the model (s. sect. A.2.1). The most common types of boundary conditions are:

- A prescribed value of the function at the model boundary (e. g. $T = 0$ at $z = 0$; s. eq. A.12),
- A prescribed gradient of the function at the model boundary (e. g. the heat flow boundary condition we used in sect. 3.4.1),
- A functional relationship between *value* and *gradient* at the model boundary (e. g. the constant heat content boundary condition used on p. 114).

Boundary conditions given by higher derivatives of functions are also possible and play an important role when integrating differential equations of higher orders (s. sect. 4.2.2, eq. 4.42). In sect. A.2 we discuss how some of these boundary conditions may be implemented.

• *Initial conditions. Initial conditions* are necessary to determine the starting point of a model. For example, if we want to use the diffusion equation (eq. 3.6) to calculate the evolution of a diffusive zoning profile over time, then we must use a function $T = f(z)$ at the time $t = 0$ from which we can start calculating. The nature of this function $T = f(z)$ must be determined by a known initial condition.

A.2 The Finite Difference Method

The finite difference method makes use of the discretization of the derivative from eq. A.2. Instead of describing the differential dy/dx by the limiting value $\Delta x \to 0$, a finite value of Δx is used (i.e. eq. A.1 is retained). For our explanation on the next pages we use Fig. A.3 showing the function $T = f(x)$ and assume that this function is a temperature profile across a metamorphic terrain along the spatial axis x. Thus, we will use the variable T instead of the more abstract y that we have used up to now in this chapter. At the point x_i (labeled in Fig. A.3a by the dotted line) the function has the slope $\mathrm{d}T/\mathrm{d}x$. When using the method of finite differences, this slope is approximated by the discrete temperature difference at two different places with a *finite* distance to each other (a bit as we have already implicitly shown in Fig. A.1). There is many ways to formulate such a difference. In Fig. A.3b we can see that one way to formulate such a difference is:

$$\frac{\mathrm{d}T}{\mathrm{d}x} \approx \frac{T_{i+1} - T_i}{x_{i+1} - x_i} = \frac{T_{i+1} - T_i}{\Delta x} \quad . \tag{A.13}$$

The index i is just a description of the number of the grid point chosen here. T_i is the temperature at the "i^{th}" point of a discrete grid of points. T_{i+1} is the temperature at the *next* point of the grid, T_{i-1} at the *previous* point. The finite difference method used in eq. A.13 is called *forward differencing method* as we have calculated the temperature gradient at x_i using the temperature *at* x_i as well as the temperature at the next *forward* point on the grid (Fig. A.4). Some other simple examples of differencing schemes have the form:

$$\frac{\mathrm{d}T}{\mathrm{d}x} \approx \frac{T_i - T_{i-1}}{\Delta x} \qquad \text{or} \qquad \frac{\mathrm{d}T}{\mathrm{d}x} \approx \frac{T_{i+1} - T_{i-1}}{2\Delta x} \quad . \tag{A.14}$$

For reasons that should now be obvious, these two methods are called *backward differencing* and *central differencing* schemes (Fig. A.4).

• *Differentiating with respect to time.* All information we have discussed so far is generally applicable, regardless of what variable is described by x, y or T. However, the use of x has suggested that we imply *spatial* differentials. In order to discriminate between the numbering of grid nodes of spatial and temporal grids, the symbols "+" and "−" are common to describe the *next*

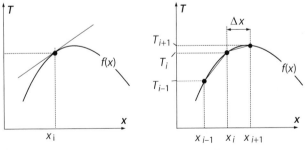

a slope at a point **b** finite difference approximation

Figure A.3. Graphical illustration of the method of finite differences. In **a** the slope of the function $f(x)$ at point x_i is accurately described by the touching tangent. Mathematically this slope is described by the differential $\mathrm{d}T/\mathrm{d}x$. In **b** the slope is approximated by the ratio of the differences of two temperature and two x values

time step and the *previous time step* while i and $i + 1$ is used for the spatial grid stepping. Thus we can write:

$$\frac{dT}{dt} \approx \frac{T^+ - T^-}{\Delta t} \ . \tag{A.15}$$

In some books "j" and "$j+1$" are used to denote time steps. However, this should not be confused with spatially two-dimensional problems in which "i" subscripts are used for grid numbering in x direction and "j" numbering of grid steps in y direction.

• *Approximations of derivatives of higher order.* For the approximation of derivatives of the second or higher order we can use the same scheme as that for the first derivative (eq. A.13, A.14). For the second derivative we must form the ratio of the difference in slope at two different grid points with the distance Δx:

$$\frac{d^2T}{dx^2} = \frac{d\left(\frac{dT}{dx}\right)}{dx} \approx \frac{\left(\frac{T_{i+1}-T_i}{\Delta x}\right) - \left(\frac{T_i-T_{i-1}}{\Delta x}\right)}{\Delta x} = \frac{T_{i+1} - 2T_i + T_{i-1}}{\Delta x^2} \ . \tag{A.16}$$

From eq. A.16 we can see that, in order to formulate the difference between slopes at two point, the slope at point i was approximated once by forward differencing and once by backward differencing. This is necessary, as we want to calculate the curvature at point i from the differences between the slopes of the curve as near as possible to it (i.e. in front of it and behind it). We can see that the curvature is described by the difference of slopes, just like we describe the slope by the differences between two function values.

• *Solution of the diffusion equation using finite differences.* If differential equations contain more than one variable (e. g. the diffusion equation (eq. 3.6), in which both spatial and temporal derivatives occur) it is necessary to combine several indices with each other. This may lead to apparently quite complicated formulations. Here we will follow the use of temporal and spatial

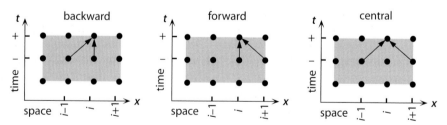

Figure A.4. Schematic illustration of three simple methods of discretization in the finite difference method. The x axis of each diagram shows four discrete points of a one-dimensional spatial grid. Each dot is a temperature value at this point in space. The y-axis shows three different time steps of the calculation. In the three diagrams, the temperature at the third grid point (labeled by subscript i) is calculated by *backward*, *forward* and *central* differencing. In each diagram this calculation is for the values at the (as yet) unknown time step "+" from known information at time "-"

indices as discussed above, i.e. the temperature at the spatial grid point i at a future time step is called T_i^+ and at a previous time step it is called T_i^- (Fig. A.4). Once we are familiar with this notation it should be straight forward to understand the following finite difference approximation of the diffusion equation (eq. 3.6) by using eqs. A.15 and A.16:

$$\frac{\partial T}{\partial t} = \kappa \frac{\partial^2 T}{\partial x^2} \approx \frac{T_i^+ - T_i^-}{\Delta t} = \kappa \frac{T_{i+1}^- - 2T_i^- + T_{i-1}^-}{\Delta x^2} \quad . \tag{A.17}$$

Solved for T at the new time step of interest this gives:

$$T_i^+ = T_i^- + \left(\frac{\kappa \Delta t}{\Delta x^2}\right) (T_{i+1}^- - 2T_i^- + T_{i-1}^-) \quad . \tag{A.18}$$

Now we can insert into eq. A.18 known temperatures at known points in space from a previous time step (starting from the known temperature profile of our initial condition) to determine the temporal evolution of the temperature profile.

The last few pages have given you an overview over the basic principles of the finite difference method. All other finite difference approximations are refinements of the above aiming at higher accuracy, higher stability and higher speeds of calculation on computers.

Why better methods may be necessary can be illustrated with eq. A.18. There, the magnitude of the constant $(\kappa \Delta t / \Delta x^2)$ is critical for the stability and accuracy of the approximation (sect. A.2.2). This number is called the *Fourier cell number* and must be smaller than 0.25 so that the solution of eq. A.18 retains stability. Since the magnitude of κ depends on material constants, we cannot change it arbitrarily. Thus, in order to fulfill the stability criterion we must make a corresponding choice with the time and space stepping. With many simple finite difference approximations this leads to unsurmountable problems: If a given problem requires high spatial resolution (small Δx) it requires a correspondingly small choice of Δt. However, if this problem should now be solved over long geological times, then we may have to iterate through too many time steps for the problem to be solvable in realistic computer time. This is one of the reasons why large computers (and more refined finite difference methods) are required for many geological questions.

A.2.1 Grids and Boundary Conditions

If we want to use the finite difference method to solve a differential equation (e. g. eq. A.18), we need to build a discrete grid on which the function is evaluated (Fig. A.5). A regular grid with n grid points has $n-1$ grid spaces. If the total length of the grid is L and the grid spacing is regular, then the distance between any two grid points will be $\Delta x = L/(n-1)$ (Fig. A.5)a. However, grid spacings need not be regular. For example, if a function is of particular interest in a special region it may be useful to make the grid especially fine in this region. On the other hand, it may not be wise to make the grid everywhere this fine as this may enlarge the time of calculation enormously. A spatially variable grid is the best solution for this. On such spatially variable grids we must substitute Δx by $(x_{i+1} - x_i)$ (see: eq. A.13)

• *Boundary conditions.* Closer consideration of eq. A.18 indicates that this equation may not be evaluated at the points $i = 1$ and $i = n$, because no grid points "$i - 1$" and "$n + 1$" exist there for which we could insert the temperatures T_{i-1} and T_{i+1} into the right hand side of the equation. These two temperatures must be determined by the boundary conditions. These boundary conditions are equivalent to the integration limits of a definite integral that are required to determine the integration constants. Thus, it is no coincidence that there is *two* grid nodes in finite difference approximations of *second* order derivatives, where the functional values can on only be determined with the aid of boundary conditions.

A.2.2 Stability and Accuracy

Finite difference solutions of differential equations have two important disadvantages:

1. They are only approximations.
2. They are often unstable.

Criteria for accuracy and stability are extensively discussed in the literature (e. g. Smith 1985, Fletcher 1991, Anderson et al. 1984). However, both problems can be reduced to a minimum by some very simple checks:

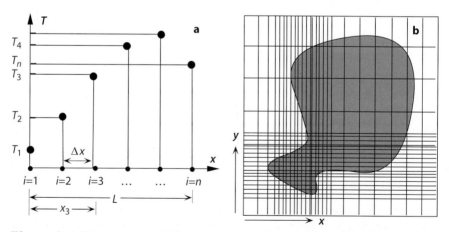

Figure A.5. Two examples of discrete grids. **a** Discrete form of the function from Fig. A.3 on a regular one-dimensional spatial grid. Different points are numbered from $i = 1$ to $i = n$. **b** An irregular two-dimensional orthogonal grid. The grid serves the description of the dark shaded region. Thus, a finer grid spacing was used for the grid in the lower left hand portion of the grid

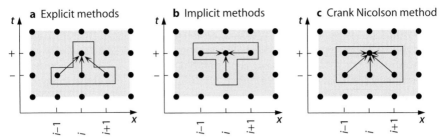

Figure A.6. The finite differencing scheme in implicit and explicit methods. The vertical axis is time, the horizontal space. Time step "+" is the time step to be calculated. Time step "-" denotes the time for which information is already available. The *Crank Nicholson* method is a mixed method. It consists of implicit and explicit parts

● *Accuracy.* The accuracy of finite difference approximations can easily be checked by successively decreasing the time or spatial stepping (for a discussion of accuracy versus precision s. p. 5). If the result does not change, the exact solution has probably been approximated well enough. A second test can be performed by simplifying the initial and boundary conditions of a giving problem enough so that analytical integration of the descriptive equations is possible. Then the numerical solution may be compared directly with the analytical results. Time and space stepping can then be relaxed and finally the initial and boundary conditions readjusted to describe the problem in the required detail.

● *Stability.* A finite difference solution is called *stable* if it converges to the correct solution. Unstable solutions diverge with progressive calculation more and more. Most unstable solutions "explode" within a few time steps. Thus, stability problems are often relatively easy to recognize as all functional values trend towards infinity (s. Fig. A.8). Stability problems can often be brought under control by decreasing the discrete stepping in the approximation.

A.2.3 Implicit and Explicit Finite Difference Methods

There are two fundamentally different types of finite difference methods that may be used to solve (approximate) differential equations:

1. explicit methods,
2. implicit methods.

There is also *mixed* methods that are partially implicit and partially explicit. Fig. A.6 illustrates what is meant with *implicit* and *explicit*. Both methods will be discussed briefly below using the example of temperature calculation with the diffusion equation. However, the principal difference between implicit and explicit solutions are the same regardless of the variables or the equations.

• *Explicit methods.* The idea behind explicit finite difference methods is illustrated in Fig. A.6a. This figure corresponds to the way the diffusion equation was solved in eq. A.18 (Fig. A.4). It may be seen that the temperature at point i at the new time step T_i^+ is calculated from the *known* temperatures (those from the previous time step) at the points $i-1$, i and $i+1$. As these temperatures $(T_i^-, T_{i-1}^-$ and $T_{i+1}^-)$ are known, the application of eq. A.18 is no problem. All methods that use schemes where *new* information is calculated exclusively from known information are called *explicit* methods.

• *Implicit methods.* Implicit finite difference methods calculate the unknown temperatures T_i^- from other unknown temperatures at the same time step (Fig. A.6b). This sounds a bit counter intuitive if not impossible, but is possible if all temperatures are calculated *simultaneously.* Remember that we have boundary conditions that tell us the new temperatures at the two ends of the grid. Thus, in a grid with n points, there is only $n-2$ points where the temperature is unknown. It is therefore possible to formulate a set of $n-1$ equations with $n-2$ unknowns. This may be solved for all unknown variables. An example of an implicit approximation of eq. 3.6, (corresponding to Fig. A.6b) is:

$$\frac{\partial T}{\partial t} = \kappa \frac{\partial^2 T}{\partial x^2} \approx \frac{T_i^+ - T_i^-}{\Delta t} = \kappa \frac{T_{i+1}^+ - 2T_i^+ + T_{i-1}^+}{\Delta x^2} \quad . \tag{A.19}$$

Solved for the temperature of interest this is:

$$T_i^+ = \frac{T_i^- + R(T_{i+1}^+ + T_{i-1}^+)}{1 + 2R} \qquad \text{where}: \qquad R = \frac{\kappa \Delta t}{\Delta x^2} \quad . \tag{A.20}$$

• *Mixed methods.* Mixed methods use explicit as well as implicit information to calculate the new data (Fig. A.6c). Mixed methods have the best accuracy and stability characteristics and are therefore commonly used. The most famous of all mixed methods is the *Crank-Nicolson-method* which is used to describe second order differentials, as they occur in the diffusion equation. The Crank Nicholson method describes this with:

$$\frac{T_i^+ - T_i^-}{\Delta t} = \frac{\kappa}{2} \left(\frac{T_{i+1}^+ - 2T_i^+ + T_{i-1}^+}{\Delta x^2} + \frac{T_{i+1}^- - 2T_i^- + T_{i-1}^-}{\Delta x^2} \right) \quad . \tag{A.21}$$

It may be seen that the expression inside the brackets is the sum of the right hand sides of eq. A.17 and eq. A.20 and that the mean of these expressions is formed. Ways to implement eq. A.21 are discussed in most books on numerical mathematics. The *Thomas algorithm* is an elegant method that can be used to implement the simultaneous solution of equations as is necessary to solve eq. A.21.

Two-dimensional Derivatives. In the following we illustrate the approximation of two-dimensional differential equations using the two-dimensional form of the heat conduction equation as an example. This equation is:

$$\frac{\partial T}{\partial t} = \kappa \left(\frac{\partial^2 T}{\partial x^2} + \frac{\partial^2 T}{\partial y^2} \right) \quad . \tag{A.22}$$

The most commonly used finite difference method to approximate eq. A.22 is the *Alternating Direction Implicit Method*, in short: ADI method. In this method, each time step is divided into two. The first half time step is *explicit* in one spatial direction and *implicit* in the other, the second half time step is done in reverse. Thus, for each step in time, two steps of calculation are required. The ADI method has a lot of similarities with the Crank Nicholson method. A discrete version of eq. A.22 looks like this:

1. step:

$$\frac{T_{i,j}^{+/2} - T_{i,j}^{-}}{\Delta t/2} = \frac{\kappa}{\Delta x^2} (T_{i+1,j}^{+/2} - 2T_{i,j}^{+/2} + T_{i-1,j}^{+/2})$$

$$+ \frac{\kappa}{\Delta y^2} (T_{i,j+1}^{-} - 2T_{i,j}^{-} + T_{i,j-1}^{-}) \quad .$$

2. step:

$$\frac{T_{i,j}^{+} - T_{i,j}^{+/2}}{\Delta t/2} = \frac{\kappa}{\Delta x^2} (T_{i+1,j}^{+/2} - 2T_{i,j}^{+/2} + T_{i-1,j}^{+/2})$$

$$+ \frac{\kappa}{\Delta y^2} (T_{i,j+1}^{+} - 2T_{i,j}^{+} + T_{i,j-1}^{+}) \quad . \tag{A.23}$$

In this equation we have used the subscripts i and j to label the grid points in the x- and y-directions and we used "$+/2$" as a description of half a time step.

A.2.4 Approximation of the Transport Equation

In sect. 3.3 we introduced a simple equation that can be used to describe the advection of material, for example the advection of rocks to the surface by erosion or the advection of fluids through a marble (eq. 3.43). This equation require the use of very different finite difference approximations from those discussed above for the diffusion equation. For example, *backward* and *central* differencing schemes are unstable when applied to approximate eq. 3.43. Thus *forward* differencing schemes must be used. A simple forward differencing scheme that can be used to approximate this transport equation is:

$$T^{+} = T^{-} + u \frac{\Delta t}{\Delta x} \left(T_{i+1}^{-} - T_{i}^{-} \right) \quad . \tag{A.24}$$

However, eq. A.24 is associated with *numerical diffusion*, a problem which makes the solution increasingly inaccurate if used over many time steps.

The Problem of Numerical Diffusion. The description of one-dimensional advection with forward finite differencing schemes causes a phenomenon known by the name *numerical diffusion*. As this is a common problem with many finite difference approximations, we explain this problem here. During numerical diffusion progressive steps of the calculation cause a rounding of parts of the function that are strongly curved. This rounding is akin to the smoothing of curves by diffusion, but is only an artifact of the numerical approximation and has nothing to do with any real diffusion process. Fig. A.7 illustrates why this numerical diffusion occurs. There, the original temperature profile is shown by the continuous line. The black dots label discrete values of this function. Using eq. A.24 and positive values for u caused advection (motion) of the temperature profile towards the origin (against the x direction). If the transport velocity is exactly $u = \Delta x/\Delta t$, then the temperature profile will be moved by exactly one grid node with every time step (dashed line and white dots). If, however, the transport rate is $u < \Delta x/\Delta t$ then the temperature profile is shifted accordingly, but also rounded off (dotted line and gray dots). If we look at eq. A.24 carefully we can see that the numerical diffusion occurs because of interpolations between grid points when the transport is not a full grid interval per time step.

Unfortunately, many geological transport rates are too small so that the number of time and space steps required to fulfill $u = \Delta x/\Delta t$ is much too large to be sensibly used. It is often necessary to use a space stepping that is $\Delta x > u\Delta t$. As a consequence, we often have to deal with numerical diffusion when describing the transport of rocks using eq. 3.43. Fig. A.8b,c illustrates how much information of a step-shaped temperature profile is lost, if the transport rate is 90 % or 30 % of $u = \Delta x/\Delta t$. It may be seen that after about 50 time steps the shape of the initial and the final temperature profile have little in common. Fig. A.8d illustrates how eq. A.24 becomes unstable if the

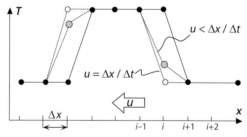

Figure A.7. Illustration of the origin of numerical diffusion in the one-dimensional transport equation (eq. 3.43). The black dots mark discrete values of the temperature profile shown by the continuous line. The dashed line (and white dots) shows the temperature profile as calculated with eq. A.24 after $1\Delta t$ time step if the transport velocity u is exactly $u = \Delta x/\Delta t$. It may be seen that the original temperature profile is advected towards the left without a change in shape. However, if $u < \Delta x/\Delta t$, then the temperature profile after $1\Delta t$, is not exactly like the original profile (dotted line, shaded dots). It has been victim to *numerical diffusion*

transport rate is $u > \Delta x/\Delta t$. The temperature profile gets jagged edges and quickly blows up meaninglessly.

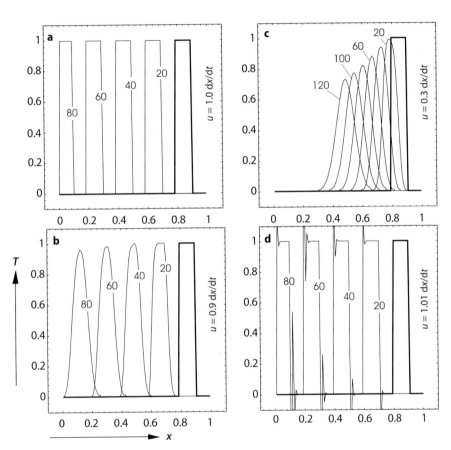

Figure A.8. Graphical illustration of numerical diffusion and instabilities using the example of the advection of a step shaped temperature profile originally at the position indicated by the thick line. The advection of a step shaped temperature profile is shown as described with the finite difference approximation of eq. 3.43 given in eq. A.24. (However, note that we use x instead of z for the spatial coordinate here). u is positive so that advection is against the x-direction. The grid consists of 100 points between 0 and 1 in the x-direction. Labels are in numbers of time steps. **a** $u = \Delta x/\Delta t$; **b, c** $u < \Delta x/\Delta t$; **d** $u > \Delta x/\Delta t$

A.2.5 Dealing with Irregular Grid Boundaries

When dealing with two-dimensional geological problems, model boundaries are often not straight lines, but are curved. For example, if we want to use the Moho heat flow as a boundary condition and the Moho is curved underneath a mountain range. Another example was discussed in sect. 3.7.3 where the irregular boundary was given by the irregular shape of the earths surface. When dealing with such problems it is often difficult to find simple finite difference approximations of the descriptive differential equations. The type of problem that we may encounter is illustrated in Fig. A.9. This figure shows an irregular body (shaded region) for which a two-dimensional heat conduction problem is to be solved. If we want to discretize the region with a rectangular grid (as for example the ADI-method in eq. A.23 would require), then this problem can only be solved for the part of the grid high lighted by the thick black dots in Fig. A.9a. The marginal points (shown in white on Fig. A.9a) must be defined by the boundary conditions. Clearly this is highly unsatisfactory. If we want the model boundaries to correspond to the *real* boundaries of the problem, then we would need to introduce all kinds of new grid lines (thin lines on Fig. A.9a) and the grid would become irregular and the entire numerical approximation full of interpolations and, in fact, quite messy.

Such problems may be avoided with one of the two following alternatives:

1. A grid may be defined that has just as many points in the x- and y directions, but in which the grid lines are not on an orthogonal grid but follow the problem boundaries (Fig. A.9b). Such grids can still be dealt with using finite difference methods, but correction terms for the changes in direction at the grid nodes must be added (s. e. g. Fletcher 1988). Calculations on such non-orthogonal grids become quite inaccurate if the direction changes at individual nodes become too large.

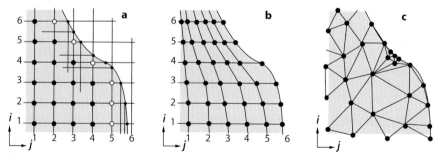

Figure A.9. Some examples for the discretization of a two-dimensional region with an irregular boundary. **a** The interpolation of the model boundary on an orthogonal grid is difficult and messy. **b** A non-orthogonal grid with a constant number of grid points in x- and y-direction. **c** The most elegant way to handle an irregular boundary is by using a triangulation of the area of interest

2. The most elegant method to deal with irregular boundaries is by using triangulations of the region of interest (Fig. A.9c). Triangles are the most simple of all geometric forms that can be used to subdivide two-dimensional regions and triangles are therefore with many respects superior to polygons with four corners. When triangulating a two-dimensional region, there are many ways to do this. For example, one could assume a certain allowed triangle size, a minimum triangle angle, or assume the number of triangles that are allowed to meet in one point. For many purposes the *Delaunay triangulation* is the best choice of *how* to triangulate a region (Fig. A.10) (e. g. Sambridge et al. 1995). The biggest disadvantage of triangulated grids is that finite difference approximations often become quite a lot more complicated than those for orthogonal grids.

A.2.6 Recommended Reading

– Anderson, Tannehill and Pletcher (1984) Computational Fluid Mechanics and Heat Transfer.
– Fletcher (1991) Computational Techniques for Fluid Dynamics Volume 1. Fundamental and General Techniques.
– Smith (1985) Numerical Solutions of Partial Differential Equations.
– Reece (1986) Microcomputer Modeling by Finite Differences.
– O'Rouke (1993) Computational geometry in C.
– Zill (1986) A First Course in Differential Equations with Applications.

A.3 Scalars, Vectors and Tensors

• *Scalars.* Geological parameters that are described by their magnitude only are *scalar quantities.* The temperature at the Moho, the elevation of a moun-

Figure A.10. A section of the grid from Fig. A.9c for the illustration of the Delaunay triangulation. In a Delaunay triangulation every circle that contains all three corners of any given triangle does *not* include any other grid nodes. Thus, all shaded triangles, except the dark shaded one, are Delaunay triangles

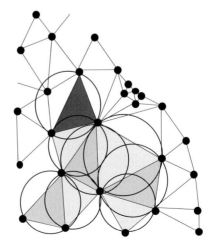

tain, density of a rock or pressure are examples. (According to Oertel 1996 pressure should be referred to as an *isotropic tensor of second rank* but for all intents and purposes of this book it is sufficient to treat it as a scalar). Variables that are scalar quantities are commonly denoted with italics, as most variables in this book.

- *Vectors.* Geological parameters that have both a *magnitude* and a *direction* are described by *vectors*. An example is the force with which India and Asia collide or the rate of intrusion of a magmatic body. The former is roughly 10^{13} N m^{-1} and is directed northwards; the latter might be some meters per year and directed upwards in the crust. Vectors are commonly represented by bold roman letters, although we refrain from this use in this book.

- *Tensors.* Parameters that are characterized by not only their magnitude and their direction, but also by a spatial dependence of this direction are described by *tensors* (s. Twiss and Moores 1992; Oertel 1996). The state of stress at a point or strain rate are the most familiar examples of tensor quantities to a geologist. It is easy to see that magnitude and direction alone are insufficient to describe stress. For example, the tensor components σ_{xx} and σ_{yx} both act in the x direction and they also may both be of the same magnitude. However, σ_{xx} is a normal stress and σ_{xy} is a shear stress, i.e. they are exerted onto planes of different orientation. Tensors are represented as matrices and are commonly abbreviated with italics. As for vectors, we do not use this notation in the present book as the tensorial quantities occurring herein (e.g. strain rate or stress) are usually simplified enough so that they reduce to simple scalar quantities (e.g by considering one-dimensional cases only).

Scalars, vectors and tensors are often called tensorial quantities of the 0., 1. and 2nd rank. In this book we treat many quantities that are actually described by vectors or tensors as if they were scalars (except in sect. 5.1.1). We have done so by making our problems so simple, so that they may be treated one-dimensionally. Only then we were able to treat many of these parameters as scalars. In fact, it is better to call them "pseudoscalars" because it is always implicit that their direction is known. Regardless, we introduce some of the basic principals of vector calculations on the following pages.

- *Common confusions.* Quantities described by scalars, vectors and tensors are often confused in the literature. Even in this book – while we try not to confuse them – we often treat tensorial quantities as if they were scalars. When doing so, we always need to remember that our considerations remain one-dimensional. For the correct consideration of two and three-dimensional problems, the full tensor quantities must be considered. Products and sums of tensors are *not* described by the sum of the one-dimensional descriptions in several spatial directions alone (e. g. Oertel 1996; Strang 1988) and it is therefore often not trivial to understand the results of two- and three-dimensional models in comparison to their one-dimensional equivalents.

- *Vectors.* Vectors describe *direction* and *magnitude* of a parameter. Thus, in Cartesian coordinates, they are described by three components:

$$\mathbf{u} = u_x \mathbf{i} + u_y \mathbf{j} + u_z \mathbf{k} \quad . \tag{A.25}$$

u_x, u_y and u_z are called the vector components of the vector $\mathbf{u}$ and $\mathbf{i}$, $\mathbf{j}$ and $\mathbf{k}$ are called the unit vectors in the three orthogonal spatial directions. The unit vectors are often omitted and vectors are usually just written as a list of three scalar components. In the literature, these are variably named u_x, u_y, u_z or u, v, w or u_1, u_2, u_3. In the following we use the first of these three notation rules. Note that vectors are commonly represented with bold characters.

The sum of two vectors $\mathbf{u}$ and $\mathbf{v}$ is given by the sum of the vector components:

$$\mathbf{w} = \mathbf{u} + \mathbf{v} = (u_x + v_x)\mathbf{i} + (u_y + v_y)\mathbf{j} + (u_z + v_z)\mathbf{k} \quad . \tag{A.26}$$

This sum is often written as:

$$\mathbf{w} = (u_x + v_x, u_y + v_y, u_z + v_z) \quad . \tag{A.27}$$

The *magnitude* (or length) of a vector is given by:

$$|\mathbf{u}| = \sqrt{u_x^2 + u_y^2 + u_z^2} \quad . \tag{A.28}$$

Eqs A.25, A.26 and A.28 may be intuitively or graphically followed using the Pythagoras theorem.

The *scalar-* or *dot product* of two vectors is a scalar quantity which is defined as the sum of the products of two vector components:

$$\mathbf{u} \bullet \mathbf{v} = u_x v_x + u_y v_y + u_z v_z \quad . \tag{A.29}$$

This is equivalent to the product of the magnitudes of the two vectors and the cos of the angle ϕ between them:

$$\mathbf{u} \bullet \mathbf{v} = |\mathbf{u}|\,|\mathbf{v}| \cos\phi \quad . \tag{A.30}$$

The scalar product has its name because the result is a scalar quantity. A nice example for a scalar product is the work required to move a plate with the force $\mathbf{F}$ (being a vector) for the distance $\mathbf{l}$ (having a length and a direction).

The *cross product* or *vector product* is denoted with $\times$ or $\wedge$. The result is a vector. It is defined as follows:

$$\mathbf{w} = \mathbf{u} \times \mathbf{v} = \mathbf{u} \wedge \mathbf{v} = (u_y v_z - u_z v_y, u_z v_x - u_x v_z, u_x v_y - u_y v_x) \quad . \tag{A.31}$$

The three values on the right hand side of eq. A.31 have the form of the determinants of matrices. Thus, vector products may be solved using the methods of matrix calculations which are not treated here. A good example for a vector product is the velocity vector. It is described by the product of the angular velocity vector $\mathbf{w}$ and the position vector $\mathbf{r}$.

• *Grad, Div and Curl.* The *gradient* of a scalar valued function (denoted with "Grad" or "Del" or: ∇; s. sect. 3.1.1) is a vector describing the spatial change of this function. It is defined as

$$\nabla \equiv \left(\frac{\partial}{\partial x}, \frac{\partial}{\partial y}, \frac{\partial}{\partial z} \right) \quad . \tag{A.32}$$

Thus, the spatial change of temperature (as a function of x, y and z: $T = T(x, y, z)$), may be described as follows:

$$\text{Grad } T \equiv \nabla T = \left(\frac{\partial T}{\partial x}, \frac{\partial T}{\partial y}, \frac{\partial T}{\partial z} \right) \quad . \tag{A.33}$$

The vector "Grad T" is normal to surfaces of constant temperature just like the dip direction of a surface is always normal to the contour lines. "Grad" is a handy tool for the description of the topography of any potential surface.

The *divergence* of a vector is a *scalar*. In the earth sciences it often describes the transfer rate of mass or energy. The divergence of a vector is defined as follows:

$$\text{Div } \mathbf{v} \equiv \nabla \bullet \mathbf{v} = \left(\frac{\partial v_x}{\partial x} + \frac{\partial v_y}{\partial y} + \frac{\partial v_z}{\partial z} \right) \quad . \tag{A.34}$$

Let us illustrate the *divergence* of a vector valued function dependent on the spatial coordinates x, y and z with an example. Assume that $\mathbf{v}$ is the rate of mass or energy transfer. The flow of mass is $q_f = \rho \mathbf{v}$ and the flow of energy is: $q = H\mathbf{v}$. There, ρ is density in kg m^{-3} and H is the volumetric energy content in J m^{-3}. Thus, flow has the units of $\text{kg m}^{-2}\,\text{s}^{-1}$ or W m^{-2}, respectively. The divergence of these flows is the sum of the change in flow in the three spatial directions (eq. A.34). If the flow of energy or mass *into* a unity cube is just as large as the flow *out of it* (general criterion for the conservation of mass), then Div $\mathbf{v} = 0$ (s. also sect. 3.1.1, 3.3).

The *Curl* or *Rot* of a vector field is a vector describing the rotation of a vector. A vector with Curl $\mathbf{u} = 0$ is called non rotating. The Curl is defined by the relationship:

$$\text{Curl } \mathbf{v} \equiv \text{Rot } \mathbf{v} \equiv \nabla \times \mathbf{v}$$

$$= \left(\frac{\partial v_z}{\partial y} - \frac{\partial v_y}{\partial z}, \frac{\partial v_x}{\partial z} - \frac{\partial v_z}{\partial x}, \frac{\partial v_y}{\partial x} - \frac{\partial v_x}{\partial y} \right) \quad . \tag{A.35}$$

A.4 An Example for Using Fourier Series

In sections 3.1.1, 3.6.1 and 3.4.1 we have been introduced to two different types of solutions of the diffusion equation (eq. 3.6). They are:

1. Solutions that may be found by integration. These include mainly problems for which the descriptive equations may be so much simplified so that it is straight forward to integrate them. Very often, these are *steady state* problems in which it is possible to assume $dT/dt = 0$.

2. Solutions containing an error function. These may be found for problems that have their boundary condition at infinity. For example, when describing the thermal evolution of intrusions that are much smaller than the thickness of the crust or their distance to the earths surface, it is possible to make this assumption (e.g. p. 102 or p. 174).

A third type of solution is necessary for time dependent problems with spatially fixed boundary conditions. We have encountered such examples when describing the erosion of mature landscapes between incising drainages with the diffusion equation, for example on p. 175. Such examples may be solved using *Fourier series*. As the diffusion equation is such a classic example where Fourier series find an important application, we will continue to use this equation as an example. The now well familiar equation that we want to use again (eq. 3.6) is:

$$\frac{\partial T}{\partial t} = \kappa \frac{\partial^2 T}{\partial x^2} \quad , \tag{A.36}$$

with T being a function of both space x and time t: $T = T(x, t)$. Let us assume that this equation is subject to zero temperature boundary conditions at $x = 0$ and $x = l$ which may be formulated as:

- $T = 0$ at $x = 0$ at time $t \geq 0$.
- $T = 0$ at $x = l$ at time $t \geq 0$.

With these boundary conditions, this problem corresponds to that discussed on page 175. There, D and H correspond to what is here κ and T and the spatial extent of the problem was there measured between $-l$ and l, while it is here only from 0 to l. On page 175 we just gave the solution of this problem in eq. 4.63 without detailing the methods of solution.

In order to understand the process of solution here in some more detail, consider the following: Eq. A.36 is satisfied if we find a term for which the first time derivative is directly proportional to the second spatial derivative. The proportionality constant is κ. A general function that satisfies this condition and the boundary conditions has the form:

$$T = \sum_{n=0}^{\infty} a_n e^{b_n t} \sin\left(\frac{n\pi x}{l}\right) \quad . \tag{A.37}$$

There, a_n and b_n are constants. Let us discuss why this solution satisfies eq. A.36 and how it may be derived:

1. We can see that the solution above contains an exponential function of time and a sine-function of x. This can be understood as follows: Differentiating an exponential function will always return an exponential function. Correspondingly, the second derivation of a sine-function is a negative sine function. This negative will result in the exponential function also being negative (as shown below), which gives a function that decays with time. Thus, the first derivative of eq. A.37 with respect to t, will always be proportional to its second derivative with respect to x. Thus, the condition of the diffusion equation is met, if the correct constants are found.
2. It may be seen that the boundary conditions at $x = 0$ and $x = l$ are always satisfied as the sine-function is always zero at these two values of x. Thus, temperature there is also always zero.
3. The fact that the solution contains an infinite sum is a generalization. If a single term of the infinite sum satisfies eq. A.36, so will the infinite sum of a series of terms.

Let us check if eq. A.37 actually satisfies eq. A.36. For clarity, we perform this check only for a single term of the infinite sum. For our check we differentiate this term with respect to time as well as space. The time derivative gives:

$$\frac{\partial T}{\partial t} = abe^{bt}\sin\left(\frac{n\pi x}{l}\right) \quad . \tag{A.38}$$

The spatial derivatives are:

$$\frac{\partial T}{\partial x} = \frac{n\pi a}{l}e^{bt}\cos\left(\frac{n\pi x}{l}\right) \tag{A.39}$$

as well as:

$$\frac{\partial^2 T}{\partial x^2} = -\frac{n^2\pi^2 a}{l^2}e^{bt}\sin\left(\frac{n\pi x}{l}\right) \quad . \tag{A.40}$$

Comparing eq. A.38 and eq. A.40 shows that eq. A.36 is satisfied if the constant b has the following value:

$$b = -\kappa\frac{n^2\pi^2}{l^2} \quad . \tag{A.41}$$

If we insert b from eq. A.41 in eq. A.37, we have an equation that satisfies all conditions of eq. A.36. The values for the constants a_n can be determined from the initial conditions. At time $t = 0$, $e^{bt} = 1$ and thus from eq. A.37 it is true that:

$$T(x,0) = f(x) = \sum_{n=0}^{\infty} a_n\sin\left(\frac{n\pi x}{l}\right) \quad . \tag{A.42}$$

Eq. A.42 is an example of a Fourier series. The coefficients a_n can be determined from the integral:

$$a_n = \frac{2}{l} \int_0^l f(x)\sin\left(\frac{n\pi x}{l}\right) \mathrm{d}x \quad , \tag{A.43}$$

the derivation of which does not follow directly from eq. A.42 and will not be discussed here. However, it may be found in any book on Fourier series. The coefficients may be evaluated from this integral if the initial condition $T(x,0) = f(x)$ is known. However, this integral is only easily evaluated for certain functions of $f(x)$. For more general functions, solutions to this integral may be obtained from either math tables or numerically.

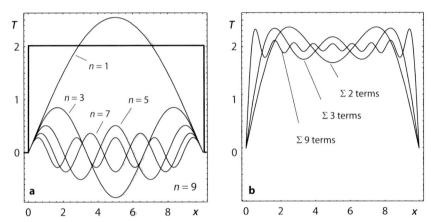

Figure A.11. a The function $f(x) = 2$ (thick line) and the first five terms of an infinite sum of sine functions from eq. A.42 at time $t = 0$. **b** The sum of the first two, three and nine terms of the function shown in **a**. It may be seen that the sum of only few terms is sufficient to approximate the thick drawn function in **a** quite good

• *Solving eq. A.36 for non-zero boundary conditions.* We can take this approach one step further to solve the diffusion equation eq. A.36 for non-zero boundary conditions:

- $T = T_1$ at $x = 0$ at time $t \geq 0$.
- $T = T_2$ at $x = l$ at time $t \geq 0$.

and initial conditions $T = f(x)$ at $t = 0$. In this case, the temperature T should evolve with time to the steady-state solution that satisfies the boundary conditions. We will denote the steady-state solution as $g(x)$. It can be shown that $g(x) = T_1 + (T_2 - T_1)x/l$. The solution for T has the form:

$$T = g(x) + \sum_{n=0}^{\infty} a_n e^{b_n t}\sin\left(\frac{n\pi x}{l}\right) \quad . \tag{A.44}$$

where b_n is given by equation eq. A.41 and a_n has the form

$$a_n = \frac{2}{l} \int_0^l (f(x) - g(x))\sin\left(\frac{n\pi x}{l}\right) \mathrm{d}x \quad . \tag{A.45}$$

A.5 Selected Numerical Tricks

A.5.1 Integrating Differential Equations

For many differential equations there are mathematical reference books containing their solutions and it certainly goes beyond the scope of this book to go into details of complicated integration methods. However, one simple example which illustrates the type of thinking that lies behind integrations like the one we used in sect. 4.1.2 will be introduced here. Eq. 4.9 has the form

$$\frac{\mathrm{d}y}{\mathrm{d}x} = ay + b \quad , \tag{A.46}$$

where a and b are constants. The equation states that the differential of the variable y is proportional to y. This information is sufficient to be able to guess that the solution will contain an exponential function of the form e^x, because exponential functions always remain exponential functions when they are differentiated (s. Table B.2). Thus, we may guess that the solution will have the form:

$$y = qe^{(cx)} + d \tag{A.47}$$

and thus:

$$\frac{\mathrm{d}y}{\mathrm{d}x} = qce^{(cx)} \quad . \tag{A.48}$$

Inserting eq. A.47 and eq. A.48 in eq. A.46 shows that $c = a$ and $d = -b/a$. It follows that:

$$y = qe^{(ax)} - \frac{b}{a} \quad , \tag{A.49}$$

for a fixed scalar q.

A.5.2 Analytically Unsolvable Equations

Many equations cannot be solved analytically. However, they often may be evaluated numerically by separating them into two parts. We illustrate this using the transcendental eq. 6.3 as an example. This equation has the form

$$cx = ae^{-bx} + d \quad . \tag{A.50}$$

All other parameters that occur in eq. 6.3 are summarized in eq. A.50 into the constants a, b, c and d. Eq. A.50 cannot be solved for x. In order to solve it numerically it is useful to split the right hand side and the left hand side of the equation into two new equations. For the left side we write:

$$z = cx \tag{A.51}$$

and for the right side we write:

$$z = ae^{-bx} + d \quad \text{or}: \quad x = \frac{\ln\left(\frac{z-d}{a}\right)}{-b} \ . \qquad (A.52)$$

These two functions are plotted in Fig. A.12. With the constants $a = 1$, $b = 2$, $c = 3$ and $d = 3$, the steep linear curve is eq. A.51 and the curve with a negative slope is eq. A.52. Their intersection is the solution of eq. A.50. This point may be found by alternating solution of eq. A.51 and A.52. For this we guess a value for x, insert this into eq. A.51 to calculate z and then insert this value for z into eq. A.52 to obtain a new x. For the example illustrated in Fig. A.24 an initial guess of $x = 0$ leads to the series: $z = 4$, $x = 1.333$, $z = 3.069$, $x = 1.023$, $z = 3.129$, $x = 1.043$, $z = 3.124$ and so forth. The result converges to a solution of approximately $x \approx 1.04$ and $z \approx 3.12$. The exact solution may be approximated as closely as desired. While the method is very simple, it may also lead to wrong results, for example if one of the two functions has local minima or maxima.

A.5.3 The Least Squares Method

A common problem in science occurs when a curve should be fitted to a number of data points and the fit of the data to this curve should be quantified. The most common method for this is to find the smallest sum of the squares of the deviations of the data from the curve, in short, the *least squares*. In the following section we explain how this is done with the example of a linear fit, i.e. the curve that is fitted to the data is a straight line. However, for more complicated choices of functions, the same rules apply. We assume that the data consist of n values for y and just as many for x. We label the data with y_i and x_i from $i = 1$ to n. The straight line we will to fit through the data cloud has the form $y = ax + b$ where a and b are unknown. If we insert our data pair for x and y into this linear equation, we obtain:

Figure A.12. Illustration of the numerical solution of eq. A.50. The constants are $a = 1$, $b = 2$, $c = 3$ and $d = 3$. The straight line respresents eq. A.51, the curved line is eq. A.52. The dashed line shows the iterative approximation of their intersection

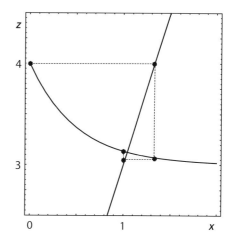

$$y_i = ax_i + b - e \ , \tag{A.53}$$

where e is the deviation of a given data point from the fitted line. In order to minimize the sum of the squares of this deviation, the sum of all e must be minimized. This means that:

$$\sum_{i=1}^{n} e^2 = \sum_{i=1}^{n} (ax_i + b - y_i)^2 \tag{A.54}$$

must be minimized. In order to do so, eq. A.54 may be partially differentiated with respect to a and b, set to zero and solved for a or b. Using the simple differentiation rules from Table B.1 the derivative with respect to a may be found to be:

$$0 = \frac{\partial \left(\sum_{i=1}^{n} (ax_i + b - y_i)^2 \right)}{\partial a} = \sum_{i=1}^{n} 2x_i (ax_i + b - y_i)^2 \ . \tag{A.55}$$

The derivative with respect to b is:

$$0 = \frac{\partial \left(\sum_{i=1}^{n} (ax_i + b - y_i)^2 \right)}{\partial b} = \sum_{i=1}^{n} 2(ax_i + b - y_i)^2 \ . \tag{A.56}$$

Eqs. A.55 and A.56 may be simplified and solved simultaneously for a and b. We get:

$$a = \frac{n \left(\sum_{i=1}^{n} x_i y_i \right) - \left(\sum_{i=1}^{n} x_i \right) \left(\sum_{i=1}^{n} y_i \right)}{n \left(\sum_{i=1}^{n} x_i^2 \right) - \left(\sum_{i=1}^{n} x_i \right)^2} \ , \tag{A.57}$$

$$b = \frac{\left(\sum_{i=1}^{n} y_i \right) \left(\sum_{i=1}^{n} x_i^2 \right) - \left(\sum_{i=1}^{n} x_i \right) \left(\sum_{i=1}^{n} x_i y_i \right)}{n \left(\sum_{i=1}^{n} x_i^2 \right) - \left(\sum_{i=1}^{n} x_i \right)^2} \ . \tag{A.58}$$

These are the coefficients of the *best fitting straight line* of eq. A.53. These two equations (eq. A.57 and eq. A.58) are straight forward to implement on a computer.

The errors of these values, as given by their standard deviations are simply given by:

$$(\partial a)^2 = \frac{n \sum_{i=1}^{n} e_i^2}{(n-2)\left(n\left(\sum_{i=1}^{n} x_i^2\right) - \left(\sum_{i=1}^{n} x_i\right)^2\right)} \tag{A.59}$$

and:

$$(\partial b)^2 = \frac{\left(\sum_{i=1}^{n} x_i^2\right)\left(\sum_{i=1}^{n} e_i^2\right)}{(n-2)\left(n\left(\sum_{i=1}^{n} x_i^2\right) - \left(\sum_{i=1}^{n} x_i\right)^2\right)} \quad . \tag{A.60}$$

A.5.4 Basic Statistical Parameters

A fundamental task of geologists is the characterization of the location and the variability of a data set, for example P-T data determined with thermobarometry, geochronological data, numbers from a whole rock analysis, digital data during image processing or dip and strike data measured in the field. Because such measurements may be very precise, but they are never perfectly accurate it is necessary to evaluate them statistically (s. p. 5). The most important parameters for such an evaluation are summarized here.

• *Normal distribution.* The statistical interpretation of many geological data is based on the assumption that the data have a normal (also called Gaussian) distribution around the exact value. A distribution is said to be normal if its probability density function is given by:

$$f(x) = \left(\frac{1}{\sigma\sqrt{2\pi}}\right) e^{\left(-\frac{(x-\mu)^2}{2\sigma^2}\right)} \quad . \tag{A.61}$$

This function is characterized by two parameters called the mean, μ, and the standard deviation σ. Eq. A.61 is plotted in Fig. A.13 and may be interpreted as the enveloping curve of a histogram. If the data are centered around a mean of $\mu = 0$ and the standard deviation is $\sigma = 1$, then eq. A.61 simplifies to: $f(x) = e^{-(x^2/2)}/\sqrt{2\pi}$. The distribution is said to be a *standard normal distribution.*

• *Mean.* The mean of a data set indicates the most probable location of the exact value. For a data set S containing n data S_i it is defined as:

$$\mu = \frac{1}{n} \sum_{i=1}^{n} S_i \quad . \tag{A.62}$$

the mean gives themost probable location of the exact value.

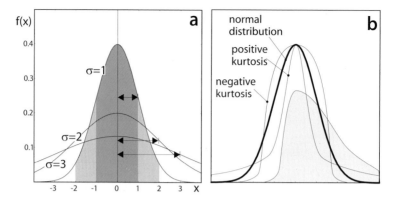

Figure A.13. a Normal distribution of data points calculated with eq. A.61. The curve for $\sigma = 1$ is a standard normal distribution. The area underneath all three curves is 1. The 1σ confidence interval is shaded dark, the 2σ confidence interval is shaded light. These two areas contain 68.26% and 95.45% of the data, respectively. **b** shows data distributions with high and low kurtosis as well as one with a significant skewness (shaded), relative to a normal distribution

- *Variance and standard deviation.* The variance is a measure of the variability of the data around the mean. It is defined as

$$Var = \frac{1}{n} \sum_{i=1}^{n} (S_i - \mu)^2 \quad . \tag{A.63}$$

If the data set is incomplete (univariat data set), where only a finite number of data points were collected, an unbiased estimate of the population variance is given by:

$$Var' = \frac{1}{n-1} \sum_{i=1}^{n} (S_i - \mu)^2 \quad . \tag{A.64}$$

The *standard deviation* is defined as the square root of the variance. If the data distribution is normal (Gaussian), then the standard deviation is abbreviated with $\sigma = \sqrt{Var}$. From eq. A.61 it may be calculated that 68.26% of all data of a normal distribution will fall within $\pm 1\sigma$ of the mean and 95.45% of the data within $\pm 2\sigma$. The latter is therefore often loosely referred to as a "95% confidence interval".

- *Skewness and kurtosis.* Skewness is a measure of the lack of symmetry of data in a histogram. Kurtosis is a measure of whether the data are peaked or flat relative to a normal distribution (Fig. A.13b). Data with a high kurtosis have a distinct peak near the mean, decline rather rapidly and have heavy tails. Data with a low kurtosis have a flat top. A uniform distribution is an extreme case of kurtosis. Skewness (for a univariate data set s. eq. A.64) is defined as:

$$skew = \frac{1}{\sigma^3(n-1)} \sum_{i=1}^{n}(S_i - \mu)^3 \ . \tag{A.65}$$

Kurtosis is defined as:

$$kurt = \left(\frac{1}{\sigma^4(n-1)} \sum_{i=1}^{n}(S_i - \mu)^4 \right) - 3 \ . \tag{A.66}$$

Many classical statistical tests depend on the assumption that the data have a normal distribution. Significant skewness and kurtosis indicate that the data are not normal. In fission track analysis the skewness and kurtosis of track length distributions bear significant information on the cooling history of the rocks.

A.6 Problems

Problem A.1. *Finite difference approximations (p. 355)*:
Eq. 3.45 describes simultaneous diffusion and one-dimensional mass transport. Write an explicit finite difference approximation for this equation. Use a forward differencing scheme to approximate the transport term. Follow the schemes introduced in eq. A.14 and eq. A.15

Problem A.2. *Finite difference approximations (p. 356)*:
Eq. 4.46 describes the elastic bending of oceanic lithosphere under applied loads. There, w is the variable for which we want to solve the equation. It is the vertical deflection of the bent plate as a function of distance x. Write an explicit finite difference approximation for this equation. Hint: it is easiest to just expand the scheme we have followed in eq. A.16

Problem A.3. *Finite difference approximations (p. 362)*:
Redraw Fig. A.7 carefully to convince yourself why *backward* finite differencing schemes will be unstable when describing the transport equation. For your considerations look at the scheme of eq. A.24, and a corresponding backward differencing scheme for $u = \Delta x / \Delta t$.

Problem A.4. *Mean and standard deviation (p. 362)*:
Determine the mean and the standard deviation of the following two data sets $S1 = \{10, 10, 10, 10, 6, 6, 6, 6\}$ and $S2 = \{1, 5, 10, 20, 15, 9, 4, 0\}$, assuming that both data sets are samples from a normal distibution.

B. Maths Refresher

Recommended Reading:

- Abramowitz and Stegun (1972) Handbook of Mathematical Functions.
- Press, Flannery, Teukolsky and Vetterling (1989) Numerical Recipes.
- Strang (1988) Linear Algebra and Applications.

Table B.1. General rules of differential calculus using the example of the function $y = f(x)$. u and v are also functions of x. a is a constant. $f'(x)$ or y' is the first derivative of y with respect to x

$f(x)$	$f'(x)$
$y = au$	$y' = a(du/dx)$
$y = x^a$	$y' = ax^{a-1}$
$y = a^x$	$y' = a^x \ln(a)$
$y = x^x$	$y' = (1 + \ln(x))x^x$
$y = u + v$	$y' = (du/dx) + (dv/dx)$
$y = uv$	$y' = u(dv/dx) + v(du/dx)$
$y = \frac{u}{v}$	$y' = (vdu/dx - udv/dx)/v^2$
$y = u^v$	$y' = u^v ((v/u)(du/dx) + \ln u(dv/dx))$

Table B.2. Special derivatives of the function $y = f(x)$. $f'(x)$ or y' is the first derivative of y with respect to x. The table may be read both ways: $f(x)$ may also be interpreted as the integral of the function $f'(x)$

$f(x)$		$f'(x)$
$y = \frac{1}{x}$	$= x^{-1}$	$y' = -x^{-2}$
$y = \ln(x)$		$y' = \frac{1}{x\ln(10)}$
$y = \log(x)$	$= \frac{\ln(x)}{\ln(10)}$	$y' = \frac{1}{x}$
$y = e^x$		$y' = e^x$
$y = e^{ax}$		$y' = ae^x$
$y = x\ln(x) - x$		$y' = \ln(x)$
$y = \sin(x)$		$y' = \cos(x)$
$y = \cos(x)$		$y' = -\sin(x)$
$y = \tan(x)$		$y' = \sec(x)$

Table B.3. Conversions between different logarithms. In this book we use "ln" for the natural logarithm (to the base of e) and "log" for the decimal logarithm (to the base of 10)

$\ln(xy)$	$= \ln(x) + \ln(y)$	
$\ln(x/y)$	$= \ln(x) - \ln(y)$	
$\ln(x^y)$	$= y\ln(x)$	
$\ln(e)$	$= 1$	
$\ln(1)$	$= 0$	
$\ln(0)$	$= -\infty$	
$\log(x)$	$= \ln(x)/\ln(10)$	$= \log(e)\ln(x)$
$\ln(10)\log(e)$	$= 1$	

Table B.4. Trigonometry of a triangle with the side lengths a, b and c. The angles α, β and γ are those opposite to the sides a, b and c, respectively

---

plane triangles with a right angle ($\gamma = 90°$)

$a^2 + b^2 = c^2$ Pythagoras

$\sin(\alpha) = a/c$

$\cos(\alpha) = b/c$

$\tan(\alpha) = a/b$

$\mathrm{ctg}(\alpha) = b/a$

$\mathrm{area} = (a \times b)/2$

general plane triangles

$a/\sin(\alpha) = b/\sin(\beta) = c/\sin(\gamma)$

$a^2 = b^2 + c^2 - 2bc \times \cos(\alpha)$

$\mathrm{area} = (bc\sin(\alpha))/2 = [s(s-a)(s-b)(s-c)]^{1/2}$ $s = 1/2(a+b$

spherical triangles

$\sin(a)/\sin(\alpha) = \sin(b)/\sin(\beta) = \sin(c)/\sin(\gamma)$

$\cos(a) = \cos(b)\cos(c) + \sin(b)\sin(c)\cos(\alpha)$

angular separation α of 2 points on a sphere (lats. λ_1, λ_2 and longs. ϕ_1, ϕ_2)

$\cos(\alpha) = \sin(\lambda_1)\sin(\lambda_2) + \cos(\lambda_1)\cos(\lambda_2)\cos(\phi_2 - \phi_1)$ ...for any α

$\sin(\alpha/2) = \left(\sin^2((\lambda_2 - \lambda_1)/2) + \cos(\lambda_2)\cos(\lambda_1)\sin^2((\phi_2 - \phi_1)/2)\right)^{0.5}$ ...for small α

transformation between spherical and Cartesian coordinates (R is sphere radius)

$x = R\cos(\phi)\cos(\lambda)$

$y = R\sin(\phi)\cos(\lambda)$

$z = R\sin(\lambda)$

spatial distance a between two points x_1, y_1, z_1 and x_2, y_2, z_2

$a^2 = (x_2 - x_1)^2 + (y_2 - y_1)^2 + (z_2 - z_1)^2$

Table B.5. Volume, surface and other important data of some geometrical bodies

---

cube with side length a

surface	$= 6a^2$
volume	$= a^3$
length of side diagonal	$= a\sqrt{2}$
length of space diagonal	$= a\sqrt{3}$

tetrahedron with side length a

surface	$= a^2\sqrt{3}$
volume	$= \frac{1}{12}a^3\sqrt{2}$

sphere with radius r

surface	$= 4\pi r^2$
volume	$= \frac{4}{3}\pi r^3$

cone with radius r and height H

surface of the mantle	$= r\pi\sqrt{r^2 + H^2}$
volume	$= \frac{1}{3}\pi r^2 H$

cylinder with radius r and height H

surface of the mantle	$= 2r\pi H$
volume	$= r^2\pi H$

Table B.6. Definitions and conversions between trigonometric functions

$$csc(x) = 1/\sin(x)$$
$$sec(x) = 1/\cos(x)$$
$$tg(x) = \sin(x)/\cos(x)$$
$$ctg(x) = \cos(x)/\sin(x) = 1/\tan(x)$$
$$\sin^2(x) + \cos^2(x) = 1$$

Table B.7. Special values of trigonometric functions (Fig. B.1)

Angle	$\alpha = 0°$	$\alpha = 30°$	$\alpha = 45°$	$\alpha = 60°$	$\alpha = 90°$
rad	0	$\pi/6$	$\pi/4$	$\pi/3$	$\pi/2$
$\sin(\alpha)$	0	0.5	$\sqrt{2}/2$	$\sqrt{3}/2$	1
$\cos(\alpha)$	1	$\sqrt{3}/2$	$\sqrt{2}/2$	0.5	0
$tg(\alpha)$	0	$1/\sqrt{3}$	1	$\sqrt{3}$	∞
$ctg(\alpha)$	∞	$\sqrt{3}\ 1$	$1/\sqrt{3}$		0

Figure B.1. The unity circle and definitions of the trigonometric functions. The four trigonometric functions are labeled for an angle of 45°

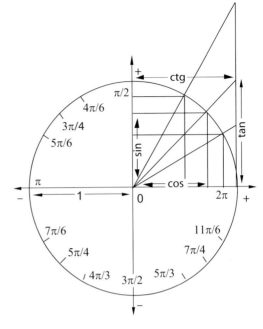

Table B.8. Solutions of quadratic equations of the form $ax^2 + bx + c = 0$

$$x_{1,2} = -\frac{b}{2a} \pm \frac{1}{2a}\sqrt{b^2 - 4ac}$$

Table B.9. Approximation of the error function using the constants $a = 0.3480242$; $b = -0.0958798$ and $c = 0.7478556$ and $y = 1/(1 + 0.47047x)$

$$\mathrm{erf}(x) \approx 1 - \left(ay + by^2 + cy^3\right)e^{-x^2}$$

C. Symbols and Units

Table C.1. Symbols and units of the variables used in this book. Physical constants of the earth are listed in Table C.3 and C.4. Variables abbreviated with Greek letters are explained in Table C.2. Variables used in the text are often specified in more detail by adding a subscript. For example q is used for heat flow, and q_s for *surface* heat flow. The most commonly used subscripts in this book are l for lithosphere; c for crust; m for mantle; i, j and n for numbering; x, y and z for spatial directions and 0 for initial values

Symbol	Variable	Unit	1. Occurrence
Ar	Argand number	–	Eq. 6.27
A	area	m^2	Eq. 3.2
A	pre exponential constant	$MPa^{-n}s^{-1}$	Eq. 5.38
A_q	. . . of quartz creep	$MPa^{-n}\,s^{-1}$	Table 5.3
A_o	. . . of olivine creep	$MPa^{-n}s^{-1}$	Table 5.3
a	general constant	variable	Eq. 3.20
b	general constant	variable	Eq. 3.20
c_p	heat capacity (of rocks)	$\approx 1\,000\ J\,kg^{-1}\,K^{-1}$	Eq. 3.4
c_{pf}	. . . of fluids	$J\,kg^{-1}\,K^{-1}$	Eq. 3.50
C	constant of integration	variable	Eq. 3.58
D	angular momentum	$kg\,m^2\,s^{-1}$	Sect. 2.2.4
D	rigidity of elastic plates	Nm	Eq. 4.41
D	deformation	–	Sect. 7
D	diffusivity of mass	$m^2\,s^{-1}$	Eq. 4.54
D_0	pre exponential diffusivity	$m^2\,s^{-1}$	Eq. 7.4
e	elongation	–	Eq. 5.1
e	error	–	Eq. A.53
E	Young's Modul	Pa	Eq. 4.43
E	energy	J	–
E_p	potential energy (per area)	$J\,m^{-2}$	Eq. 5.44
E_k	kinetic energy	J	Eq. 5.58
f	frequency	s^{-1}	Eq. 3.103
f	ellipsidity	–	Eq. 4.1
f_c	vertical strain of crust	–	Eq. 4.3
f_l	vertical strain of lithosphere	–	Eq. 4.3
F	force	N	Sect. 2.2.4
F_b	buoyancy force (per length)	$N\,m^{-1}$	Eq. 5.44
F_d	tectonic driving force (per length)	$N\,m^{-1}$	Eq. 6.22
F_{eff}	effective driving force (per length)	$N\,m^{-1}$	Eq. 6.22
F_l	integrated strength	$N\,m^{-1}$	Eq. 5.43
g	geothermal gradient	$°C\,m^{-1}$	Eq. 3.48
g	acceleration (gravitational)	$m\,s^{-2}$	Eq. 4.16

Table C.1. ... *continuation*

Symbol	Variable	Unit	1. Occurrence
G	geometrical factor for T_c	–	Eq. 7.9
G	Gibb's energy	J	Eq. 7.5
h	elastic thickness	m	Eq. 4.43
h	dimensionless height	–	Eq. 1.2
h_r	exponential drop off	m	Eq. 3.67
h_s	erodable thickness	m	Eq. 4.55
H	heat content (volumetric)	$J\,m^{-3}$	Fig. 3.10; Eq. 3.99
H	elevation, height	m	Eq. 1.2
I	momentum	$kg\,m\,s^{-1}$	Sect. 2.2.4
I	tensor invariant	–	Eq. 5.5
J	moment of inertia	$kg\,m^2$	Sect. 2.2.4
k	thermal conductivity	$J\,s^{-1}\,m^{-1}\,K^{-1}$	Eq. 3.1
K	bulk modulus	Pa	Sect. 5.1.2
K	equilibrium coefficient	–	Eq. 7.5
l	thickness, length scale	m	Eq. 3.18
L	latent heat of fusion	$J\,kg^{-1}$	Eq. 3.34
L	sediment thickness	m	Eq. 6.2
L_0	... of decompacted layer	m	Eq. 6.2
L^*	... of decompacted pile	m	Eq. 6.2
m	mass	kg	Eq. 1.1
M	bending moment	N	Eq. 4.41
N	number, counter	–	Eq. 3.93
n	power law-exponent	–	Eq. 5.39
n	general counter	–	Eq. 3.16
Pe	Peclet number	–	Eq. 3.51
P	pressure	Pa	Eq. 5.7
Q	activation energy (diffusion)	$J\,mol^{-1}$	Eq. 7.4
Q	activation energy (creep)	$J\,mol^{-1}$	Eq. 5.38
Q_q	... of quartz creep	$J\,mol^{-1}$	Table 5.3
Q_o	... of olivine creep	$J\,mol^{-1}$	Table 5.3
Q_D	... of Dorn law creep	$J\,mol^{-1}$	Eq. 5.42
q	load on a plate	Pa	Eq. 4.42
q	heat flow	$W\,m^{-2}$	Eq. 3.1
q_s	... at the surface	$W\,m^{-2}$	Eq. 3.61
q_m	... at the Moho	$W\,m^{-2}$	Eq. 3.61
q_{rad}	... caused by radioactivity	$W\,m^{-2}$	Eq. 6.13
q_r	water flux	$m\,s^{-1}$	Eq. 4.71
q_f	sediment flux in rivers	$m\,s^{-1}$	Eq. 4.71
r	radius	m	Eq. 1.1, 3.12
r	distance in polar coordinates	m	Eq. 1.1, 3.12
R	radius of earth	m	Eq. 2.1
R_A	... at the equator	6 378.139 km	Eq. 4.1
R_P	... at the pole	6 356.75 km	Eq. 4.1

Table C.1. *continuation*

Symbol	Variable	Unit	1. Occurrence
s	stretch	–	Eq. 5.1
s	cooling rate	$°C\,s^{-1}$	Eq. 3.90
S	rate of heat production	$J\,s^{-1}\,m^{-3}$	Eq. 3.24
S_{chem}	. . . chemical	$J\,s^{-1}\,m^{-3}$	Eq. 3.23
S_{mec}	. . . mechanical	$J\,s^{-1}\,m^{-3}$	Eq. 3.23
S_{rad}	. . . radioactive	$J\,s^{-1}\,m^{-3}$	Eq. 3.23
S_0	. . . radioactive at surface	$J\,s^{-1}\,m^{-3}$	Eq. 3.67
S	entropy	$J\,K^{-1}$	Eq. 3.33
SL	sea level	m	Eq. 6.5
t	time	s	Eq. 3.4
t_a	degradation coefficient	m^2	Eq. 4.66
t_E	erosional time constant	s	Eq. 4.10
T	temperature	$°C$; K	Eq. 3.1
T_A	. . . at the start	$°C$; K	Eq. 7.8
T_b	. . . of the host rock	$°C$; K	Eq. 3.85
T_c	. . . closure temperature	$°C$; K	Eq. 7.9
T_E	. . . at the end	$°C$; K	Eq. 7.8
T_i	. . . intrusion temperature	$°C$; K	Eq. 3.85
T_l	. . . at base of lithosphere	$\approx 1\,200\text{–}1\,300\,°C$	Eq. 3.60
T_l	. . . liquidus	$°C$; K	Eq. 3.41
T_s	. . . solidus	$°C$; K	Eq. 3.41
T_s	. . . at the surface	$°C$; K	Eq. 3.81
T_0	. . . initial temperature	$°C$; K	Eq. 3.103
u	velocity (often: in x direction)	$m\,s^{-1}$	Eq. 3.43
U	circumference	m	Sect. 4.0.1
v	velocity (often: in y direction)	$m\,s^{-1}$	Eq. 4.4
v_f	. . . of fluids	$m\,s^{-1}$	Eq. 3.50
v_{ex}	exhumation rate	$m\,s^{-1}$	Eq. 4.4
v_{er}	erosion rate	$m\,s^{-1}$	Eq. 4.7
v_{ro}	uplift rate of rocks	$m\,s^{-1}$	Eq. 4.4
v_{up}	uplift rate of the surface	$m\,s^{-1}$	Eq. 4.4
V	volume	m^3	Eq. 3.2
w	angular velocity	s^{-1}	Sect. 2.2.4
w	water depth	m	Eq. 4.30
w	plate deflection	m	Eq. 4.41
x	spatial coordinate (horizontal)	m	Sect. 1.2
X	mole fraction	–	Eq. 7.6
y	spatial coordinate (horizontal)	m	Sect. 1.2
z	spatial coordinate (vertical)	m	Sect. 1.2
z_c	thickness of crust	m	Sect. 2.4.1; Eq. 3.63
z_i	initial depth	m	Eq. 3.49
z_l	thickness of lithosphere	m	Sect. 2.4.1
z_{rad}	thickness of radioactive crust	$\approx 7\text{–}10$ km	Eq. 3.61

Table C.2. Greek symbols

Symbol	Variable	Unit	1. Occurrence
α	coefficient of thermal expansion	$\approx 3 \cdot 10^{-5}\,^{\circ}\mathrm{C}^{-1}$	Eq. 3.32
α	general angle	radian	Eq. 3.101
α	flexural parameter	m	Eq. 4.47
β	isothermal compressibility	Pa^{-1}	Eq. 3.31; 5.24
β	stretching factor (mantle lithosph.)	–	Eq. 6.10
δ	stretching factor (crust)	–	Eq. 6.10
δ	density ratio	–	Eq. 4.25
η	viscosity	Pa s	Eq. 5.37
ϵ	strain	–	Eq. 5.20
$\dot{\epsilon}$	strain rate	s^{-1}	Eq. 1.5
κ	diffusivity	$\mathrm{m}^2\,\mathrm{s}^{-1}$	Eq. 3.6
λ	longitude	degree	Eq. 2.2; Fig. 2.7
λ	wave length	m	Eq. 3.105
λ	pore fluid pressure ratio	–	Eq. 5.29
λ_c	. . . in the crust	–	Fig. 5.17
λ_l	. . . in the lithosphere	–	Fig. 5.17
μ	coefficient of internal friction	–	Eq. 5.25
ν	Poisson ratio	–	Eq. 4.43
ϕ	latitude	degree	Fig. 2.7
ϕ	porosity	–	Eq. 3.50
ϕ_0	. . . at the surface	–	Eq. 6.1
ϕ	friction angle	radian	Eq. 5.28
ρ	density	$\mathrm{kg\,m}^{-3}$	Eq. 3.4
ρ_c	. . . of the crust	$\mathrm{kg\,m}^{-3}$	Eq. 4.18
ρ_g	. . . of sediment grains	$\mathrm{kg\,m}^{-3}$	Fig. 6.3; Eq. 6.7
ρ_0	. . . of the mantle at $0\,^{\circ}\mathrm{C}$	$\approx 3\,300\ \mathrm{kg\,m}^{-3}$	Eq. 4.22
ρ_L	. . . of a sedimentary pile	$\mathrm{kg\,m}^{-3}$	Eq. 6.5
ρ_m	. . . of the mantle at T_l	$\approx 3\,200\ \mathrm{kg\,m}^{-3}$	Eq. 4.18
ρ_w	. . . of water	$\approx 1\,000\ \mathrm{kg\,m}^{-3}$	Eq. 4.30
σ	stress	Pa	Eq. 5.2
$\sigma_1,\ \sigma_2,\ \sigma_3$	principal stress	Pa	Eq. 5.4
σ_d	differential stress	Pa	Eq. 3.25
σ_n	normal stress	Pa	Eq. 5.25
σ_D	. . . critical, Dorn law creep	Pa	Eq. 5.42
τ	thermal time constant	s	Eq. 3.18
τ	shear stress	Pa	Eq. 5.3
θ	dimensionless temperature	–	Eq. 1.3; 3.88
θ	angle of a failure surface to σ_1	radian	Fig. 5.6
ξ	expansion ratio	–	Eq. 4.25

Table C.3. Important data of the earth

equatorial radius	6 378.139 km
polar radius	6 356.750 km
diameter of core	3 468 km
volume	$1.083 \cdot 10^{21}$ m^3
mass	$5.973 \cdot 10^{24}$ kg
surface area	$5.10 \cdot 10^{14}$ m^2
area of the continents	$1.48 \cdot 10^{14}$ m^2
area of continental lithosphere	$2.0 \cdot 10^{14}$ m^2
area of oceanic lithosphere	$3.1 \cdot 10^{14}$ m^2
mean elevation of the continents	825 m
mean depth of the oceans	3 770 m
total length of mid oceanic ridges	$\approx 60\,000$ km
mean continental surface heat flow	≈ 0.056 W m^{-2}
mean oceanic surface heat flow	≈ 0.078 W m^{-2}

Table C.4. Important physical constants

Constant	Symbol	Value
gas constant	R	8.3144 J mol^{-1} K^{-1}
gravitational constant	G	$6.6732 \cdot 10^{-11}$ N m^2 kg^{-2}
speed of light	c	$2.99792 \cdot 10^8$ m s^{-1}

Table C.5. SI-units

Physical Parameter	Symbol in Text	Unit	Abbreviation
distance	x, y, z	meter	m
time	t	second	s
mass	m	kilogram	kg
temperature	T	Kelvin	K

Table C.6. Important derived units

Physical Parameter	Symbol in Text	Unit	Abbreviation	SI-Unit
force	F	Newton	N	$\mathrm{kg\,m\,s^{-2}}$
pressure	P	Pascal	$\mathrm{Pa = N\,m^{-2}}$	$\mathrm{kg\,m^{-1}\,s^{-2}}$
energy	E	Joule	$\mathrm{J = N\,m}$	$\mathrm{kg\,m^{2}\,s^{-2}}$
power	–	Watt	$\mathrm{W = J\,s^{-1}}$	$\mathrm{kg\,m^{2}\,s^{-3}}$

Table C.7. Conversions between derived units

Physical Parameter	Conversion
force	$= \text{mass} \times \text{acceleration}$ $= \text{pressure} \times \text{distance} = \text{Pa m}$
pressure	$= \text{force per area} = \text{N m}^{-2}$ $= \text{energy per volume} = \text{J m}^{-3}$
energy	$= \text{force} \times \text{distance} = \text{Nm}$ $= \text{mass} \times \text{velocity}^2 = \text{kg m}^2 \text{s}^{-2}$
power	$= \text{work per time} = \text{J s}^{-1} = \text{W}$
velocity	$= \text{distance per time} = \text{m}^{-1}$
acceleration	$= \text{velocity change per time} = \text{m s}^{-2}$

Table C.8. Important commonly used variables and their conversion into SI-units. Numbers are given to a maximum of 4 digits

Parameter	Unit	Conversion
length	1 angstrom	$= 10^{-10}$ m
	1 micrometer (μm)	$= 10^{-6}$ m
	1 millimeter (mm)	$= 10^{-3}$ m
	1 kilometer (km)	$= 10^{3}$ m
	1 foot (ft)	$= 0.3048$ m
	1 inch (in)	$= 2.54$ cm
	1 mile (mi)	$= 1.6093$ km
	1 yard (yd)	$= 0.9144$ m
	1 nautical mile (nmi)	$= 1.852$ km
	1° latitude	$= 60$ nmi $= 111.12$ km
area	1 hectar (ha)	$= 10^{4}$ m^2
	1 acre	$= 4046.9$ m^2
volume	1 liter (l)	$= 10^{-3}$ m^3
	1 gallon (US)	$= 3.7854$ l
	1 gallon (UK)	$= 4.5461$ l
	1 hectoliter (hl)	$= 100$ l
	1 barrel (US)	$= 158.98$ l
time	1 day	$= 8.64 \cdot 10^{4}$ s
	1 million years (my)	$= 3.1557 \cdot 10^{13}$ s
temperature	1 °C	$= 1$ K (0 °C $= 273.16$ K)
force	1 dyne	$= 1$ g cm s^{-2} $= 10^{-5}$ N
pressure	1 bar	$= 10^{5}$ Pa
	1 atmosphere (atm)	$= 1.0133 \cdot 10^{5}$ Pa $= 760$ mm Hg
	1 mm Hg (torr)	$= 1.3332 \cdot 10^{2}$ Pa
	1 lb in^{-2}	$= 6.8947 \cdot 10^{3}$ Pa
energy	1 cal	$= 4.184$ J
	1 erg	$= 1$ dyne cm $= 10^{-7}$ J
	1 heat flow unit (hfu)	$= 10^{-6}$ cal cm^{-2} s^{-1} $= 0.04184$ W m^{-2}
	1 horse power (PS)	$= 746$ W
viscosity	1 poise	$= 0.1$ Pa s

D. Answers to Problems

Problem 2.1. According to Fig. 2.4, both the Aleute arc and the Java trench appear to have small circle radii corresponding to roughly $25°$ latitude which is $\approx 2\,700$ km. Thus, the ping pong ball model of eq. 2.1 predicts subduction angles of the order of $25°$. The much steeper observed dip may be due to additional forces exerted onto the subducted slab by asthenospheric convection.

Problem 2.2. The experiment can be performed by starting at $1+(2\pi n)^{-1}$ km from the south pole, where n is an integer. Thus, there is not only a single point, but an infinite number of rings around the south pole from where this experiment can be started. 1 km south of each of these rings the circumference of the earth along a line going due east or west is $1/n$ km.

Problem 2.3. $(5/360) \cdot 24$ hours $= 20$ minutes. Knowledge of the latitude is not necessary.

Problem 2.4. The difference in geographic longitude is $4°\ 42' = 282'$. The improved solution of Problem 2.3 is then $282/(360 \cdot 60) \cdot 24$ hours $= 18.8$ minutes.

Figure D.1. Sketch of the solution of Problem 2.5

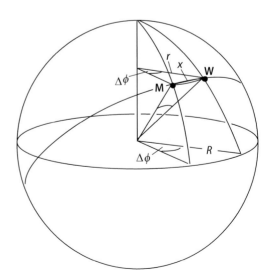

Problem 2.5. a) The difference in longitude between Munich and Vienna is: $\Delta\phi = 5°$. Therefore: $we = \cos\lambda\frac{\Delta\phi}{360}2R\pi = 371\,961.3$ m. The small circle radius r for the 48th degree of latitude is $r = R\cos\lambda = 4\,262.362$ km. The great circle angular separation (using the formula in Table B.4) is: 3.34507°. Accordingly, the great circle distance is 371 896 m. This is about 65 m shorter than the distance along the small circle. The direct (Euclidian) distance along a hypothetical tunnel is: $x = 371\,843$ m. It is only about 53 m shorter than the great circle distance.

Problem 2.6. At the given angular velocity, a full rotation takes $2\,\pi/(10^{-15}\mathrm{s}^{-1})$ ≈ 200 my. The equatorial circumference of the earth is $2\,R\pi \approx 40\,000$ km. Thus, the plate velocity at the equator is roughly 20 cm per year. At 48° north the small circle circumference is only 26 781 km (eq. 2.2). Thus, the relative plate velocity at this latitude is only 13 cm per year.

Problem 2.7. The torque at the equator is: $M = R \cdot 10^{12}$ Nm. North and south of the equator we can write: $M = r \cdot 10^{12}$ Nm. Using $r = R\cos\lambda$ (e. 2.2) we get a latitude of: $\lambda \approx 51°$ either north or south.

Problem 2.8. $x = C_1\phi$ and $y = C_2\mathrm{tg}\lambda$. The constants C_1 and C_2 can have any value for example 1. They are not really necessary, but are introduced here in order to show that the Mercator projection can be scaled to rectangles of any aspect ratio.

Problem 2.9. In thermally stabilized continental lithosphere, the base of the crust is at about $z_c = 30$ - 40 km depth, the base of the lithosphere is at about $z_l = 100$ - 200 km depth. The temperature at the Moho is about $500\,°C$ and at the base of the lithosphere it is about $T_l \approx 1\,200 - 1\,300\,°C$. Thus, at least 100 - 200 °C of the Moho-temperature may be attributed to radioactive heat. The mean density of crustal rocks is of the order of $\rho_c = 2\,800\,\mathrm{kg\ m}^{-3}$ and that of mantle rocks in the asthenosphere about $\rho_l = 3\,200\,\mathrm{kg\ m}^{-3}$. The thermal expansion coefficient of rocks is about $\alpha = 3 \times 10^{-5}\,°C^{-1}$ so that the density changes by several percent within both the crust and the mantle part of the lithosphere.

Problem 2.10. a) The relative velocities may be read from Table 2.3. b) A new plate boundaries is currently forming along the East African Rift system. However, how the world will look like in future is very much a trial and error game. Have a look at http://www.scotese.com. c) Ancient plate boundaries exist just about anywhere where there is ancient mountain belts. Fig. 2.3 gives an overview over the last 170 my.

Problem 2.11. There is a total of 10 possible triple junctions: RRR, FFF, TTT, RRT, RRF, FFT, FFR, TTF, TTR, RTF. RRR-triple junctions are always stable, FFF-triple junctions are always unstable. Most other triple junctions may occur in stable or in unstable configuration. Whether they are unstable or not does not depend on the qualitative nature of the plate motion, but rather on the angles and relative velocities.

Problem 3.1. The amount of heat in one gram of mass is 10^{-3} kg × $300\,000^2$ km^2 s^{-2} = $9 \cdot 10^{13}$ J, which is enough to light a 60 W globe for $9 \cdot 10^{13}$ J $/$ 60 W $\approx 48\,000$ years.

Problem 3.2. See Table C.8.

Problem 3.3. It may be seen intuitively that the ore body will be nearly isothermal if it has a high conductivity. As a consequence, the isotherms above and below the ore body have a closer spacing (Fig. D.2).

Problem 3.4. At the contact between two rocks of different conductivity the heat flow must be the same on both sides of the contact. Thus, for the given linear thermal gradient the temperatures at different depths will be: $T_{5km} = 100\,^\circ C$; $T_{7km} = 120\,^\circ C$ and $T_{10km} = 180\,^\circ C$.

Problem 3.5. The thermal time constant of the pile is roughly 40 my. This is significantly longer than the deformation in a), but significantly shorter than that in b). ($\dot{\epsilon} = 10^{-12}$ s^{-1} implies that the pile doubles in thickness in a mere 30 000 years). The implication of this result is that the deformation in b) occurs largely in thermal equilibrium. In contrast, in a) the crust will be out of thermal equilibrium at the end of deformation. The deformation will be followed by thermal equilibration. Metamorphic parageneses are expected to grow *across* the fabric in a) but grow syn-deformational in b).

Problem 3.6. Using eq. 3.18 it is easy to see that the thermal time constant (giving an estimate of the duration of heat conduction) of a 10 km large body is at least one order of magnitude longer than the questioned period of observation of 10^5 years. Heat conduction may therefore be neglected and the temperature is given (from eq. 3.22) by $T = t \times S_{rad}/\rho c_p$. Using the given values for density and heat capacity this implies a temperature rise of about $T \approx 100^\circ C$.

Problem 3.7. a) The product of deviatoric stress and strain rate gives minimum and maximum values for the heat production rate of $3 \cdot 10^{-5}$ and $3 \cdot 10^{-8}$ W m^{-3}, respectively (eq. 3.25). These minimum and maximum values

Figure D.2. Answer to Problem 3.3

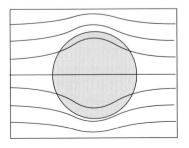

are significantly higher and smaller than typical values for radiogenic heat production rates, respectively. b) Multiplying these values with a deformation period of 1 my, the total shear heat production is 10^9 J m^{-3} for the largest stresses and strain rates and 10^6 J m^{-3} for the small stresses and strain rates. Thus, according to eq. 3.26, the maximum temperature that can be reached is given by the ratio of volumetric heat content and ρc_p. This gives 370 °C and 0.37 °C for the maximum and minimum values, respectively. Whether or not these temperatures can be reached depends on the efficiency of heat conduction away from the site of heat production and on the change in strength of the shear zone during deformation. The former depends on the thickness of the heat producing shear zone.

Problem 3.8. a) For this problem it is useful to convert energy into mass × velocity^2. b) For this estimate, you need to remember that the gravitational acceleration at the center of the earth is zero. Assuming a mean temperature of 4 000 °C and a mean acceleration of about 5 m s^{-2}, the adiabatically caused temperature change would amount to roughly 3 600 °C. c) The adiabatic compressibility is smaller than the isothermal compressibility because adiabatic compression is associated with heating. Thus, thermal expansion acts against the compressibility.

Problem 3.9. None! All heat of the fire is reaction heat (chemically produced heat). It is released by the exothermic chemical reaction: wood → CO_2 + water.

Problem 3.10. The duration for which the rock will remain at constant temperature is given by the ratio of *total amount of released fusion heat* to *rate of heat loss by conduction*. For example, if x J are produced by the crystallization and the conductive cooling causes a heat loss of x Js^{-1}, then it the cooling history will be halted for exactly 1 second. The volumetric amount of heat that is released at the solidus is $0.3 \times L\rho$. The rate of heat loss due to cooling is $dT/dt\rho c_p$ (s. eqs. 3.3, 3.99). Thus, the duration of thermal buffering is $= L/c_p(dt/dT) \approx 1$ my. The fact that the cooling history is halted for 1 my may imply that metamorphic parageneses re-equilibrate partially or fully at this temperature.

Problem 3.11. Using eq. 3.51 and a standard value for the thermal diffusivity of $\kappa = 10^{-6}$ m^{2}s^{-1}, the three advection (erosion) rates give Peclet numbers of $Pe \approx 0.1$, $Pe \approx 1$ and $Pe \approx 5$, respectively. Thus, the slow erosion rate of a) warrant description of the regional thermal evolution with consideration of conduction only (e.g. eq. 3.6), while the intermediate rates of b) require consideration of both conductive and advective processes (e.g. eq. 3.45) and the rapid rates of c) would allow to consider the thermal evolution as a solely advective process (e.g. eq. 3.43).

Problem 3.12. The answer is discussed in eqs. 6.14 and 6.15.

Figure D.3. Answer to
Problem 3.15b. Curves are
labeled in $\mathrm{mW\,m^{-2}}$

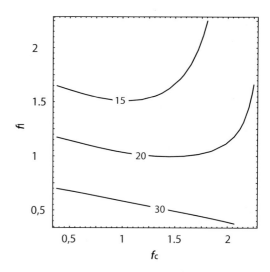

Problem 3.13. The total amount of heat produced by the crust in model
b in Fig. 3.16 is: $S_0 z_{rad}$. The question requires that the depth integrated
heat production above the depth z_w in model c (eq. 3.67) corresponds to this
value. Thus: $S_0 z_{rad} = \int_0^{z_w} S_0 e^{-z/h_r}$. Integration and inserting the boundaries
gives: $S_0 z_{rad} = S_0 h_r \left(1 - e^{(-z_w/h_r)}\right)$. Using $h_r = 2 z_{rad}$ we arrive at: $0.5 =$
$e^{-z_w/(2 z_{rad})}$ or: $z_w = 1.38 z_{rad}$.

Problem 3.14. After inserting the cosine-shaped distribution of heat sources
in the crust into eq. 3.56 a first integration gives:

$$k\frac{dT}{dz} = -S_0 z + \frac{S_0 z_c}{4\pi}\sin\left(2\pi z/z_c\right) + C_1$$

The first integration constant has the value $C_1 = q_m + S_0 z_c$, so that the
boundary condition $k dT/dz = q_m$ at $z = z_c$ is fulfilled. A second integration
gives the result:

$$T = \frac{S_0 z_c z}{2k} + \frac{q_m z}{k} - \frac{S_0 z^2}{4k} - \frac{S_0 z_c^2}{8k\pi^2}\cos\left(\frac{2\pi z}{z_c}\right) + C_2$$

The second constant of integration has the value $C_2 = S_0 z_c^2/(8k\pi^2)$, so that
$T = 0$ at $z = 0$. The given heat source distribution describes a maximum
crustal heat production in the center of the crust. This situation could arise
if a crust with a "normal" heat source distribution (decreasing with depth)
is overthrust by a thick pile of rocks with little heat production. .

Problem 3.15. a) The derivative of eq. 3.76 with respect to z is:

$$\frac{dT}{dz} = \frac{T_l}{f_l z_l} + f_c h_r S_0 e^{-z/(f_c h_r)} + \frac{f_c^2 h_r^2 S_0}{f_l z_l}\left(e^{-\frac{z_l f_l}{f_c h_r}} - 1\right)$$

The mantle heat flow may be found by multiplying this equation with k
and evaluating it at $z = z_c$. b) is solved in Fig. D.3. c) The mantle heat

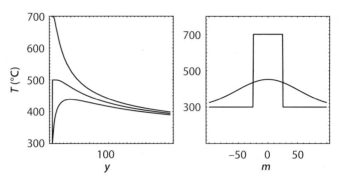

Figure D.4. Answer to Problem 3.18b and c

flow changes insignificantly if the lithosphere is thickened homogeneously. However, it changes dramatically during homogeneous thinning.

Problem 3.16. a) Eq. 3.81 shows that the error function must have a value of 0.8333 so that the temperature is $T=1\,000\,°$C. For this, Fig. 3.5 shows that the argument of the error function must be ≈ 0.98 (thus: $0.98 = z/\sqrt{4\kappa t}$). From this we get: $z = 98$ km. b) 0 km $/0\,°$C; 10 km $/371\,°$C; 20 km $/689\,°$C; 30 km $/921\,°$C; 40 km $/1\,067\,°$C; 50 km $/1\,144\,°$C; 75 km $/1\,196\,°$C; 100 km $/1\,199\,°$C.

Problem 3.17. a) Using $\tau = l^2/\kappa$ the result is ≈ 80 years. b) for symmetry reasons this temperature is half way between the intrusion and the host rock temperature if both have the same heat capacity: $500\,°$C, directly at the contact. c) 50 m · $(T_i - T_b)\rho c_p = 5.4 \cdot 10^{10}$ J $/$ m^2 dike surface. d) 40 years is roughly half of the total thermal history. The cooling of the dike is most rapid at the start. Thus, the dike will have cooled a bit more than half by then. The area underneath the T-z-profile must remain constant as no heat may be lost. This information may be used to estimate the width of the contact metamorphic halo. (s. also Problem 3.18).

Problem 3.18. a) Inserting eq. 3.91 in eq. 3.89 gives: $439\,°$C. Note that for the choice of coordinate system for which eq. 3.89 was written, z must be $z = 35$ in order to describe a point at 10 m from the contact. b) and c) are solved in Fig. D.4.

Problem 3.19. The depth of the rocks multiplied with the geothermal gradient gives the terrain temperature prior to metamorphism: $300\,°$C. If $10\,\%$ of the terrain were at $1\,100\,°$C and $30\,\%$ of the terrain area at $700\,°$C, and the remainder at $300\,°$C, then the mean temperature is $\overline{T}_{cool} = 0.1 \times 1\,100 + 0.3 \times 700 + 0.6 \times 300 = 500°$C. This is $200\,°$C above the host rock temperature. Thus, the heat of the intrusives is insufficient to explain metamorphism if latent heat of crystallization is neglected. The additional temperature contribution of the crystallization heat from 40% of the terrain is $T_{cryst} = 0.4 \times L/c_p = 128\,°$C. Thus, the potential metamorphic temperature is $\overline{T}_{cool} + T_{cryst} = 628\,°$C, as the mean potential temperature for the whole terrain. This is within $10\,\%$-error

Figure D.5. Solution of Problem 3.19

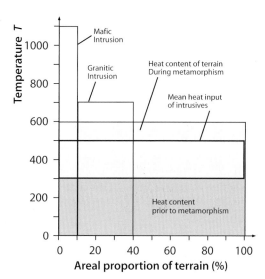

of the observed metamorphic temperature. Contact metamorphism is therefore a plausible model explanation for the observations. Fig. D.5 illustrates the problem answer graphically.

Problem 3.20. The depths and temperatures of the starting condition are: $z[1]=0$, $T[1]=0$; $z[2]=10$ km, $T[2]=125\,°C$; $z[3]=20$ km, $T[3]=250\,°C$; $z[4]=30$ km, $T[4]=375\,°C$; $z[5]=40$ km, $T[5]=500\,°C$; $z[6]=50$ km, $T[6]=0\,°C$; $z[7]=60$ km, $T[7]=125\,°C$; $z[8]=70$ km, $T[8]=250\,°C$; $z[9]=80$ km, $T[9]=375\,°C$; $z[10]=90$ km, $T[10]=500\,°C$. The time stepping must be smaller than 0.7937 my. In order to perform the calculations with a round number it is useful to use something like $\Delta t = 0.5$ my $= 1.575\cdot 10^{13}$ s, i.e. $R=0.1575$. After $1\Delta t$ the temperatures are: $T[5]=402\,°C$ and $T[6]=98\,°C$. After $2\Delta t$ the temperatures are: $T[4]=360\,°C$, $T[5]=350\,°C$, $T[6]=150\,°C$ and $T[7]=140\,°C$. All other temperatures remain up to this time constant because there is no spatial curvature of the temperature profile around them (eq. 3.6).

Problem 4.1. Using eq. 4.2 we get a distance of the summit of Mt Everest of 6 382.207 km from the center of the earth and 6 384.32 km for Chimborazzo. Thus, Chimborazzo is actually about 2 100 m higher than Everest, if measured from the enter of the earth.

Problem 4.2. In contrast to Fig. 4.5 this diagram is defined for all positive values of f_c and f_{ml}. Homogeneous thickening of the lithosphere is represented by a diagonal line through the diagram. Thickening of the mantle part of the lithosphere with no change of crustal thickness is represented by a line parallel to the f_{ml}-axis. Crustal thickening at constant thickness of the entire lithosphere must be accompanied by a corresponding thinning of the mantle part of the lithosphere by $f_{ml} = z_c/z_{ml} + 1 - f_c$ (because: $z_c f_c + z_{ml} f_{ml} = z_c + z_{ml}$).

Problem 4.3. a) There is *no* rock uplift at all. b) The rock is not exhumed at all. It is buried by 1 km. This example is actually what occurs to rocks on the Tibetan plateau at the moment where extreme surface uplift and sedimentation occur at the same time.

Problem 4.4. a) Morphological equilibrium will be reached after a long time. Thus, eq. 4.12 may be used for $t \to \infty$: $t_E = Hb/(\dot{e}(H + a)) = 1.834$ my^{-1}. b) If $z = 0$ and $t = 40$ my are inserted into eq. 4.13 and this is solved for z_i we get: $z_i = 10\,202$ m. c) This may be evaluated by finding the derivative of eq. 4.13 with respect to t, setting the result to 0 and solving for t. The result is 11.8 my.

Problem 4.5. Eq. 4.17 is solved in eq. 4.20. Adapting the variables by inserting $z_1 = 1$ for a randomly chosen thickness of the iceberg of 1 and $H = 0.1$, because 10% of this thickness are above water and using the density of water for ρ_m we get for the density of ice: $\rho_{ice} = 900$ kg m^{-3}.

Problem 4.6. The solution is shown graphically in Fig. D.6.

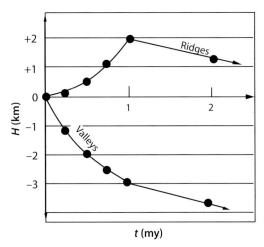

Figure D.6. Solution of Problem 4.6. The mean vertical denudation x within 1 my is given by the relationship $lx = (vt)^2$. This relationship was derived geometrically from Fig. 4.39. Thermal expansion or the influence of the mantle lithosphere need not be considered as they remain constant. The mean subsidence at any time is given by: $x(\rho_m - \rho_c)/\rho_m$. Thus, the elevation of the ridges is after 0.25 my: 132 m, after 0.5 my: 527 m, after 0.75 my: 1187 m and after 1 my: 2110 m. Thereafter, the ridges subside with a rate of 781 m my^{-1}. The subsidence of the valley floors may be easily derived from the product of incision rate and time substracted from the results for the ridge elevations

Problem 4.7. a) $\Delta H = 10$ km $(\rho_m - \rho_u)/\rho_m = 937$ m. Density and thickness of the crust do not change and therefore need not be considered. b) $\Delta H = 391$ m. c) the thinning of the crust in b) causes a subsidence of $(f_c - 1)(\rho_m - \rho_c)/\rho_m z_c = 2343$ m. This is partially balanced by an uplift of 391 m by the basaltic underplate. Relative to a) there is a total elevation difference of $937 + (2343 - 391) = 2889$ m.

Problem 4.8. The relative importance of thermal expansion and material differences is given directly by the parameters δ and ξ. With the given values these are: $\delta = 0.1563$ and $\xi = 0.018$. Thus thermal expansion accounts for roughly 12 % of the density contrast between lithosphere and asthenosphere. However, thermal expansion (or contraction) concerns the entire lithosphere, while the material difference to the asthenosphere exists only in the crust. Thus, because the lithosphere is roughly three times as thick as the crust, thermal contraction accounts for up to 40 % of the surface elevation of a mountain belts in isostatic equilibrium. In lithospheres with thin crust or very thick mantle lithosphere the importance of thermal contraction may even outweigh material differences.

Problem 4.9. a) The answer is: 2.5 cm y^{-1}. From this and the given data the water depths are: 469, 878, 1049, 1483, 1756, 1906, 2392, 2633, 2795, 2930, 3318, 3709, 3855 and 3995 m for the given distances from the mid oceanic ridge. b) These model values correspond well with the observed values up to an age of roughly 100 my. From there on, sedimentary loading of the plate, its increasing strength and changes of the lower boundary condition by mantle convection cause the divergence between model and observation.

Problem 4.10. a) The units of D are Nm. The fourth derivative has the units of m^{-4} and q has the units of stress or force per area. The load is the vertical stress. F has therefore the units of N m^{-1}, or Newton per meter length of orogen. b) $D_{h=10\text{km}} \approx 8.8 \times 10^{21}$ Nm and $D_{h=70\text{km}} \approx 3 \times 10^{24}$ Nm.

Problem 4.11. a) The given hints are required to determine the constants of integration. For the first and second integration they the constants must be: $C_1 = -qL/D$ and: $C_2 = qL^2/(2D)$, respectively. For the third and fourth integration the constants must be: $C_3 = 0$ and $C_4 = 0$, respectively. Thus: $w = qx^2/D \left(x^2/24 - Lx/6 + L^2/4\right)$. b) At $x = L$ the deflection is: $w = 3qL^4/(24D)$. Using the parameters given in the problem we get: $D = 0.0008$ Nm.

Problem 4.12. a) For the lowest point where there is a density difference between the two columns we can write: column A: $\sigma_{zz} = \rho_c g(H + z_c + w)$; column B: $\sigma_{zz} = \rho_c g z_c + \rho_m g w$. The buoyancy is given by the difference between these two values. This is the origin of eq. 4.44. b) If the load is q_a is not considered, then we can write for the downward force per unit area: $(\rho_c - \rho_w)wg$, and for the upward directed force: $(\rho_m - \rho_c)wg$. The net force per unit area is therefore: $wg(\rho_w - \rho_m)$. The total load is: $q = q_a + wg(\rho_w - \rho_m)$.

Problem 4.13. $\alpha \approx 65$ km.

Problem 4.14. At the highest point of the elastic bulge the slope of the plate is $dw/dx = 0$. The slope of the plate as a function of distance is given by the first spatial derivative of eq. 4.48. If this is set to zero: $0 = -2w_0/\alpha e^{-x/\alpha} \sin(x/\alpha)$, we get: $x = \pi\alpha$. This is the highest point of the fore bulge. In 250 km distance from this point we get from eq. 4.47: $D = 2.2 \cdot 10^{23}$ Nm, and eq. 4.43: $h \approx 33$ km. The value of w_0 does not change the distance of the forebulge from the load and its knowledge is therefore not required.

Problem 4.15. The problem is easiest solved by plotting the data and graphing eq. 4.61 iteratively for different times until a best fit for the data is found. For this a half height of the scarp $a \approx 7$m should be used, which is roughly half the elevation difference between base and top. It is also useful to subtract roughly 150m from all x values so that the inflection point of the scarp is at $x = 0$. Then, a best fit gives an age of about 30 000 years.

Problem 4.16. The two boulders have been exposed to weathering for the same time. This result is easiest arrived at by measuring l (for example in millimeters) off the photograph for the two different boulders and comparing the topographic shapes of the boulders with plots of eq. 4.63. Neither the magnitude of D, not the actual length of l need be known as we perform only a comparison between the two profiles.

Problem 4.17. In the steady state, eq. 4.64 simplifies to $Dd^2H/dx^2 =- v_{ro}$. A first integration renders $dH/dx =- v_{ro}x/D + C_1$. C_1. As $dH/dx = 0$ at $x = 0$. A second integration renders $H = -v_{ro}x^2/(2D) + C_2$, where C_2 is determined by the boundary condition $H = 0$ at $x = l$. The result is $H = v_{ro}/(2D) \times (l^2 - x^2)$ which is actually the first term of eq. 4.65. Evaluating this at $x = 0$ for the parameter values in Fig. 4.30b we can see that the steady state elevation of the crest above the streams is about 800 m, so that the plotted age of 30 000 years is already practically in steady state.

Problem 4.18. According to eq. 4.73 the fractal dimension of the object shown in Fig. 4.41 is: $D = \log 3/\log 2 = 1.585$.

Problem 5.1. The stretch is 2, the elongation is 1. The vertical strain is 100%.

Problem 5.2. Force $=$ mass $\times$ acceleration. As weight is a force and the acceleration on the surface of the earth is $g = 9.81$ m/s^2, a mass of 1 kg hat has the weight of: 1 kg $\cdot$ 9.81 m/s$^2 = 9.81$ N.

Problem 5.3. Energy is given by the product of force times distance. Thus, the collision process releases 10^{13}N $\times 10^5$ m $= 10^{18}$ J. The most important forms of energy into which this mechanical energy is transformed are frictional heat and potential energy of the mountain belt that formed in response to the collision.

Problem 5.4. As there are no shear stresses and the problem is two dimensional, we can assume that $\sigma_{zz} = \sigma_1$ and $\sigma_{xx} = \sigma_2$ (there is no σ_3). For the continent that is pulled apart by the subducting plate we can write: $\sigma_{xx} = \sigma_2 = -A$ and $\sigma_{zz} = \sigma_1 = 0$. From this, the mean stress (pressure) is: $\sigma_m = -A/2$. According to eq. 5.11 the vertical and horizontal components of deviatoric stress are: $+A/2$ and $-A/2$, respectively. For the continent that collapses under the weight of a mountain range we can write: $\sigma_{xx} = \sigma_3 = 0$ and $\sigma_{zz} = \sigma_1 = A$. From this, the mean stress is given by: $\sigma_m = A/2$. The vertical and horizontal components of deviatoric stress are also: $+A/2$ and $-A/2$. Thus, we can conclude that state of deviatoric stress is the same in both continents but the pressure is different.

Problem 5.5. As no shear stresses apply to the surfaces of the rock, all components of eq. 5.16 that describe shear stresses are zero. Eq. 5.16 simplifies to: $\mathrm{d}\sigma_{zz}/\mathrm{d}z = \rho g$. The corresponding equations for the other two spatial directions in which there is no gravitational acceleration simplify to: $\mathrm{d}\sigma_{xx}/\mathrm{d}x = 0$ and $\mathrm{d}\sigma_{yy}/\mathrm{d}y = 0$. Integration of these three equation gives: $\sigma_{xx} = C_1$, $\sigma_{yy} = C_2$ and $\sigma_{zz} = \rho g z + C_3$. C_1, C_2 and C_3 are the constants of integration. They are determined by the boundary conditions of the problem. The normal stresses can be assumed to be zero on each free surface (Fig. 5.34): $\sigma_{zz} = 0$ at $z = -H$. From this we get: $C_3 = \rho g H$ and $\sigma_{zz} = \rho g(H+z)$. Furthermore, because there is no stresses on the sides it must be true that: $C_2 = C_1 = 0$; $\sigma_{xx} = \sigma_{yy} = 0$. The mean principal stress (pressure P) as a function of depth is: $P = (\sigma_{xx} + \sigma_{yy} + \sigma_{zz})/3 = \rho g(H + z)/3$. The principal components of the deviatoric stress tensor are, according to eq. 5.9: $\sigma_{xx} - P = \sigma_{yy} - P = -\rho g(H + z)/3$ and $\sigma_{zz} - P = 2\rho g(H + z)/3$.

Problem 5.6. Similar to problem 5.5, integration of the stress balance equations gives: $\sigma_{xx} = C_1$; $\sigma_{yy} = C_2$ and $\sigma_{zz} = \rho g z + C_3$. As fluids support only negligible differential stresses it is true that: $\sigma_{xx} = \sigma_{yy} = \sigma_{zz}$. With the same boundary condition for the vertical component of stress as in Problem 5.5 we get: $\sigma_{xx} = \sigma_{yy} = \sigma_{zz} = \rho g(z + H) = P$. Thus, at any given depth, the pressure in the fluid is three times as high as it is in the rock of Problem 5.5. All components of the deviatoric stress tensor are zero.

Problem 5.7. If the rock can flow apart without resistance, then it must be true that: $\sigma_{xx} = 0$. Thus, $\sigma_{xx}=0$, $\tau_{xx} = \sigma_{xx} - P = -\rho g(H + z)/3$ and according to eq. 5.37, the horizontal strain rate is: $\dot{\epsilon}_{xx} = -\rho g(z + H)/(6\eta)$. The rock flows quickly at its base and not at all at its top surface ($z = -H$). The vertical strain rate is $\dot{\epsilon}_{zz} = \rho g(z + H)/(3\eta)$ as $\tau_{zz} = \sigma_{zz} - P = 2\rho g(H + z)/3$.

Problem 5.8. If the side wall is fixed, then the stain rate is: $\dot{\epsilon} = 0$ and $\sigma_{zz} = \sigma_{xx}$. The mean force per area is given by the vertically integrated vertical stresses $\rho g(z + H)$ using the limits of integration $-H$ and 0 from $F_b = \rho g H^2/2$. The mean horizontal stress $\overline{\sigma}_{xx}$ is therefore $\overline{\sigma}_{xx} = F_b/H = \rho g H/2$.

Problem 5.9. In the fluid: $\sigma_{xx} = \sigma_{yy} = \sigma_{zz} = P = \rho_m g z$. In the rock the horizontal stresses for $z < 0$ are: $\sigma_{xx} = \sigma_{yy} = 0$, and for $z > 0$ they are: $\sigma_{xx} = \sigma_{yy} = \rho_m g z$. The vertical stresses in the whole rock are: $\sigma_{zz} = \rho_c g(H + z)$. The pressure in the rock is given by the mean of the principal stresses. For $z < 0$ it is: $P = (\rho_c g(H + z))/3$. For $z > 0$ it is: $P = (2\rho_m g z + \rho_c g(H + z))/3$. All principal components of the deviatoric stress tensor in the fluid are 0. In the rock, for $z < 0$ they are: $\tau_{xx} = \tau_{yy} = \sigma_{xx} - P = \sigma_{yy} - P = -(\rho_c g(H + z))/3$ and: $\tau_{zz} = \sigma_{zz} - P = (2\rho_c g(H + z))/3$. For $z > 0$ they are: $\tau_{xx} = \tau_{yy} = \sigma_{xx} - P = \sigma_{yy} - P = (\rho_m g z - \rho_c g(z + H))/3$ and $\tau_{zz} = \sigma_{zz} - P = -(2\rho_m g z - 2\rho_c g(z + H))/3$.

Problem 5.10. From eq. 5.11 we get: $\epsilon = \sigma/E = 5 \cdot 10^7 \text{Pa}/5 \cdot 10^{10} \text{Pa} = 0.001 = 0.1\%$.

Problem 5.11. The vertical stress at any given depth is: $\sigma_{zz} = \rho g z$. The strain of a unity cube at depth z in the lithosphere is therefore: $\epsilon = \rho g z/E$. From this, the total strain integrated over the entire lithosphere is given by: $\int_0^{z_l} \rho g z/E \, dz = \rho g z^2/2E$. Using the given numerical values for the parameters this is: $2\,500$ m. Thus, the elastic thickness change of the total lithosphere due to its own weight is roughly 2.5% of its total thickness.

Problem 5.12. The solution is graphically shown in Fig. D.7.

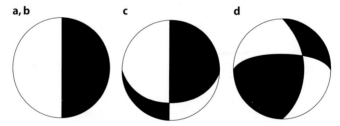

a, b   **c**   **d**

Figure D.7. Solution of Problem 5.12

Problem 5.13. For the quartz data the stresses are $\sigma_d = 117$ MPa; $\sigma_d = 24$ MPa and $\sigma_d = 11$ MPa, for $n = 2$, $n = 3$ and $n = 4$, respectively. For olivine the stresses are $\sigma_d = 1.4 \times 10^8$ MPa; $\sigma_d = 2.7 \times 10^5$ MPa and $\sigma_d = 1.2 \times 10^4$ MPa, for $n = 2$, $n = 3$ and $n = 3$, respectively.

Problem 5.14. When taking the logarithm of eq. 5.41 it becomes linear. It then has the form: $\log(\dot{\epsilon} = n\log(\sigma_d) + \log(A)-Q/RT$. Thus, if the experimental results are plotted in a diagram of $\log(\dot{\epsilon})$ against $\log(\sigma_d)$ the slope of the data gives the power law exponent n and the intercept is $(\log(A)-Q/RT)$. Replotting the data from a constant stress on a diagram of $\log(\dot{\epsilon}$ against $1/T$

Figure D.8. Graphical solution
of Problem 5.20

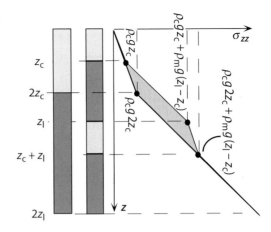

gives a line with the slope of $-Q/R$ from which the activation energy may
be determined.

Problem 5.15. It is because in oceanic lithosphere olivine dominates the
lithospheric rheology at much shallower levels (and therefore colder temperatures) than in continental lithosphere.

Problem 5.16. According to eq. 5.45 it is: $E_\mathrm{p} = \int_0^{-H} \sigma_{zz}\mathrm{d}z = \int_0^{-H} \rho g(H + z)\mathrm{d}z = \rho g H^2/2$.

Problem 5.17. The solution is $F_\mathrm{b} \approx 10.5 \times 10^{12}\,\mathrm{N}\,\mathrm{m}^{-1}$. The result is the
same as that from Problem 5.20 as can also be seen from Fig. D.8

Problem 5.18. Eq. 5.52 gives a buoyancy force of $F_\mathrm{b} \approx 6.6 \times 10^{12}\,\mathrm{N}\,\mathrm{m}^{-1}$.

Problem 5.19. Inserting the given values into eq. 5.54 give a horizontal
buoyancy force of the mountain range of $F_\mathrm{b} \approx 7.71 \times 10^{12}\,\mathrm{N}\,\mathrm{m}^{-1}$. This result
may also be read from Fig. 5.32. There, the point $f_\mathrm{c} = 2$, $f_\mathrm{l} = 1$ gives a value
of about $F_\mathrm{b} \approx 8.3 \times 10^{12}\,\mathrm{N}\,\mathrm{m}^{-1}$. The difference arises from slightly different
assumptions of the densities and thicknesses. The result is larger than that
from eq. 5.52 because the refined equation (eq. 5.54) also considers thermal
contraction of the mantle part of the lithosphere.

Problem 5.20. The graphical solution of this problem is shown in Fig. D.8.
The net force between the two columns corresponds to the shaded area on this
figure (see also Problem 5.17). Calculating this area algebraically, gives a net
force of $10.5 \cdot 10^{12}\,\mathrm{N}\,\mathrm{m}^{-1}$ towards the homogeneously thickened lithosphere.
This is comparable to the magnitude of plate tectonic driving forces, even
though there is *no* elevation difference between the two lithospheric columns.

Problem 5.21. The kinetic energy ($E_{kin} = mv^2/2$) of the plate is: $E_{kin} = 3 \cdot 10^{20}\mathrm{kg} \cdot (0.03\,\mathrm{m/y})^2 = 271\,\mathrm{J}$ (or $\mathrm{kg}\,\mathrm{m}^2\,\mathrm{s}^{-1}$). The potential energy per cubic
meter is: $E_p = \rho g h$. Thus, with a kinetic energy of 271 J we can lift one cubic
meter of rock (using $\rho = 2\,700\,\mathrm{kg}\,\mathrm{m}^{-3}$ and $g = 10\,\mathrm{m}\,\mathrm{s}^{-2}$) merely by about

1 cm. Consequently, the kinetic energy is completely insignificant to have an influence on the orogenic force balance. Plate motions are too slow!

Problem 6.1. Fig. 6.1 shows that the elevation contours have a slope of roughly $(f_c/f_l)=1.333$. Thus, during extension (when both f_c and f_l become smaller than 1) uplift will occur when $(f_c/f_l)>1.333$; for the assumed thicknesses and densities of Fig. 6.1. Eq. 4.29 shows that the elevation does not change when $\delta z_c(f_c-1)=\xi z_l(f_l-1)$ and subsidence will occur when $\delta z_c(f_c-1)<\xi z_l(f_l-1)$. This can be solved for any of the variables in that equation.

Problem 6.2. According to eq. 6.1 we can write for the cross over between sandstone and shale porosities: $\phi_0^{sand}\exp(-c^{sand}z)=\phi_0^{shale}\exp(-c^{shale}z)$. This can be solved for z by taking the logarithm of the equation giving: $z=(\log(\phi_0^{sand})-\log(\phi_0^{shale}))/(c^{sand}-c^{shale})$. Using the parameter values from Fig. 6.3 we get $z=1\,116\,\text{m}$.

Problem 6.3. Inserting eq. 6.1 into eq. 6.4 and using the parameter values from Fig. 6.3 gives $L_0=130\,\text{m}$.

Problem 6.4. According to eq. 4.17 we can formulate from Fig. 6.5: $Lg\rho_L+z_cg\rho_c=z_sg\rho_w+z_cg\rho_c+(L-z_s)g\rho_m$. Solving for z we get eq. 6.5 with a reversed sign. The different sign arises because the surface is *lower* before sedimentation, whereas we have automatically changed the sign in eq. 6.5 to deal only with positive numbers. The equation that describes the change if water depth may be derived analogously (s. Problem 6.5).

Problem 6.5. In isostatic equilibrium the mass of all vertical columns above the isostatic compensation depth must be the same. The data for density and thicknesses are redundant as they are the same for all three basins. It must be true that: $2\,\text{km}\cdot\rho_{air}+(w-2)\,\text{km}\cdot\rho_l=w\,\text{km}\cdot\rho_w$. This implies a water depth of: w of 2.9 km. The base of the lake is therefore 900 m lower than the valley floor. Analogous calculation for the sedimentary pile gives: $s=7.1\,\text{km}$. The surface of the sedimentary basin is 5.1 km lower than the valley floor. In other words: sediment filled basins are roughly 2.5 times as deep as water filled basins (as may be directly concluded by the density ratio of water to sediment).

Problem 6.6. The answer is shown in Fig. D.9. The first three columns show the field data: *a* illustrates thickness and lithology (sandstone: gray, slate: lines), *b* shows the age of the lithological boundaries and *c* the water depth. The strata are numbered from bottom to top from 1 to 5. For the following calculations the center points of each layer (shown in the 4th column) are used as reference point. For the porosities we use the data from Fig. 6.3. Using eq. 6.1 for the uppermost, $1\,000\,\text{m}$ thick layer we get a porosity of 34.4% at the mean depth of 500 m (1st column of Table *d*). The thickness of each layer is based on field measurements and their densities are given by eq. 6.7. In the second column the data for porosity, thickness and density *after* removal of the uppermost layer ($i = 5$) are plotted. For the second layer from the top ($i = 4$) that has a mean depth of $1\,250$ m, we begin with a determination of porosity. Its decompacted thickness is then calculated with eq. 6.4 using the porosities of the first column and the decompacted porosities. The porosity of the third layer after removing the first two strata is given by eq. 6.1 at a depth of 655 m $+ 500$ m and so on. The same principle is repeated for all columns using always the data from the fist column so that no methodical errors are propagated. The sum of the thicknesses and densities if the profile (in the bottom row) are calculated with eq. 6.8 and 6.9. *e* is a graphical illustration of the data from Table *d*.

Problem 6.7. To determine the ratio we take the subsidence in eq. 6.10 to be zero, i.e.: $H = 0$. Thus, after rearranging: $z_c/z_l = \rho_l \alpha T_l/(2(\rho_l - \rho_w))$. For the given numerical values of the parameters: $z_c/z_l = 13.7$ km. At smaller starting ratios of the thicknesses, the contribution of the mantle part of the lithosphere is so large, so that homogeneous stretching of the lithosphere (when thinned) will lead to surface *uplift*.

Problem 6.8. Eq. 6.11 shows that the duration of the thermal sag period is only determined by the ratio t/τ in the last term of the equation. The time scale of thermal equilibration τ is the same we have discussed in eq.3.18 and Table 3.2. It differs from that only by the factor π^2, but is in principle equivalent. We can see that when $t = \tau$, then H_{sag} has reached $(1 - e^{-1}) \times 100\%$ completion. 90% of the thermal sag is completed when: $(1 - e^{-t/\tau}) = 0.9$ or: $t = -\tau \ln(0.1)$. This is ≈ 75 my for 100 km thick lithosphere and ≈ 300 my for 200 km thick lithosphere.

Problem 6.9. For the thickened lithosphere we can use eq. 6.15. We get: $q_s = 0.06$ Wm^{-2}; $q_s = 0.075$ Wm^{-2} and $q_s = 0.09$ Wm^{-2}, for the three heat source distributions, respectively. For the thinned lithosphere we must use $q_s = 2q_m + q_{rad}/2$ and get: $q_s = 0.0525$ Wm^{-2}; $q_s = 0.075$ Wm^{-2} and $q_s = 0.0975$ Wm^{-2}, respectively. We can see that the first distribution results in an *increase* of the surface heat flow with increasing thickness, while the last distribution results in a *decrease* of the surface heat flow with increasing thickness.

Problem 6.10. (a) Fig. 5.32 shows that the buoyancy force F_b balances the driving force roughly at $f_c \approx 1.65$ (while $f_l = 1$). Fig. 4.16 shows that this corresponds to a surface elevation of about $H \approx 3.5$ km. (b) In order to use eq. 5.52, the thickness of the root w must be substituted by $w = H\rho_c/\Delta\rho$. It then can be solved for to give $H = 4\,400$ m at a buoyancy force of $F_b = F_d = 5 \cdot 10^{12}$ N m^{-1}. (c) Eq. 5.54 simplifies significantly for $f_l = 1$. It then can be used to derive $f_c = 1.7$ to give $F_b = F_d = 5 \cdot 10^{12}$ N m^{-1}. From this, eq. 4.25 gives a surface elevation of $H = 3\,800$ m. The estimate from (b) is substantially higher than those from (a) and (c) because the thermal contraction in the mantle part of the lithosphere is not considered here.

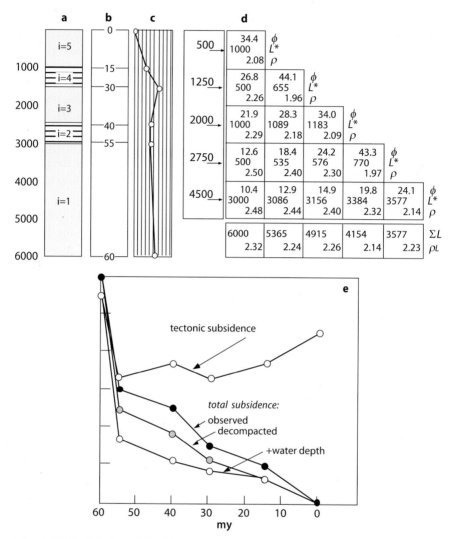

Figure D.9. Solution of Problem 6.6

Problem 6.11. Using the parameter values from Table 5.3 and eq. 6.28 gives: $Ar^{(T=400)} \approx 0.27$ and $Ar^{(T=400)} \approx 15$. Thus, deformation in the "colder" orogen will be concentrated around the collision zone establishing substantial topography and potential energy contrasts, while deformation in the "warmer" orogen will be dissipated over a large region with little topography. However, note that the Argand number is used in geodynamics in order to *avoid* the use of ill-constrained material constants and temperatures in as much detail as we do here. Thus, it is not generally meaningful to use examples like this one for any quantitative consideration of a real orogen.

Problem 6.12. The lithostatic component of pressure in 10 and 15 km depth will be $\sigma_{zz} = \rho g z = 275$ and 321 MPa, respectively. The temperature at this depth is 300 °C and 450 °C, respectively. The strain rate is: $1/5$ my^{-1}. From the power law we get differential stresses of 824 MPa and 40 MPa, respectively. According to eq. 6.30 the pressure component related to differential stress in 10 or 15 km depth 412 MPa or 20 MPa, respectively. This means that the non lithostatic component of pressure in 10 km depth may be larger than the stress exerted by burial. In 15 km depth pressure is roughly equal to the vertical normal stress.

Problem 7.1. a) We can insert into eq. 7.1 the values: $z = 10\,000$ m; $P = 1$ atm $\approx 10^5$ Pa und $g \approx 10$ m s^{-2}. This gives a mean density of $\rho = 1$ kg m^{-3}.
b) 1 atm ≈ 1 bar $= 10^{-3}$ kbar.

Problem 7.2. One mole of almandine has the volume: $V_{almandine} = 11.5$ J $/ 10^5$ Pa $= 1.15 \cdot 10^{-4}$ m$^3 = 115$ cm^3.

Problem 7.3. The cation diffusivities at the 3 temperatures are: $D_{300} \approx 1.6 \cdot 10^{-30}$ m^2 s^{-1}; $D_{500} \approx 6.9 \cdot 10^{-25}$ m^2 s^{-1}; $D_{1000} \approx 1.5 \cdot 10^{-18}$ m^2 s^{-1}. All these are many orders of magnitude smaller than the diffusivity of thermal energy. Thus, chemical zoning profiles in minerals will only develop because of the sluggishness of cation diffusion, but not because temperature gradients in a crystal.

Problem 7.4. Similar to problem 7.3 we can use the Arrhenius relationship to obtain a diffusivity of $D_{600} \approx 4.9 \cdot 10^{-23}$ m^2 s^{-1}. Inserting this diffusivity and $l = 1$ mm (being careful to use SI units!) we obtain a time scale of more than 600 million years. This is much longer than the duration of most metamorphic events. We can conclude that 1 mm sized garnets are unlikely to be equilibrated at 600 °C by diffusion. Chemically equilibrated garnets of this size must have grown at this temperature.

Problem 7.5. The mean diffusivity between 700 °C and 400 °C is: $\overline{D} \approx 1.1 \cdot 10^{-22}$ m^2 s^{-1}.

Problem 7.6. a) The thermal maximum. b) Because the activation volume in eq. 7.4 is much smaller than the activation energy Q. c) The uppermost curve in Fig. 7.4a. d) Yes, for example if crustal shortening follows intrusion and contact metamorphism.

Problem 7.7. Inserting the given values into eq. 7.9 gives a cooling rate: $s \approx 42$ °C my^{-1}.

Problem 7.8. a) The characteristic time sale of cooling (the thermal time constant) of the dike is: $4 \cdot 10^{12}$ s. The diffusivities at T_i and T_b are $D_{1200} \approx 1.3 \cdot 10^{-14}$ m^2 s^{-1} and $D_{500} \approx 7.4 \cdot 10^{-20}$ m^2 s^{-1}. The diffusive time constants of mica are $\tau_{1200} \approx 1.9 \cdot 10^9$ s and $\tau_{500} \approx 3.35 \cdot 10^{14}$ s. Thus, the chemical equilibration of the micas is much more rapid than the cooling of the dike at 1 200 °C but much slower than the cooling at 500 °C. Thus, it may be - in principal - meaningful to use the micas to determine the intrusion age geochronologically. However, 50 m from the intrusion contact the maximum contact metamorphic temperatures are much lower than 1 200 °C. Thus it is necessary to obtain better information on the contact metamorphic temperatures to investigate if the dating of these micas is meaningful. b) Eq. 3.91 shows that the maximum contact metamorphic temperature of the micas is reached $2.83 \cdot 10^{11}$ s after intrusion. Inserting this into eq. 3.89 we get the contact metamorphic peak temperature the micas experienced to be: $T_{max} = 829$ °C. Using this number as the starting temperature T_A and 500 °C as the final temperature T_E, eq. 7.8 may be used to estimate the mean diffusivity to be: $\overline{D} \approx 1.91 \cdot 10^{-17}$ m^2 s^{-1} and $\overline{\tau} = 1.3 \cdot 10^{12}$ s. This value is only a bit shorter that the characteristic time scale of cooling of the dike. Thus, equilibration of the micas is likely to be at least affected by the contact metamorphism. Whether the micas are partially or fully equilibrated can not be answered and requires careful analysis.

Problem 7.11. The T-t-curve passes following points: 700 °C; 0 my $\rightarrow$ 700 °C; 3 my $\rightarrow$ 400 °C; 6 my $\rightarrow$ 0 °C; 26 my. However, the last point will not be reached as the cooling curve intersects the stable geotherm before that. The corner points of the z-t-curve are: 20 km; 0 my $\rightarrow$ 14 km; 5 my $\rightarrow$ 0 km; 40 my. The intersection of the cooling curve with the stable geotherm may be found by finding the intersection of the linear equations $T = z \times 20$ °C / km and $T = 280$ °C - $z \times 50$ °C / km. It is at: $T = 187$ °C and $z = 9\,333$ m. According to these results, the P-T-path follows the following evolution: 5.4 kbar; 700 °C $\rightarrow$ 4.43 kbar; 700 °C $\rightarrow$ 3.78 kbar; 500 °C $\rightarrow$ 3.67 kbar; 400 °C $\rightarrow$ 2.52 kbar; 187 °C $\rightarrow$ 0 kbar; 0 °C.

Problem A.1. The solution of this problem is simply the sum of the approximations from eq. A.13 and eq. A.16:

$$T_i^+ = T_i^- \left(1 - \frac{2\kappa \Delta t}{\Delta z^2} - u\Delta t \right) + T_{i+1} \left(\frac{2\kappa \Delta t}{\Delta z^2} + u\Delta t \right) + T_{i-1} \left(\frac{2\kappa \Delta t}{\Delta z^2} \right)$$

Problem A.2. The answer is:

$$w_i = \frac{-D}{6D + \Delta x^4 (\rho_m - \rho_c)g} (w_{i+1} - 4w_{i+1} - 4w_{i-1} + w_{i-2})$$

This equation was derived by forming the difference between the curvature at neighboring grid points in an analogous way to the way eq. A.16 was formulated.

Problem A.3. Using $u = \Delta x / \Delta t$ forward and backward finite differencing schemes give the following approximations: $T_i^+ = T_{i+1}^-$ bzw. $T_i^+ = 2T_i^- -$

T^-_{i-1}. If you draw these equations according to the scheme of Fig. A.5, it becomes visible that backward differencing schemes become unstable.

Problem A.4. Both data sets have a mean of $\mu = 8$. The first data set hgas a variance of $Var = 4$, but an unbiased estimate of the population variance of $Var' = 4.57$. For the second data set the variance is $Var = 42$, but the unbiased estimate of the population variance is $Var' = 48$.

E. Internet Addresses

On the next few pages there is a random selection of internet addresses that are in some way related to geodynamics. The list was not created according to any important selective criteria and obviously claims no completeness. It is merely a number of addresses that were encountered during the writing of this book.

Software and Online Tools:

ANSYS. Commercial FEM package
- http://www.ansys.com/

DimensionMG. General purpose 3D mesh generator
- http://www.ahpcrc.org/ johnson/SOFTWARE/MESHGEN/

ESRI. widely used GIS software and data depot
- http://www.esri.com/

GeoVu. Software to simplify access to diverse data
- http://www.ngdc.noaa.gov/seg/geovu/

GMT. Software to produce maps from digital data (e.g. Fig. 1.2, 1.1)
- http://gmt.soest.hawaii.edu/

MARC. Commercial FEM package
- http://www.marc.com/

MATLAB. Software to do mathematics
- http://www.mathworks.com/products/matlab/

MICRODEM. Software for imaging of digital elevation models
- http://www.usna.edu/Users/oceano/pguth/website/microdemdown.htm

Online topographic map creation facility
- http://www.aquarius.geomar.de/omc/

Online paleogeographic map creation (Fig. 2.3)
- http://www.odsn.de/odsn/services/paleomap/paleomap.html

SCOTESE. Plate and climate reconstruction
- http://www.scotese.com/

Spectral analysis of data toolkit
- http://www.atmos.ucla.edu/tcd/ssa/

Data Centers, Downloads and Support:

Bathymetry, topography
- http://www.ngdc.noaa.gov/mgg/bathymetry/relief.html

Bathymetric chart of the arctic ocean
- http://www.ngdc.noaa.gov/mgg/bathymetry/arctic/arctic.html

Bathymetric chart of the oceans
- http://www.ngdc.noaa.gov/mgg/gebco/gebco.html

Digital elevation models for the US
- http://mcmcweb.er.usgs.gov

Digital elevation model Etopo5
- http://www.ngdc.noaa.gov/mgg/global/seltopo.html
- http://www.ncgia.ucsb.edu/pubs/spherekit/platforms.html

Digital elevation model Gtopo30
- http://edcdaac.usgs.gov/gtopo30/gtopo30.html

Digital elevation model Global Topo 2min
- ftp://topex.ucsd.edu/pub/global_topo_2min/

Digital elevation model of the US
- http://edcwww.cr.usgs.gov/Webglis/glisbin/guide.pl/glis/hyper/guide/7_min_c

Earth sciences websites
- http://www.websites.noaa.gov/guide/sciences/earth/earth.html

Earthquake catalogue
- http://quake.geo.berkeley.edu/cnss/

Earthquake Database
- http://www.ngdc.noaa.gov/seg/hazard/sig_srch.shtml

Geographic data download of USGS.
- http://edcwww.cr.usgs.gov/doc/edchome/ndcdb/ndcdb.html

Geological time table
- http://www.dinosauria.com/dml/history.htm

Geological time table
- http://www.ucmp.berkeley.edu/help/timeform.html

Geomagnetics
- http://www.noaa.gov/geomagnetics.html

Geophysical data base of the USGS
- http://crustal.cr.usgs.gov/crustal/geophysics/index.html

Geothermal heat flow data
- http://www.noaa.gov/geothermal.html

GIS data depot
- http://www.gisdatadepot.com/

Gravity land
- http://www.noaa.gov/landgravity.html

Gravity data marine (Sandwell)
- http://topex.ucsd.edu/marine_grav/mar_grav.html

Gravity data global
- ftp://topex.ucsd.edu/pub/global_grav_2min/

Gravimetric data base
- http://bgi.cnes.fr:8110/bgi_service_a.html

Heavens above cities
- http://www.heavens-above.com/countries.asp

IERS. International earth rotation service
- http://maia.usno.navy.mil/

IGS. International GPS service
- http://tonga.unavco.ucar.edu/

International GPS support association
- http://igscb.jpl.nasa.gov/

Landform atlas of the US
- http://fermi.jhuapl.edu/states/states.html

Marine geology and geophysics data bases
- http://www.ngdc.noaa.gov/mgg/mggonline.html

National geophysical data center of the US
- http://www.ngdc.noaa.gov/ngdc.html

Natural hazards
- http://www.noaa.gov/hazards.html

Satellite images
- http://earth.jsc.nasa.gov/
- http://visibleearth.nasa.gov/
- http://nix.nasa.gov/
- http://www.earthkam.ucsd.edu/
- http://www.spaceimaging.com/index_text.html

Seafloor topography from satellite
- http://www.ngdc.noaa.gov/mgg/announcements/announce_predict.html

Sediment thicknesses around the world
- http://www.ngdc.noaa.gov/mgg/sedthick/sedthick.html

Seismologic data bases
- http://www.seismology.harvard.edu/CMTsearch.html

Shoreline data base
- http://www.ngdc.noaa.gov/mgg/shorelines/gshhs.html

Shoreline of the United States
- http://www.csc.noaa.gov/products/shorelines/digdata.htm

Statistic data of the world
- http://www.un.org/Depts/unsd/

Topography analysis tools
- http://www.bell-labs.com/project/topo/

World stress map project (Fig. 2.2)
- http http://www-wsm.physik.uni-karlsruhe.de/

World fact book of the CIA
- http://www.odci.gov/cia/publications/factbook/index.html

World maps
- http://ortelius.maproom.psu.edu/dcw/

Important Earth Science Journals:

Links to a range of journals
- http://www.earth.monash.edu.au/Journals.html

Australian Journal of Earth Science (Blackwell)
- http://www.blacksci.co.uk/products/journals/xajes.htm

Computers and Geoscience (Elsevier)
- http://www.elsevier.com/inca/publications/store/3/9/8/

Earth and Planetary Science Letters (Elsevier)
- http://www.elsevier.nl:80/inca/publications/store/5/0/3/3/2/8/

Earth Science Reviews (Elsevier)
- http://www.elsevier.com/inca/publications/store/5/0/3/3/2/9/

Electronic Geosciences
- http://link.springer.de/link/service/journals/10069/index.htm

EOS
- http://www.agu.org/pubs/eos.html

Geochemistry, Geophysics, Geosystems (electronic journal)
- http://146.201.254.53/

Geophysical Journal International (Blackwell)
- http://www.blacksci.co.uk/products/journals/gji.htm

Geophysical Research Letters
- http://www.agu.org/grl/

GSA Bulletin, Geology, GSA today
- http://www.geosociety.org/pubs/

Journal of Applied Geophysics (Elsevier)
- http://www.elsevier.com/inca/publications/store/5/0/3/3/3/3/

Journal of Geophysical Research
- http://www.agu.org/journals/jb/

Journal of Metamorphic Geology (Australian site)
- http://www.es.mq.edu.au/jmg/jmg.html

Journal of African Earth Science (Elsevier)
- http://www.elsevier.com/inca/publications/store/6/9/1/

Journal of Geodynamics (Elsevier)
- http://www.elsevier.com/inca/publications/store/8/7/4/

Journal of Petrology (Elsevier)
- http://petrology.oupjournals.org/

Journal of Structural Geology (Elsevier)
- http://www.elsevier.nl:80/inca/publications/store/5/3/9/

Journal of the Virtual Explorer (electronic journal)
- http://www.virtualexplorer.com.au/

Mineralogy and Petrology
- http://link.springer.de/link/service/journals/00710/index.htm

Physics of the Earth and Planetrary Interiors
- http://www.elsevier.com/inca/publications/store/5/0/3/3/5/6/

Precambiran Research (Elsevier)
- http://www.elsevier.com/inca/publications/store/5/0/3/3/5/7/

Reviews in Geophysics
- http://www.agu.org/rog/

Tectonics
- http://www.agu.org/journals/tc/

Important Geological Organisations:

American Geophysical Union (AGU)
- http://www.agu.org/

Austrian Geological Survey
- http://www.geolba.ac.at/

Austrian portal to academic institutions home pages
- http://www.portal.ac.at/

Australian earth science departments
- http://www.earth.monash.edu.au/aus_depts.html

Australian Geological Survey
- http://www.agso.gov.au/

British Geological Society
- http://www.geolsoc.org.uk

Geological Society of America
- http://www.geosociety.org/

German portal to academic Institutions Home pages
- http://www.bildungsserver.de/

International Earth Science Departments and Organsiations
- http://dir.yahoo.com/Science/Earth_Sciences/Geology_and_Geophysics/Institutes/

NASA
- http://www.nasa.gov/NASA_homepage.html/

Swiss portal to academic institutions home pages
- http://www.switch.ch/

European geoscience department listing
- http://www.uni-mainz.de/FB/Geo/Geologie/GeoInst/Europa.html

References

Abramowitz M, Stegun IA (1972) Handbook of mathematical functions with formulas, graphs mathematical tables. Dover Publications, New York, 1045 p

Ahnert F (1970) Functional relationships between denudation, relief and uplift in large mid-latitude drainage basins. Am J Sci 268:243–263

Ahnert F (1976) Brief description of a comprehensive three-dimensional process-response model of landform development. Z Geomorphol Suppl 25:29–49

Ahnert F (1984) Local relief and the height limits of mountain ranges. Am J Sci 284:1035–1055

Albers HC (1805) Beschreibung einer neuen Kegelprojektion. Zachs monatliche Korrespondenz zur Beförderung der Erd- und Himmelskunde, Nov:450–459

Allen PA, Allen JR (1990) Basin analysis. Principles and applications. Blackwell Scientific Publication, Oxford, 450 p

Anderson DA, Tannehill JC, Pletcher RH (1984) Computational fluid mechanics and heat transfer. Series in computational methods in mechanics and thermal sciences. McGraw Hill, New York, 599 p

Anderson EM (1951) The dynamics of faulting. Oliver and Boyd, Edinburgh, 206 p

Anderson GM, Crerar DA (1993) Thermodynamics in geochemistry, the equilibrium model. Oxford University Press, Oxford, 588 p

Andrews DJ, Bucknam RG (1987) Finite degradation of shoreline scarps by a nonlinear diffusion model. J Geophys Res 92:12857–12867

Angelier J (1984) Tectonic analysis of fault slip data sets. J Geophys Res 89:5835–5848

Angelier J (1994) Fault slip analysis and palaeo-stress reconstruction. In: Hancock P (ed) Continental deformation. Pergamon Press, Oxford, pp 53–101

Angevine CL, Heller PL, Paola C (1990) Quantitative sedimentary basin modelling. AAPG Continuing Education Course Note Series 32, 133 p

Argus DF, Heflin MB (1995) Plate motion and crustal deformation estimated with geodetic data from the Global Positioning System. Geophys Res Lett 22:1973–1976

Armstrong AC (1980) Soils and slopes in a humid environment. Catena 7:327–338

Artyushkov EV (1973) Stresses in the lithosphere caused by crustal thickness inhomogenities. J Geophys Res 78:7675–7708

Atkins PW (1994) Physical chemistry, 5th edition. Oxford University Press, Oxford, 1031 p

Avigard D (1992) On the exhumation of coesite-bearing rocks in the Dora-Maira massif (Western Alps, Italy). Geology 18:466–469

Avouac JP (1993) Analysis of scarp profiles: evaluation of errors in morphological dating. J Geophys Res 98:6745–6754

Avouac JP, Peltzer G (1993) Active tectopnics in southern Xinjiang, China: Analysis of terrace riser and normal fault scarp degradation along the Hotan-Qira Fault System. J Geophys Res 98:21773–21807

Barr TD, Dahlen FA (1989) Brittle frictional mountain building 2nd thermal structure and heat budget. J Geophys Res 94:3923–3947

Barr TD, Houseman GA (1996) Deformation fields around a fault embedded in a non-linear ductile medium. Geophys J Int 125:473–490

Barrell J (1914) The strength of the Earth's crust. J Geol 22:441–468

Barton CM, England PC (1979) Shear heating at the Olympos (Greece) thrust and the deformation properties of carbonates at geological strain rates. Geol Soc Am Bull 90:483–492

Barton MD, Hanson RB (1989) Magmatism and the development of low pressure metamorphic belts: implications from the western United States and thermal modelling. Geol Soc Am Bull 101:1051–1065

Bassi G (1991) Factors controlling the style of continental rifting: insights from numerical modelling. Earth Planet Sci Lett 105:430–452

Bassi G, Keen CE, Potter P (1993) Contrasting styles of rifting: models and examples from the eastern Canadian margin. Tectonics 12:639–655

Batt GE, Braun J (1997) On the thermomechanical evolution of compressional orogens. Gephys J Int 128:364–382

Beaumont C (1981) Foreland basins. Geophys J Roy Astron Soc 65:291–329

Beaumont C, Fullsack P, Hamilton J (1992) Erosional control of active compressional orogens. In: McClay KR (ed) Thrust tectonics. Chapman and Hall, New York, pp 1–18

Beaumont C, Ellis S, Hamilton J, Fullsack P (1996) Mechanical model for subduction-collision tectonics of Alpine-type compressional orogens. Geology 24: 675–678

Begin SB, Meyer DF, Schumm SA (1981) Development of longitudinal profiles of alluvial channels in response to base level lowering. Earth Surface Processes and Landforms 6:49–68

Bell JS, Gough DI (1979) Northwest-southeast compressive stress in Alberta: evidence from oil wells. Earth Planet Sci Lett 45:475–482

Benfield AE (1949) The effect of uplift and denudation on underground temperatures. J Appl Phys 20: 66–70

Berman RG (1988) Internally-consistent thermodynamic data for minerals in the system Na_2O-K_2O-CaO-MgO-FeO-Fe_2O_3-Al_2O_3-SiO_2-TiO_2-H_2O-CO_2. J Pet 29: 445–522

Bickle MJ, McKenzie D (1987) The transport of heat and matter by fluids during metamorphism. Contrib Mineral Petrol 95:384–392

Bickle MJ, Hawkesworth CJ, England PC, Athey D (1975) A preliminary thermal model for regional metamorphism in the eastern Alps. Earth Planet Sci Lett 26:13–28

Bird P (1979) Continental delamination and the Colorado Plateau. J Gephys Res 84:7561–7571

Bird P, Piper K (1980) Plane stress finite element model of tectonic flow in southern California. Phys Earth Planet Int 21:158–195

Blanckenburg F von, Davies JH (1995) Slab breakoff: a model for syncollisional magmatism and tectonics in the Alps. Tectonics 14:120–131

Bohlen SR (1987) Pressure temperature time paths and a tectonic model for the evolution of granulites. J Geol 95:617–632

Bond GC, Kominz MA, Devlin WJ (1983) Thermal subsidence and eustasy in the lower Paleozoic miogeocline of western North America. Nature 306:775–779

Bons PD, Barr TD, ten Brink CE (1997) The development of delta-clasts in non-linear viscous materials: a numerical approach. Tectonophys 270:29–42

Bott MHP (1993) Modelling the plate-driving mechanism. J Geol Soc London 150:941–951

Bott MHP, Waghorn GD, Whittaker A (1989) Plate boundary forces at subduction zones and trench-arc compression. Tectonophys 170:1–15

Brace WF, Kohlstedt DL (1980) Limits on lithospheric stress imposed by laboratory experiments. J Geophys Res 94:3967–3990

Braun J (1992) Dynamics of compressional orogens; beyond thin sheet and plane strain approximations. EOS tranctions 73:292

Braun J, Beaumont C (1995) Three dimensional numerical experiments of strain partitioning at oblique plate boundaries: Implications for contrasting tectonic styles in the southern coast ranges, California central south island, New Zealand. J Geophys Res 100:18059–18074

Braun J, Sambridge M (1997) Modelling landscape evolution on geological time scales: a new method based on irregular spatial discretization. Basin Res. 9:27–52

Brown RW (1991) Backstacking apatite fission track "stratigraphy": a method for resolving the erosional and isostatic rebound components of tectonic uplift histories. Geology 19:74–77

Brun JP, Cobbold PR (1980) Strain heating and thermal softening in continental shear zones: a review. J Struct Geol 2:149–158

Buck WR (1991) Modes of continental lithosphere extension. J Geophys Res 96:20161–20178

Buck WR, Martinez F, Steckler MS, Cochran JR (1988) Thermal consequences of lithosphere extension: pure and simple. Tectonics 7:213–234

Burbank DW, Anderson RS (2001) Tectonic geomorphology. Blackwell Science, Malden, Mass 274 p

Byerlee JD (1968) Brittle ductile transition in rocks. J Geophys Res 73:4741–4750

Byerlee JD (1970) Friction of rocks. In: Everden JF (ed) Experimental studies of rock friction with application to earthquake prediction. USGS, Menlo Park, pp 55–77

Carey SW (1976) The expanding Earth. Elsevier, Amsterdam, 488 p

Carslaw HS, Jaeger JC (1959) Conduction of heat in solids. Oxford Science Publications, Oxford University Press, Oxford, 510 p

Carson CJ, Powell R, Wilson CJL, Dirks PHMG (1997) Partial melting during tectonic exhumation of a granulite terrain: an example from the Larsemann Hills, East Antarctica. J Met Geol 15:105–127

Carson MA, Kirkby MJ (1972) Hillslope form and processes, Cambridge University Press, Cambridge, 475 p

Chamberlain CP, Sonder LJ (1990) Heat-producing elements and the thermal and baric patterns of metamorphic belts. Science 250:763–769

Chapple WM (1978) Mechanics of thin-skinned fold-and-thrust belts. Geol Soc Am Bull 89:1189–1198

Chase CG (1992) Fluvial landsculpting and the fractal dimension of topography. Geomorphology 5:39–57

Chemenda A, Mattauer M, Bokun AN (1996) Continental subduction and a mechanism for exhumation of high-pressure metamorphic rocks: new modelling and field data from Oman. Earth Planet Sci Lett 143:173–182

Christensen UR, Yuen DA (1984) The interaction of a subducting lithospheric slab with a chemical or phase boundary. J Geophys Res 89:4389–4402

Christie JM, Ord A (1980) Flow stress from microstructures of mylonites: example and current assessment. J Geophys Res 85:6253–6262

Cliff RA, Droop GTR, Rex DC (1985) Alpine metamorphism in the south-east Tauern Window, Austria: 2nd Rates of heating, cooling and uplift. J Met Geol 3:403–415

Cloetingh S, van Wees JD, van der Beek PA, Spadini G (1995) Role of pre-rift rheology in kinematics of extensional basin formation: constraints from thermo-mechanical models of Mediterranean and intracratonic basins. Mar Petrol Geol 12:793–807

Cloos M (1982) Flow melanges: numerical modelling and geologic constraints on their origin in the Franciscan subduction complex, California. Bull Geol Soc Am 93:330–345

Cloos M (1984) Flow melanges and the structural evolution of accretionary wedges. Geol Soc Am Spec Pap 198:71–79

Cloos M, Shreve RL (1988) Subduction channel model of accretion, melange formation, sediment subduction and subduction erosion at convergent plate margins 1. Background and description. Pure and Applied Geophys 128:455–500

Coblentz D, Richardson RM, Sandiford M (1994) On the potential energy of the Earth's lithosphere. Tectonics 13:929–945

Coblentz D, Sandiford M (1994) Tectonic stress in the African plate: constraints on the ambient lithospheric stress state. Geology 22:831–834

Cochran JR (1982) The magnetic quite zone in the eastern gulf of Aden: implications for the early development of the continental margin. Geophys J Roy Astron Soc 68:171–202

Cochran JR (1983) Effects of finite extension times on the development of sedimentary basins. Earth Planet Sci Lett 66:289–302

Coffin MF, Eldholm O (1994) Large igneous provinces: crustal strucutre, dimensions, and external consequences. Rev Geophys 32:1–36

Coffin MF (1997) Models for the emplacement of large igenous provinces; a review. In: International lithosphere program on volcanic margins, abstract volume, Geoforschungszentrum Potsdam.

Coffin MF, Eldholm O (1993a) Scratching the surface: Estimating dimensions of large igeneous provinces. Geology 21:515–518

Coffin MF, Eldholm O (1993b) Large igneous provinces. Scientific American 269: 42–49

Connolly JAD, Thompson AB (1989) Fluid and enthalpy production during regional metamorphism. Contrib Mineral Petrol 102:347–366

Conrad CP, Molnar P (1999) Convective instability of a boundary layer with temperature- and strain-rate-dependent viscosity in terms of "abvailable buoyancy". Geophys J Int 139:51–68

Cooke RA, OBrian PJ, Carswell DA (2000) Garnet zoning and the identification of equilibrium mineral compositions in high pressure-temperature granulites from the Moldanubian zone, Austria. J Metam Geol 18:551–569

Coulomb CA (1773) Sur une application des régles de maximis et minimis a quelques problémes de statique relatifs a l'árchitecture. Acad Roy Sci Mem de Math et de Phys 7:343–382

Cowan DS, Silling RM (1978) A dynamic scaled model of accretion and trenches and its implications for the tectonic evolution of subduction complexes. J Geophys Res 83:5389–5396

Cox A (1972) Plate tectonics and geomagnetic reversals. Freeman and Company, San Francisco, 702 p

Crank J (1975) The mathematics of diffusion, 2nd edn. Oxford Science Publications. Clarendon Press, Oxford, 414 p

Creager KC, Jordan TH (1984) Slab penetration into the lower mantle. J Geophys Res 89:3031–3050

Crough ST (1983) Hot spot swells. Ann Rev Earth Planet Sci 11:165–193

Cull JP (1976) The measurement of thermal parameters at high pressures. Pageoph 114:301–307

Culling WEH (1960) Analytical theory of erosion. J Geol 68:336–344

Cygan RT, Lasaga AC (1985) Self-diffusion of magnesium in garnet at 750°C to 900°C. Am J Sci 285:328–350

Dahlen FA (1984) Non-cohesive critical Coulomb wedges: an exact solution. J Geophys Res 89:10125–10133

Dahlen FA, Barr TD (1989) Brittle frictional mountain building 1st Deformation and mechanical energy budget. J Geophys Res 94:3906–3922

Dahlen FA, Suppe J, Davis D (1984) Mechanics of fold-and-thrust belts and accretionary wedges: cohesive coulomb theory. J Geophys Res 89:10087–10101

Dalmayrac B, Molnar P (1981) Parallel thrust and normal faulting in Peru and constraints on the state of stress. Earth Planet Sci Lett 55:473–481

Davis D, Suppe J, Dahlen FA (1983) Mechanics of fold-and-thrust belts and accretionary wedges. J Geophys Res 88:1153–1172

DeMets C, Gordon RG, Argus DF, Stein S (1990) Current plate motions. Geophys J Int 101:425–478

Dewey JF (1988) Extensional collapse of orogens. Tectonics 7:1123–1139

DeYoreo JJ, Lux DR, Guidotti CV (1991) Thermal modelling in low-pressure/high-temperature metamorphic belts. Tectonophys 188:209–238

Dickinson WR (1976) Plate tectonic evolution of sedimentary basins. AAPG Continuing Education Course Note Series 1:1–62

Dodson MH (1973) Closure temperature in cooling geochronological and petrological systems. Contrib Mineral Petrol 40:259–274

Doin, Fleitout (1996) Thermal evolution of the oceanic lithosphere: an alternative view. Earth Planet Sci Lett 142:121–136

Doglioni C (1993) Some remarks on the origin of foredeeps. Tectonophys 228:1–20

Dunning GH, Etheridge MA, Hobbs BE (1982) On the stress dependence of subgrain size. Textures and Microstructures 5:127–152

Edmond JM, Damm K von (1983) Heiße Quellen am Grund der Ozeane. Spektrum der Wissenschaft 6:74–87

Ehlers K, Powell R (1994) An empirical modification of Dodson's equation for closure temperature in binary systems. Geochim Cosmochim Acta 58:241–248

Ehlers K, Powell R, Stüwe K (1994a) The determination of cooling histories from garnet-biotite equilibrium. Am Min 79:737–744

Ehlers K, Stüwe K, Powell R, Sandiford M, Frank W (1994b) Thermometrically inferred cooling rates from the Plattengneiss, Koralm region, eastern Alps. Earth Planet Sci Lett 125:307–321

Elison MW (1991) Intracontinental contraction in western North America: continuity and episodicity. Geol Soc Am Bull 103:1226–1238

Engelder T (1993) Stress regimes in the lithosphere. Princeton University Press, Princeton, 451 p

Engelder T (1994) Deviatoric stressitis: a virus infecting the Earth science community. EOS 75:18, 210–213

England PC (1981) Metamorphic pressure estimates and sediment volumes for the Alpine orogeny: an independent control on geobarometers? Earth Planet Sci Lett 56:387–397

England PC (1987) Diffusive continental deformation: length scales, rates and metamorphic evolution. Phil Trans R Soc Lon 321:3–22

England PC (1996) The mountains will flow. Nature 381:23–24

England PC, Holland TJB (1979) Archimedes and the Tauern eclogites: the role of buoyancy in the preservation of exotic tectonic blocks. Earth Planet Sci Lett 44:287–294

England PC, Houseman G (1986) Finite strain calculations of continental deformation 2nd Comparison with the India-Asia collision zone. J Geophys Res 91:3664–3676

England PC, Houseman GA (1988) The mechanics of the Tibetan plateau. Phil Trans Roy Soc Lon A326:301–319

England PC, Houseman G (1989) Extension during continental convergence, with application to the Tibetan plataeu. J Geophys Res 94:17561–17579

England PC, Jackson J (1989) Active deformation of the continents. Ann Rev Earth Planet Sci 17:197–226

England PC, McKenzie D (1982) A thin viscous sheet model for continental deformation. Geophys J Roy Astron Soc 70:295–321

England PC, Molnar P (1990) Surface uplift, uplift of rocks and exhumation of rocks. Geology 18:1173–1177

England PC, Molnar P (1991) Inferences of deviatoric stress in actively deforming belts from simple physical models. Phil Trans R Soc Lond 337:151–164

England PC, Molnar P (1993) The interpretation of inverted metamorphic isogrades using simple physical calculations. Tectonics 12:145–157

England PC, Richardson SW (1977) The influence of erosion upon the mineral facies of rocks from different metamorphic environments. J Geol Soc London 134:201–213

England PC, Thompson A (1984) Pressure-temperature-time paths of regional metamorphism I. Heat transfer during the evolution of regions of thickened continental crust. J Pet 25:894–928

England PC, Houseman GA, Sonder LJ (1985) Length scales for continental deformation in convergent, divergent and strike-slip environments: analytical and approximate solutions for a thin viscous sheet model. J Geophys Res 90:3551–3557

Ernst WG (1971) Do mineral paragenesis reflect unusually high-pressure conditions of Franciscan metamorphism? Am J Sci 270:81–108

Evenden G (1990) Cartographic projection procedures for the UNIX environment – a user's manual. USGS Open File Report 90–284

Fischer GW (1973) Non-equilibrium thermodynamics as a model for diffusion controlled metamorphic processes. Am J Sci 273:897–924

Fleitout L, Froidvaux C (1982) Tectonics and topography for a lithosphere containing density heterogeneities. Tectonics 1:21–57

Fletcher CAJ (1991) Computational techniques for fluid dynamics 1st Fundamental and general techniques, 2nd edn. Springer series in computational physics. Springer, Berlin Heidelberg New York, 401 p

Forsyth DW (1985) Subsurface loading and estimates of the flexural rigidity of continental lithosphere. J Geophys Res 90:12623–12632

Forsyth DW, Uyenda S (1975) On the relative importance of the driving forces of plate motion. Geophys J Roy Astron Soc 43:163–200

Fortier SM, Giletti BJ (1991) Volume self diffusion of oxygen in biotite, muscovite and phlogopite micas. Geochim Cosmochim Acta 55:1319–1330

Fourier JBJ (1816) Theorie de la chaleur. Annales de Chimie et de Physique 3:350–376

Fourier JBJ (1820) Extrait d'un memoire sur le refroidissement seculaire du globe terrestre. Annales Chimie et de Physique 13:418–437

Fowler CMR (1990) The solid Earth. An introduction to global geodynamics. Cambridge University Press, Cambridge, 472 p

Fowler CMR, Nisbet EG (1982) The thermal background to metamorphism II. Simple two-dimensional conductive models. Geoscience Canada 9:208–214

Frohlich C, Coffin MF, Massell C, Mann P, Schuur CL, Davis SD, Jones T, Karner G (1997) Constraints on Macquarie Ridge tectonics provided by Harvard focal mechanisms and teleseismic earthquake locations. J Geophys Res 102:5029–5041

Frost HJ, Ashby MF (1982) Deformation-mechanism maps. Pergamon Press, Cambridge, 260 p

Frottier LG, Buttles J, Olson P (1995) Laboratory experiments on the structure of subducted lithosphere. Earth Planet Sci Lett 133:19–34

Gacia-Castellanos D, Fernandez M, Torne M (1997) Numerical modeling of foreland basin formation: a program relating subsidence, thrusting, sediment geometry and depth dependent lithosphere rheology. Computers and Geoscience 23:993–1003

Genser J, Neubauer F (1989) Low angle normal faults at the eastern margin of the Tauern window (Eastern Alps). Mitt Österr Geol Ges 81:233–243

Gilchrist AR, Kooi H, Beaumont C (1994) Post-Gondwana geomorphic evolution of southwestern Africa: implications for the controls on landscape development from observations and numerical experiments. J Geophys Res 99:12211–12228

Gleason GC, Tullis J (1995) A flow law for dislocation creep of quartz aggregates determined with the molton salt cell. Tectonophys 247:1–23

Goetze C (1978) The mechanisms of creep in olivine. Phil Trans Roy Soc London 288:99–119

Goetze C, Evans B (1979) Stress and temperature in the bending lithosphere as constrained by experimental rock mechanics. Geophys J Roy Astron Soc 59:463–478

Gordon RB (1965) Diffusion creep in the Earth's mantle. J Geophys Res 70:56–61

Graham CM, England PC (1976) Thermal regimes and regional metamorphism in the vicinity of overthrust faults: an example of shear heating and inverted metamorphic zonation from southern California. Earth Planet Sci Lett 31:142–152

Grasemann B, Mancktelow NS (1993) Two-dimensional thermal modelling of normal faulting: the Simplon Fault Zone, Central Alps, Switzerland. Tectonophys 225:155–165

Greenfield JE, Clarke GL, White RW (1998) A sequence of partial melting reactions at Mt Stafford, central Australia. J Metam Geol 16:363–378

Greenwood HJ (1989) On models and modelling. Canad Mineral 27:1–14

Griggs D (1939) A theory of mountain building. Am J Sci 237:611–650

Haack U (1983) On the content and vertical distribution of K, Th and U in the continental crust. Earth Planet Sci Lett 62:360–366

Harker A (1939) Metamorphism. Lethuer S. Co., London, 462 p

Harley SL (1989) The origin of granulites: a metamorphic perspective. Geol Mag 126:215–247

Harrison CGA (1994) Rates of continental erosion and mountain building. Geol Rundsch 83:431–447

Harrison TM, Clark GK (1979) A model of the thermal effects of igneous intrusion and uplift as applied to Quottoon pluton, British Columbia. Canad J Earth Sci 16:410–420

Harrison TM, Copeland P, Kidd WSF, Yin A (1992) Raising Tibet. Science 255:1663–1670

Hawkesworth C, Turner S, Gallagher K, Hunter A, Bradshaw T, Rogers N (1995) Calc-alkaline magmatism, lithosphere thinning and extension in the basin and Range. . J Geophys Res 100:10271–10286

Hawkins SW (1988) A brief history of time. Space Time Publications. Guild Publishing, 213 p

Heezen BC (1962) The deep sea floor. In: Runcorn SK (ed) Continental drift. Academic Publishers, New York, pp 235–268

Hess H (1961) History of ocean basins. In: Engle AEJ, Petrologic Studies. A volume in honour of AE Buddington, Geol Soc Am pp 599–620

Hilst R van der, Engdahl ER, Sparkman W, Nolet G (1991) Tomographic imaging of subducted lithosphere below north-west Pacific island arcs. Nature 357:37–43

Hodges KV (1996) Self-organization and the metamorphic evolution of mountain ranges. MSG meeting Kingston University abs

Hoffman PF, Grotzinger JP (1993) Orographic precipitation, erosional unloading and tectonic style. Geology, 21:195–198

Hoffman PF, Kaufman AJ, Halverson GP, Schrag P (1998) A Neoproterozoic snowball Earth. Science 281:1342–1346

Hoke L, Hilton DR, Lamb SH, Hammerschmidt K, Friedrichsen H (1994) ^{3}He evidence for a wide zone of active mantle melting beneath the Central Andes. Earth Planet Sci Lett 128:341–355

Holland TJB, Powell R (1990) An enlarged and updated internally consistent thermodynamic dataset with uncertainties and correlations: the system K_2O-Na_2O-CaO-MgO-FeO-Fe_2O_3-Al_2O_3-TiO_2-SiO_2-C-H_2-O_2. J Met Geol 8:89–124

Holmes A (1929) Radioactivity and Earth movements. Trans Geol Soc Glasgow 18:559–607

Horton RE (1945) Erosional development of streams and their drainage basins: hydrophysical approach to quantitative geomorphology. Bull Geol Soc Am 56:275–370

House MA, Wernicke BP, Farley KA (1998) Dating topography of the Sierra Nevada, California, using apatite (U-Th)/He ages. Nature 396:66–69

House MA, Farley KA, Kohn BP (2000) an empirical test of helium diffusion in apatite:borehole data from the Otway basin, Australia. Earth Planet Sci Lett 170:463–474

Houseman G, England P (1986a) Finite strain calculations of continental deformation 1st method and general results for convergent zones. J Geophys Res 91:3651–3663

Houseman G, England P (1986b) A dynamical model of lithosphere extension and sedimentary basin formation. J Geophys Res 91:719–729

Houseman G, Gubbins D (1997) Deformation of subducted oceanic lithosphere. Geophys J Int 131:535–551

Houseman G, Molnar P (1997) Gravitational (Rayleigh-Taylor) instability of a layer with non-linear viscosity and convective thinning of continental lithosphere. Geophys J Int 128:125–150

Houseman G, McKenzie DP, Molnar P (1981) Convective instability of a thickened boundary layer and its relevance for the thermal evolution of continental convergent belts. J Geophys Res 86:6115–6132

Hubbert MK (1948) A line-integral method for computing the gravimetric effects of two-dimensional masses. Geophysics 13:215–225

Huppert HE, Sparks RSJ (1988) The generation of granitic magmas by intrusion of basalt into continental crust. J Pet 29:599–624

Isacks BL, Barazangi M (1977) Geometry of Benioff zones: lateral segmentation and downward bending of the subducted lithosphere. In: Talwani M, Pitman III WC (ed) Island Arcs, deap-sea trenches and back arc basins. Am Geophys U, Maurice Ewing Series Washington 1:99–114

Isacks BL, Oliver J, Sykes LR (1968) Seismology and the new global tectonics. J Geophys Res 73:5855–5899

Issler D, McQueen H, Beaumont C (1989) Thermal consequences of simple shear extension of the continental lithosphere. Earth Planet Sci Lett 91:341–358

Itayama K, Stüwe HP (1974) Mittlere Reaktionsgeschwindigkeit bei zeitlich veränderter Temperatur. Zeitschr Metallk 65:70–72

Jackson ED, Shaw HR (1975) Stress fields in the central portions of the pacific plate:delineated in time by linear volcanic chains. J Geophys Res 80:1861–1874

Jackson JA, McKenzie DP (1988) The relationship between plate motions and seismic moment tensors, and the rates of active deformation in the Mediterranean and the Middle East. Geophys J Roy Astron Soc 93:45–73

Jaeger JC (1964) Thermal effects of intrusions. Rev Geophys 2:443–466

Jaeger JC, Cook NGW (1979) Fundamentals of rock mechanics, 3rd edn. Science Paperbacks, Chapman and Hall, London, 585 p

Jarvis GT, McKenzie DP (1980) Sedimentary basin formation with finite extension rates. Earth Planet Sci Lett 48:42–52

Jaupart C, Provost A (1985) Heat focusing, granite genesis and inverted metamorphic gradients in continental collision zones. Earth Planet Sci Lett 73:385–397

Jeanloz R (1988) High-pressure experiments and the Earth's deep interior. Physics Today 41:44–45

Jeanloz R, Richter FM (1979) Convection, composition and the thermal state of the lower mantle. J Geophys Res 84:5497–5504

Jessel MW, Lister GS (1990) A simulation of the temperature dependence of quartz fabrics. In: Knipe RJ, Rutter EH (eds) Deformation mechanisms, rheology and tectonics. Geol Soc Spec Publ 54:353–362

Joesten R (1977) Evolution of mineral assemblage zoning in diffusion metasomatism. Geochim Cosmochim Acta 41:649–670

Jones RR (1997) Lateral extrusion in transpression zones; the importance of boundary conditions. J Struct Geol 19:1201–1217

Jordan TH (1981) Continents as a chemical boundary layer. Phil Trans Roy Soc Lond A301:359–373

Jordan TE (1981) Thrust loads and foreland basin evolution. Cretaceous, western United States. American Assoc Pet Geol 65:2506–2520

Jordan TH (1989) Chemical boundary layers of the mantle and core. Phil Trans Roy Soc Lond A328:441–442

Julien PY (1998) Erosion and Sedimentation Cambridge University Press, Cambridge UK, 298 p.

Jull M, Keleman PB (2001) On the conditions for lower crustal convective instability. J Geophys Res 106:6423–6446

Karner GD, Watts AB (1983) Gravity anomalies and flexure of the lithosphere at mountain ranges. J Geophys Res 88:10449–10477

Keen CE (1980) The dynamics of rifting: deformation of the lithosphere by active and passive driving mechanisms, Geophys. J Roy Astron Soc 62:631–647

Keen C, Peddy C, de Voogd B, Mathew D (1989) Conjugate margins of Canada and Europe: results from deep reflection profiling. Geology 17:173–176

Keller EA, Pinter N (1996) Active Tectonics - Earthquakes, Uplift and Landscape. Prentice Hall, Upper Saddle River, 338 p.

Kelvin Lord (1864) The secular cooling of the Earth. Trans Roy Soc Edinburgh 23:157

Kincaid C, Silver P (1996) The Role of Viscous Dissipation in the Orogenic Process. Earth Planet Sci Lett 142:271–288

King LC (1953) Canons of landscape evolution. Geol Soc Am Bull 64:721–753

King LC (1983) Wandering continents and spreading sea floors on an expanding Earth. John Wiley and Sons, Chichester, 232 p

King SD (1996) You don't need plumes to make flood basalts; receipes that don't require a plume. EOS 77:768

Kirchner JW (1993) Statistical inevitability of Horton's laws and the apparent randomness of stream channel networks. Geology 21:591–594

Kirkby MJ (ed.) (1994) Process Models and Theoretical Geomorphology. (British Geomorphological research group symposia) John Wiley and Sons. 432p.

Kooi J, Beaumont C (1994) Escarpment evolution on high-elevation rifted margins insights derived from a surface processes model that combines diffusion, advection and reaction. J Geophys Res 99:12191–12209

Koons PO (1990) Two-sided orogen: Collision and erosion from the sandbox to the southern Alps, New Zealand. Geology 18:679–682

Kuznir NJ, Park RG (1986) Continental lithosphere strength: the critical role of lower crustal deformation. In: Dawson JB, Carswell DA, Hall J, Wedepohl KH (ed) The nature of the lower continental crust. Geol Soc Spec Pub 24:79–93

Lachenbruch AH (1968) Preliminary geothermal model for the Sierra Nevada. J Geophys Res 73:6977–6989

Lachenbruch AH (1970) Crusal temperature and heat production: implications of the linear heat flow relationship. J Geophys Res 75:3291–3300

Lachenbruch AH, Bunker CM (1971) Vertical gradients of heat production in the continental crust 2nd Some estimates from borehole data. J Geophys Res 76:3852–3860

Lachenbruch AH (1980) Frictional heating, fluid pressure and the resistance to fault motion. J Geophys Res 85:6097–6112

Lambeck K (1991) Glacial rebound and sea level change in the British Isles. Terra Nova 3:379–389

Lambeck K (1993) Glacial rebound and sea level change, an example of a relationship between mantle and surface processes. Tectonophys 223:15–37

Lambert JH (1772) Beiträge zum Gebrauch der Mathematik und deren Anwendung, Teil III/Abschn. 6: Anmerkungen und Zusätze zur Entwerfung der Land- und Himmelscharten. Berlin

Larson RL (1991) Geological consequences of superplumes. Geology 19:963–966

Lasaga AC (1983) Geospeedometry: an extension of geothermometry. In: Saxena SK (ed) Kinetics and equilibrium in mineral reactions, Springer, Berlin Heidelberg New York, 82–114 p

Le Pichon X (1983) Land-locked oceanic basins and continental collision: the eastern Mediterranean as a case example. In: Hsu KJ (ed) Mountain building processes. Academic, Orlando USA, pp 201–211

Le Pichon X, Francheteau J, Bonnin J (1976) Plate tectonics. Developments in geotectonics (2nd edn. New York, Amsterdam, 311 p

Le Pichon X, Angelieru J, Sibuet JC (1982) Plate boundaries and extensional tectonics. Tectonophys 81:239–256

Lister GS, Etheridge MA (1989) Towards a general model. Detachment models for uplift and volcanism in the eastern highlands and their application to the origin of passive margin mountains. In: Johnson RW, Knutson J, Taylor SR (eds) Intraplate volcanism in eastern Australia and New Zealand. Cambridge Univ. Press. Cambridge, pp 297–313

Lister GA, Etheridge MA, Symonds PA (1986) Detachment faulting and the evolution of passive continental margins. Geology 14:246–250

Lister GA, Etheridge MA, Symonds PA (1991) Detachment models for the formation of passive continental margins. Tectonics 10:1038–1064

Lux DR, DeYoreo JJ, Guidotti CV and Decker ER (1986) Role of plutonism in low pressure metamorphic belt formation. Nature 323:794–796

Lyon-Caen H, Molnar P (1983) Constraints on the structure of the Himalaya from an analysis of gravity anomalies and a fexural model of the lithosphere. J Geophys Res 88:8171–8191

Lyon-Caen H, Molnar P (1989) Constraints on the deep structure and dynamic processes beneath the Alps and adjacent regions from an analysis of gravity anomalies. Geophys J Int 99:19–32

Mackin JH (1948) Concept of the graded river. Bull Geol Soc Am 59:463–512

Mahoney JJ, Coffin MF (1997) (eds) Large igneous provinces; continental oceanic, and planetary flood volcanicsm. Geophysical Monograph 100: 438 p

Malanson GP, Butler DR, Georgakakos KP (1992) Nonequilibrium geomorphic processes and deterministic chaos. Geomorphology 5:311–322

Malniverno A, Pockalny RA (1990) Abyssal hill topography as an indicator of episodicity in crustal accretion and deformation. Earth Planet Sci Lett 99:154–169

Mancktelow NS (1993) Tectonic overpressure in competent mafic layers and the development of isolated eclogites. J Met Geol 11:801–812

Mancktelow NS (1995) Nonlithostatic pressure during sediment subduction and the development and exhumation of high pressure metamorphic rocks. J Geophys Res 100:571–583

Mancktelow NS, Grasemann B (1997) Time-dependent effects of heat advection and topography on cooling histories during erosion. Tectonophys 270:167–195

Mancktelow NS (1999) Finite-element modelling of single-layer folding in elasto-viscous materials: the effect of initial perturbation geometry. J Struct Geol 21:161–177

Mandelbrot BB (1967) How long is the coast of Briotain? Statisitcal self similarity and fractional dimension. Science 156:636-638

Mandelbrot B (1975) Stochastic models for the Earth's relief, the shape and fractal dimension of the coastlines and the number-area rule for islands. Proc Nat Acad Sci, USA 72:3825–3828

Mandelbrot B (1982) The fractal geometry of nature. Freeman, New York, 460 p

Mastin L (1988) Effect of borehole deviation on breakout orientations. J Geophys Res 93:9187–9195

McClay KR (1992) Thrust tectonics. Chapman and Hall, London, 447 p

McDonough WF, Rudnick RL (1998) Mineralogy and composition of the upper mantle. Reviews in Mineralogy 37:139–164

McKenzie DP (1967) Some remarks on heat-flow and gravity anomalies. J Geophys Res 72:6261–6273

McKenzie DP (1969a) The relationship between fault plane solutions for earthquakes and the directions of the principle stresses. Bull Seismol Soc Am 59:591–601

McKenzie DP (1969b) Speculations on the consequences and causes of plate motions. Geophys J Roy Astron Soc 18:1–32

McKenzie DP (1972) Active tectonics in the Mediterranean region. Geophys Roy Astron Soc 30:109–185

McKenzie DP (1977a) Surface deformation, gravity anomalies and convection. Geophys J Roy Astron Soc 48:211–238

McKenzie DP (1977b) The initiation of trenches: a finite amplitude instability. In: Talwani M, Pitman III WC (eds) Island arcs, deap-sea trenches and back arc basins. Am Geophys U, Maurice Ewing Series 1:57–61

McKenzie DP (1978) Some remarks on the development of sedimentary basins. Earth Planet Sci Lett 40:25–32

McKenzie DP (1984) The generation and compaction of partially molten rock. J Pet 25:713–765

McKenzie DP, Bickle MJ (1988) The volume and composition of melt generated by extension of the lithosphere. J Pet 29:625–679

McKenzie DP, Morgan WJ (1969) Evolution of triple junctions. Nature 224:125–133

McKenzie DP, Roberts JM, Weiss NO (1974) Convection in the Earth's mantle: towards a numerical simulation. J Fluid Mech 62:465–538

McLaren S, Sandiford M, Hand M (1999) High radiogenic heat producing granites and metamorphism - an example from the western Mt Isa Inlier, Australia. Geology 27:679–682

Means WD (1976) Stress and strain – basic concepts of continuum mechanics for geologists. Springer, Berlin Heidelberg New York, 339 p

Menard HW (1964) Marine Geology of the Pacific. McGraw Hill, New York, 271 p

Meyerhoff AA (1995) Surge tectonic evolution of southeastern Asia: A geohydrodynamics approach. Journal of SE Asian Earth Science 12:145–247

Michael AJ (1987) The use of focal mechanisms to determine stress: a control study. J Geophys Res 92:357–368

Miyashiro A (1973) Metamorphism and metamorphic belts. Allen and Unwin, New York, 492 p

Mohr O (1900) Welche Umstände bedingen die Elastizitätsgrenze und den Bruch eines Materials? Z Ver dt Ing 44:1524–1530, 1572–1577

Mollweide C (1805) Über die von Prof. Schmidt in Giessen in der zweyten Abteilung seines Handbuches der Naturlehre angegebene Projektion der Halbkugelfläche: Zachs monatliche Korrespondenz 13:152–163

Molnar P (1992) Brace-Goetze strength profile, the partitioning of strike-slip and thrust faulting at zones of oblique convergence and the stress heat flow paradox of the San Andreas Fault. In: Evans B, Wong T (eds) Fault mechanics and transport properties of rocks. A festschrift in honour of WF Brace. Acad. Press, San Diego, pp 435–459

Molnar P, England PC (1990a) Temperatures, heat flux and frictional stress near major thrust faults. J Geophys Res 95:4833–4856

Molnar P, England PC (1990b) Late Cenozoic uplift of mountain ranges and global climatic change: chicken or egg? Nature 346:29–34

Molnar P, England PC (1995) Temperatures in zones of steady-state underthrusting of young oceanic lithosphere. Earth Planet Sci Lett 131:57–70

Molnar P, Gibson JM (1996) A bound on the rheology of continental lithosphere using very long baseline interferometry: the velocity of South China with respect to Eurasia. J Geophys Res 101:545–553

Molnar P, Lyon-Caen H (1988) Some simple physical aspects of the support, structure and evolution of mountain belts. Geol Soc Am Spec Pap 218:179–207

Molnar P, Lyon-Caen H (1989) Fault plane solutions of earthquakes and active tectonics of the Tibetan plateau and its margins. Geophys J Roy Astron Soc 99:123–153

Molnar P, Tapponier P (1975) Cenozoic tectonics of Asia: effects of a continental collision. Science 189:419–426

Molnar P, Tapponier P (1978) Active tectonics of Tibet. J Geophys Res 83:5361–5374

Molnar P, England P, Martinod J (1993) Mantle dynamics, uplift of the Tibetan plateau the Indian monsoon. Rev Geophys 31:357–396

Molnar P, Houseman G, Conrad C (1998) Rayleigh-Taylor instability and convective thinning of mechanically thickened lithosphere: effects of non-linear viscosity decreasing exponentially with depth and of horizontal shortening of the layer. Geophys J Int 133:568–584

Montgomery DR (1994) Valley incision and the uplift of mountain peaks. J Geophys Res 99:13913–13921

Morgan J (1968) Rises, trenches, great faults and crustal blocks. J Geophys Res 73:1959–1982

Morgan WJ (1971) Convection plumes in the lower mantle. Nature 230:42–42

Müller B, Reinecker J, Fuchs K (2000) The 2000 release of the world stress map (available online at http://www-wsm.physik.uni-karlsruhe.de)

Newman WI (1983) Nonlinear diffusion: Self-similarity and travelling waves. Pure Appl Geophys 121:417–441

North GR, Cahalan RF, Coakley JA (1981) Energy balance climate models. Rev Geophys Scace Phys 19:91-121

Oertel G (1996) Stress and deformation. A handbook on tensors in geology. Oxford University Press, New York, 292 p

Ollier CD (1985) Morphotectonics of continental margins with great escarpments. In: Morrison M, Hack JT (eds) Tectonic geomorphology. Allan & Unwin, Boston, 390 p

Onsager L (1931) Reciprocal relations in irreversible processes. I Phys Rev 37:405–426

O'Rouke J (1993) Computational Geometry in C. Cambridge University Press, 346 p

Oxburgh ER (1980) Heat flow and magma genesis. In: Hargraves RB (eds) Physics of magmatic processes. Princeton University Press, Princeton NJ, pp 161–199

Oxburgh ER (1982) Heterogeneous lithospheric stretching in early history of orogenic belts. In: Hsü K (ed) Mountain building processes. Academic Press, London, pp 85–94

Oxburgh ER, Turcotte DL (1974) Thermal gradients and regional metamorphism in overthrust terrains with special reference to the eastern Alps. Schweiz Mineral Petrograph Mitt 54:641–662

Parsons B, McKenzie D (1978) Mantle convection and the thermal structure of the plates. J Geophys Res 83:4485–4496

Parsons B, Richter FM (1980) A relationship between the driving force and the geoid anomaly associated with the mid-ocean ridges. Earth Planet Sci Lett 51:445–450

Parsons B, Sclater JG (1977) An analysis of the variation of ocean floor bathymetry with age. J Geophys Res 82:803–827

Passchier CW, Myers JS, Kröner A (1990) Field geology of high-grade gneiss terrains. Springer Verlag, Berlin 150p.

Pavlis TL (1986) The role of strain heating in the evolution of megathrusts. J Geophys Res 91:12407–12422

Peacock SM (1989) Numerical constraints on rates of metamorphism and fluid production fluid flux during regional metamorphism. Geol Soc Am Bull 101:476–485

Perry J (1895) On the age of the Earth. 51:224–227 & 341–342 & 582–585

Pfiffner OA, Lehner P, Heitzmann P, Mueller S, Steck A (eds) (1997) Deep structure of the Swiss Alps. Results of NRP20. Birkhäuser, Basel, 380 p

Pierce KL, Coleman SM (1986) Effect of height and orientation (microclimate) on geomorphic degradation rates and processes, late glacial terrace scarps in central Idaho. Geol Soc Am Bull 97:869–885

Pinter N, Brandon MT (1997) How erosion builds mountains. Scientific American, April 1997, pp 74–79

Pitman WC, Andrews JA (1985) Subsidence and thermal history of small pull-apart basins. In: Biddle K, Christie-Blick N (eds) Strike-slip deformation, basin formation and sedimentation. Spec Publ Soc Econ Palaeont Mineral 37:45–119

Platt JP (1990) Thrust mechanics in highly overpressured accretionary wedges. J Geophys Res 95:9025–9034

Platt JP (1993a) Mechanics of oblique convergence. J Geophys Res 98:16239–16256

Platt JP (1993b) Exhumation of high-pressure rocks: a review of concepts and processes. Terra Nova 5:119–133

Platt JP, England PC (1994) Convective removal of lithosphere beneath mountain belts: thermal and mechanical consequences. Am J Sci 294:307–336

Pluijm B, Marshack S (1997) Earth structure – an introduction to structural geology and tectonics. McGraw Hill, New York, 495 p

Pollack HN, Chapman DS (1977) On the regional variation of heat flow, geotherms and lithosphere thickness. Tectonophys 38:279–296

Powell R (1978) Equilibrium thermodynamics in petrology. Harper and Row, London, 284 p

Press WH, Flannery BP, Teukolsky SA, Vetterling WT (1989) Numerical receipes in Pascal. The art of scientific computing. Cambridge University Press, Cambridge, 759 p

Price PH, Slack MR (1954) The effect of latent heat on numerical solutions of the heat flow equation. Br J Appl Phys 3:379–384

Putnis A, McConnell JDC (1980) Principles of mineral behaviour. Geoscience Texts 1, Elsevier, New York, 257 p

Ramsay JG, Huber M (1983) The techniques of modern structural geology, Strain analysis. Academic Press, San Diego, 1:1-300

Ramsay JG, Huber M (1987) The techniques of modern structural geology, Folds and fractures 2:301-700.

Ramsay JG, Lisle RJ (2000) The techniques of modern structural geology, Applications of continuum mechanics in structural geology. Academic Press, San Diego, 3:701-1061

Ranalli G (1987) Rheology of the Earth, 2nd edn. Chapman and Hall, London, 413 p

Ranalli G (1994) Nonlinear flexure and equivalent mechanical thickness of the lithosphere. Tectonophys 240:107–114

Ratschbacher L, Frisch W, Linzer HG, Merle O (1991) Lateral extrusion in the Eastern Alps, part 2: structural analysis. Tectonics 10:245–256

Reece G (1986) Microcomputer modelling by finite differences. MacMillian. Basingstoke UK, 125 p

Reiner M (1964) The Deborah number. Physics Today 17:62

Reiner M (1969) Deformation, strain and flow. HK Lewis, London, 360 p

Richardson RM (1992) Ridge forces, absolute plate motions and the intra-plate stress field. J Geophys Res 97:11739–11748

Ridley J (1989) Vertical movement in orogenic belts and the timing of metamorphism relative to deformation. In: Daly JS, Cliff RA, Yardley BWD (eds) Evolution of metamorphic belts. Geol Soc Lond Spec Pub 43: 103–115.

Ringwood (1988) Phase transformations and their bearing on the constitution and dynamics of the mantle. Geochim Cosmochim Acta 55:2083–2110

Robinson P (1990) The eye of the petrographer, the mind of the petrologist. Am Mineral 76:1781–1810

Robinson AH, Sale RD, Morrison JL, Muehrke PC (1984) Elements of Cartography, 5th edn. John Wiley and Sons, New York, 544 p

Roering JJ, Kirchner JW, Dietrich E (2001) Hillslope evolution by nonlinear, slope dependent transport: Steady state morphology and equilibrium adjustment time scales. J Geophys Res 106:16499–16513

Roy RF, Decker ER, Blackwell DD, Birch F (1968) Heat flow in the United States. J Geophys Res 73:5207–5221

Royden L, Keen CE (1980) Rifting processes and thermal evolution of the continental margin of eastern Canada determined from subsidence curves. Earth Planet Sci Lett 51:343–361

Royden LH (1993a) The tectonic expression of slab pull at continental convergent boundaries. Tectonics 12:303–325

Royden LH (1993b) The steady-state thermal structure of eroding orogenic belts and accretionary prisms. J Geophys Res 98:4487–4507

Ruppel C (1995) Extensional processes in continental lithosphere. J Geophys Res 100:24187–24216

Rutland RWR (1965) Tectonic overpressures. In: Pitcher WS, Flinn GW (eds) Controls of metamorphism. Oliver and Boyd, New York, pp 119–139

Sabodini R, Lambeck K, Boschi E (eds) (1991) Glacial isostasy, sea level and mantle rheology. NATO ASI Series Serie C: Mathematical and physical sciences, 334 p

Sahagian DL, Holland SM (1993) On the thermomechanical evolution of continental lithosphere. J Geophys Res 98:8261–8274

Sambridge M, Braun J, McQueen H (1995) Geophysical parameterization and interpolation of irregular data using natural neighbours. Geophys J Int 122:837–857

Sandiford M, Coblentz DD (1994) Plate-scale potential-energy distribution and the fragmentation of ageing plates. Earth Planet Sci Lett 126:143–159

Sandiford M, Powell, R (1990) Some isostatic and thermal consequences of the vertical strain geometry in convergent orogens. Earth and Planet Sci Lett 98:154–165

Sandiford M, Martin N, Zhou S, Fraser G (1991) Mechanical consequences of granite emplacement during high-T, low-P metamorphism and the origin of anticlockwise PT paths. Earth Planet Sci Lett 107:164–172

Sandiford M, Coblentz DD, Richardson RM (1995) Ridge-torques and continental collision in the Indo-Australian plate, Geology 23:653–656

Sandiford M, Hand M (1998a) Controls on the locus of intraplate deformation in central Australia. Earth Planet Sci Lett 162:97–110

Sandiford M, Hand M (1998b) Austalian Proterozoic high-temperature, low pressure metamorphism in the conductive limit. In: Treolar P, O'Brian P (eds.) What controls metamorphism. Gological Society of London special Publicaion 138: 103–114

Sandiford M, Hand M, McLaren S, (1998) High geothermal gradient metamorphism during thermal subsidence. Earth Planet Sci Lett 163:149–165

Sandiford M, McLaren S, Neumann N (2002) Long term thermal consequences of the redistribution of heat producing elements associated with large scale granitic complexes. J Metam Geol in press

Saunders AD, Tarney J, Kerr AC, Kent RW (1996) The formation and fate of large oceanic igneous provinces. Lithos 37:81–95

Sawyer DS (1985) Brittle failure in the upper mantle during extension of continental lithosphere. J Geophys Res 90:3021–3025

Schatz JF, Simmons G (1972) Thermal conductivity of Earth materials at high temperatures. J Geophys Res 77:6966–6983

Scheidegger AE (1961) Theoretical geomorphology. Prentice Hall, Englewood Cliffs, N. J., 333 p

Schmalholz SM, Podladchikov Y (1999) Buckling versus folding: Importance of Viscoelasticity. Geophys Res Lett 26:2641–2644

Scholz CH (1980) Shear heating and the state of stress on faults. J Geophys Res 85:6174–6184

Schumm SA (1956) The evolution of drainge basins and slopes in badlands at Perth Amboy, New Jersey. Bull Geol Soc Am 67:597–646

Schumm SA, Dumont JF, Holbrook JM (2000) Active Tectonics and Allivial Rivers. Cambridge University Press. 276p.

Sclater JG, Christie PA (1980) Continental stretching: an explanation of the post-mid Cretaceous subsidence of the central North Sea basin. J Geophys Res 85:3711–3739

Sclater JG, Jaupart C, Galson D (1980) The heat flow through oceanic and continental crust and the heat loss of the Earth. Rev Geophys Space Phys 18:269–311

Selverstone J (1988) Evidence for east-west crustal extension in the Eastern Alps: implications for the unroofing history of the Tauern Window. Tectonics 7:87–105

Seth HC (1999) Flood basalts and large igneous provinces from deep mantle plumes: fact, fiction and fallacy. Tectonophysics 311:1–29

Shemenda AI (1994) Subduction: Insights from physical modelling. Kluwer Academic Ser. Modern Approaches in Geophysics. Nethelands pp 215

Shreve RL (1967) Infinite topologically random networks. J Geol 75:178–186

Sleep NH (1971) Thermal effects of formation of Atlantic continental margins by continental breakup. Geophys J Roy Astron Soc 24:325–350

Sleep NH (1979) A thermal constraint on the duration of folding with reference to Acadian geology, New England (USA). J Geol 87:583–589

Sleep NJ (1992) Hotspot volcanism and mantle plumes. Annual Reviews of Earth and Planetary Science 20:19–43

Smith GD (1985) Numerical solutions of partial differential equations: finite difference methods, 3rd edn. Oxford Applied Mathematics and Computing Science Series. Clarendon Press, Oxford, 337 p

Smoot NC (1997) Magma floods, micro plates and orthogonal intersections In Choi DR, Dickins JM (eds) New concepts in global tectonics Newsletter 5:8–13

Snyder JP (1987) Map projections – a working manual. USGS Prof Pap 1395, 383 p

Snyder JP, Voxland RM (1989) An album of map projections: USGS Prof Pap 1453, 249 p

Sonder LJ, Chamberlain CP (1992) Tectonic controls of metamorphic field gradients. Earth Planet Sci Lett 111:517–535

Sonder LJ, England P (1986) Vertical averages of rheology of the continental lithosphere: relation to „thin sheet"-parameters. Earth Planet Sci Lett 77:81–90

Spear FS (1993) Metamorphic phase equilibria and pressure-temperature-time paths. Min Soc Am Monograph, Book Crafters, Chelsea, Michigan, 799 p

Spear FS, Florence FP (1992) Thermobarometry in granulites: pitfalls and new approaches. Precamb Res 55:209–241

Spear FS, Peacock SM (1989) Metamorphic pressure-temperature-time paths. Am Geophys Union Short Course in Geology AGU press 7, 102 p

Spiegel MR (1968) Mathematical handbook of formulas and tables. Schaums outline series in mathematics. McGraw Hill, New York, 271 p

Spiegelman M, McKenzie D (1987) Simple 2-D models for melt extraction at mid-ocean ridges and island arcs. Earth Planet Sci Lett 83:137–152

Steckler MS, Watts AB (1978) Subsidence of Atlantik-type continental margin off New York. Earth Planet Sci Lett 41:1–13

Steckler MS, Watts AB (1981) Subsidence history and tectonic evolution of Atlantic-type continental margins. In: Scrutton, RA (ed) Dynamics of passive margins. Am Geophys U, Geodynamics Series 6:184–196

Stefan J (1891) Über die Theorie der Eisbildung, insbesondere über die Eisbildung im Polarmeere. Annalen der Physik und Chemie 42:269–286

Stephansson O (1974) Stress-induced diffusion during folding. Tectonophys 22:233–251

Stockmal GS (1983) Modelling of large scale accretionary wedge deformation. J Geophys Res 88:8271–8287

Strahler AN (1964) Quantitative geomorphology of drainage basins and channel networks. In: Chow VT (ed) Handbook of applied hydrology, McGraw-Hill, New York Section 4-II

Strang G (1988) Linear algebra and its applications, 3rd edn. Harcourt Brace Jovanovich International Edition, 505 p

Strömgard KE (1973) Stress distribution during formation of boudinage and pressure shadows. Tectonophys 16:215–248

Stüwe K (1991) Flexural constraints on the denudation of asymmetric mountain belts. J Geophys Res 96:10401–10408

Stüwe K (1994) Process and age constraints for the formation of Ayers Rock / Australia – An example for two-dimensional mass diffusion with pinned boundaries. Z Geomorph 38:435–455

Stüwe K (1995) The buffering effects of latent heat of fusion on the equilibration of partially melted metamorphic rocks. Tectonophys 248:39–51

Stüwe K (1997) Effective bulk composition changes due to cooling: a model predicting complexities in retrograde reaction textures. Contrib Mineral Petrol 129:43–52

Stüwe K (1998a) Heat sources for Eoalpine metamorphism in the Eastern Alps. A discussion. Tectonophys 287:251–269

Stüwe K (1998b) Keyword: Diffusion. In: Marshall CP, Fairbridge RW (eds.) Encyclopedia of Geochemistry. Chapman and Hall, London 133–140

Stüwe K, Sandiford M (1994) On the contribution of deviatoric stresses to metamorphic PT paths: an example appropriate to low-P, high-T metamorphism. J Met Geol 12:445–454

Stüwe K, Sandiford M (1995) Mantle-lithospheric deformation and crustal metamorphism with some speculations on the thermal and mechanical significance of the Tauern Event, Eastern Alps. Tectonophys 242:115–132

Stüwe K, Sandiford M, Powell R (1993a) On the origin of repeated metamorphic-deformation events in low-P high-T terranes. Geology 21:829–832

Stüwe K, Will TM, Zhou S (1993b) On the timing relationships between fluid production and metamorphism in metamorphic piles. Some implications for the origin of post-metamorphic gold mineralisation. Earth Planet Sci Lett 114:417–430

Stüwe K, White L, Brown R (1994) The influence of eroding topography on steady-state isotherms. Application to fission track analysis. Earth Planet Sci Lett 124:63–74

Stüwe K, Ehlers K (1997) Multiple metamorphic events at Broken Hill, Australia. Evidence from chloritoid-bearing parageneses in the Nine-Mile region. J Pet 38:1167–1186

Stüwe K, Barr T (1998) On uplift and exhumation during convergence. Tectonics 17:80–88

Stüwe K, Barr T (2000) On the relationship between surface uplift and gravitational extension. Tectonics 19:1056–1064

Stüwe K, Hintermüller M (2000) Topography and isotherms evisited: the influence of laterally migrating drainage divides. Earth Planet Sci Lett 184:287–303

Summerfield MA (1991) Global Geomorphology. Longman House, Burnt Mill, England, 537 p

Summerfield MA, Hutton NJ (1994) Natural controls of fluvial denudation rates in major world drainage basins. J Geophys Res 99:13871–13883

Suppe J (1981) Mechanics of mountain building and metamorphism in Taiwan. Mem Geol Soc China 4:67–89

Suppe J (1985) Principles of structural geology. Prentice Hall, Englewood Cliffs, N. J., 537 p

Suppe J (1987) The active Taiwan mountain belt. In: Schaer JP, Rodgers J (eds) The anatomy of mountain ranges. Princeton University Press, Princeton N. Y., pp 277–293

Talwani M, Worzel JL, Landisman M, (1959) Rapid gravity computations for two-dimensional bodies with application to the Mendocino Submarine fracture zone. J Geophys Res 64:49–61

Tapponier P, Molnar P (1976) Slip-line field theory and large scale continental tectonics. Nature 264:319–324

Tapponier P, Peltzer G, Le Dain AY, Armijo R, Cobbolt P (1982) Propagating extrusion tectonics in Asia: new insights from simple experiments with plasticine. Geology 10:611–616

Tapponier P, Meyer B, Avouac JP Peltzer G, Gaudemer Y, Shunmin G, Hongfa X, Kelun Y, Zhitai C, Shuahua C, Huagang D (1990) Active thrusting and folding in the Qilian Shan, and decoupling between upper crust and mantle in northeeastern Tibet. Earth Planet Sci Lett 97:382–403

Tarboton DG (1996) Fractal rivers networks, Horton's laws and Tokunga cyclicity. J Hydrology 187:105–117

Taylor FB (1910) Bearing of the Tertiary mountain belt on the origin of the Earth's plan. Bull Geol Soc Am 21:179–226

Telford WM, Geldart LP, Sheriff RE (1990) Applied geophysics, 2nd edn. Cambridge University Press, Cambridge, 770 p

Tex E den (1963) A commentary on the correlation of metamorphism and deformation in space and time. Geol Mijnbouw 42:17176

Thompson AB, England PC (1984) Pressure-temperature-time paths of regional metamorphism Part II: some petrological constraints from mineral assemblages in metamorphic rocks. J Pet 25:929–955

DuToit A (1937) Our wandering continents. Oliver and Boyd, Edinburgh, 366 p

Tong H (1983) Threshold models in non-linear time series analysis. Lecture notes in statistics, Springer, Berlin Heidelberg New York, 323 p

Tucker GE, Slingerland R (1994) Erosional dynamics, flexural isostasy and long-lived escarpments: A numerical modeling study. J Geophys Res 99:12229–12243

Tucker GE, Slingerland R (1996) Predicting sediment flux from fold and thrust belts. Basin Res 8:329–349

Turcotte DL (1979) Flexure. Advances in Geophys 21:51–86

Turcotte DL (1983) Mechanisms of crustal deformation. J Geol Soc London 140:701–724

Turcotte DL (1997) Fractals and chaos in geology and geophysics, 2nd edn. Cambridge University Press, Cambridge, 398 p

Turcotte DL, Oxburgh ER (1973) Mid-plate tectonics. Nature, 244:337–339

Turcotte DL, Schubert G (1982) Geodynamics. Applications of continuum physics to geological problems. John Wiley and Sons, New York, 450 p

Turcotte DL, Haxby WF, Ockendon JR (1977) Lithospheric instabilities. In: Talwani M, Pitman III WC (eds) Island arcs, deep sea trenches and back-arc basins. Am Geophys U, Maurice Ewing Series Washington 1:63–69

Turner S, Arnaud N, Liu J, Rogers N, Hawkesworth C, Harris N, Kelley S, Van Calsteren P, Deng W (1995) Post collision shoshonitic volcanism in the Tibetan plateau: Implications for convective thinning of the lithosphere and the source of ocean island basalts. J Pet 37:45-71

Twiss RJ, Moores EM (1992) Structural geology. Freeman and Company, New York 532 p

Vilotte JP, Daignieres M, Madariaga R (1982) Numerical modeling of intraplate deformation: simple mechanical models of continental collision. J Geophys Res 87:10709–10728

Vine F, Mathews D (1963) Magnetic anomalies over oceanic ridges, Nature 199:947–949

Voorhoeve H, Houseman G (1988) The thermal evolution of lithosphere extending on a low angle detachment zone. Basin Res 1:1–9

Waschbusch PJ, Royden LH (1992) Episodicity in fordeep basins. Geology 20:915–918

Watts AB, Ryan WBF (1976) Flexure of the lithosphere and continental margin basins. Tectonophys 36: 24–44

Watts AB (1976) Gravity and bachymetry in the central Pacific ocean. J Geophys Res 81:1533–1553

Wees JD, Jong KC, Cloetingh S (1992) Two dimensional *P-T-t*-modelling and the dynamics of extension and inversion in the Beltic Zone (SE Spain). Tectonophys 302:305–324

Wegener A (1912a) Die Entstehung der Kontinente. Petermanns Geographische Mitteilungen 9127:185–308

Wegener A (1912b) Die Entstehung der Kontinente. Geol Rundsch 3:276–292

Wegener A (1915) Die Entstehung der Kontinente und Ozeane. Sammlung Vieweg 23, 94 p

Weijermars R (1997) Principles of rock mechanics. Alboran Science Publishing, Amsterdam, 359 p

Wells PRA (1980) Thermal models for the magmatic accretion and subsequent metamorphism of continental crust. Earth Planet Sci Lett 46:253–265

Wernicke B (1985) Uniform sense normal simple shear of the continental lithosphere. Canad J Earth Sci 22:108–125

Wessel P, Smith WHF (1995) New version of the generic mapping tool released, EOS, Amer. Geophys U 76, 329

Wheeler J (1991) Strucutral evolution of a subducted continental sliver: the northern Dora Maira massif, Italian Alps. J Geol Soc London 148:1103–1113

White R, McKenzie D (1989) Magmatism at rift zones: the generation of volcanic continental margins and flood basalts. J Geophys Res 94:7685–7729

Wickham SM (1987) The segregation and emplacement of granitic magmas. J Geol Soc London 144:281–297

White RW, Powell R, Holland TJB (2001) Calculation of partial melting equilibria in the system Na_2O-CaO-K_2O-FeO-MgO-Al_2O_3-SiO_2-H_2O. J Metam Geol 19:139–153

Wignall PB (2001) Large igneous provinces and mass extinctions. Earth Science Reviews 53:1–33

Wilde M, Stock J (1997) Compression directions in southern California (from Santa Barbara to Los Angeles basin) obtained from borehole breakouts. J Geophys Res 102:4969–4983

Will TM (1998) Phase equilibrian in metamorphic rocks. Lecture notes in Earth Science, Springer 315 pp.

Willet S (1992) Dynamic and kinematic growth and change of a Coulomb wedge. In: McClay KR (ed) Thrust tectonics. Chapman and Hall, pp 19–31

Willet S, Beaumont C, Fullsack P (1993) Mechanical model for the tectonics of doubly vergent compressional orogens. Geology 21:371–374

Willgoose G, Bras RL, Rodriguez-Iturbe I (1991) Results from a new model of river basin evolution. Earth Surface Processes and Landforms 16:237–254

Wilson JT (1963) A possible origin of the Hawaiian islands. Canad J of Physics 41:863–868

Wilson JT (1972) Continents adrift. Readings from Scientific American Freeman WH and Company, San Francisco, 172 p

Wilson M (1993) Plate-moving mechanisms: constraints and controversies. J Geol
    Soc London 150:923–926

Winter J (2001) Introduction to igneous and metamorphic petrology. Prentice Hall,
    699 pp.

Wintsch RP, Andrews MS (1988) Deformation induced growth of sillimanite:
    "stress"minerals revisited. J Geol 96:143–161

Won IJ, Bevis M (1987) Computing the gravitational and magnetic anomalies due
    to a polygon: algorithms and Fortram subroutines. Geophys 52:232–238

Xu G, Will TM, Powell R (1994) A calculated petrogenetic grid for the system K_2-
    FeO-MgO-Al_2O_3-SiO_2-H_2O, with particular refeence to contact metamorphosed
    pelites. J Metam Geol 12:99–119

Yale LB, Carpenter SJ (1998) Large igneous provinces and giant dike swarms;
    proxies for supercontinent cyclicity and mantle convection. Earth Planet Sci
    Lett 163:109–122

Yardley BWD (1989) An introduction to metamorphic petrology. Longman Harlow
    UK, 248 p

Zen E-an (1966) Construction of pressure-temperature diagrams for multicompo-
    nent systems after the method of Schreinemakers: a geometric approach. USGS
    Bull 1225, 56 p

Zhou S, Sandiford M (1992) On the stability of isostatically compensated mountain
    belts. J Geophys Res 97:14207–14221

Zhou S, Stüwe K (1994) Modeling of dynamic uplift, denudation rates thermome-
    chanical consequences of erosion in isostatically compensated mountain belts.
    J Geophys Res 99:13923–13939

Ziegler PA (1992) Plate tectonics, plate moving mechanisms and rifting. Tectono-
    phys 215:9–34

Ziegler PA (1993) Plate-moving mechanisms: their relative importance. J Geol Soc
    London 150:927–940

Zill DG (1986) A first course in differential equations with applications (3rd edition).
    PWS-Kent Publishing, Boston, 520 p

Zoback ML (1992) First- and second-order patterns of stress in the lithosphere: the
    world stress map project. J Geophys Res 97:11703–11728

Zoback ML, Haimson BC (1983) Hydraulic fracturing stress measurements US Nat
    Comm for Rock Mechanics, Nat Acad Press, Washington D. C., 270 p

Index

Printing (Computer to Film): Saladruck Berlin
Binding: Stürtz AG, Würzburg